AF580876

Geometric Properties of Natural Operators Defined by the Riemann Curvature Tensor

Geometric Properties of Natural Operators Defined by the Riemann Curvature Tensor

Peter B. Gilkey

University of Oregon, USA

Published by

World Scientific Publishing Co. Pte. Ltd.

P O Box 128, Farrer Road, Singapore 912805

USA office: Suite 1B, 1060 Main Street, River Edge, NJ 07661

UK office: 57 Shelton Street, Covent Garden, London WC2H 9HE

British Library Cataloguing-in-Publication Data
A catalogue record for this book is available from the British Library.

GEOMETRIC PROPERTIES OF NATURAL OPERATORS DEFINED BY THE RIEMANN CURVATURE TENSOR

ISBN 981-02-4752-4

Printed in Singapore by World Scientific Printers

Preface

A central problem in differential geometry is to relate algebraic properties of the Riemann curvature tensor to the underlying geometry of the manifold. The full curvature tensor is in general quite difficult to deal with. We will use the curvature tensor to define several natural associated operators; the Jacobi operator, the Szabó operator, and the skew-symmetric curvature operator are all natural operators of differential geometry which are defined in terms of the curvature tensor and its covariant derivative. We also consider other related operators. We shall discuss the geometric conditions which are imposed when we assume that one of these operators has constant eigenvalues.

Chapter 1 of this book is devoted to algebraic preliminaries.

Chapter 2 deals with the skew-symmetric curvature operator.

Chapter 3 deals with the Jacobi and Szabó operators.

Chapter 4 discusses results from algebraic topology which are needed previously.

The first section in Chapters 1-4 contains a lengthy introduction to the material contained therein. An extensive bibliography is provided at the end of the book.

It is an honor to acknowledge the service of Academic Vice President Lorraine Davis to the University of Oregon and to acknowledge useful conversations with Dr. G. Steigelman on many matters.

Professor R. Ivanova and Ms. I. Stavrov offered invaluable assistance by reading the manuscript carefully and by providing helpful comments, both stylistic and mathematical.

This book is dedicated to James Gordon Gilkey Jr (my father) and to Arnie Zweig (Philosopher and friend).

The research of this book has been partially supported by the Max Planck Institute for Mathematical Sciences (Leipzig, Germany) and by the National Science Foundation (USA).

Contents

Chapter 1
Algebraic Curvature Tensors

1.1 Introduction

In Chapter 1.1, we present introductory material we shall need subsequently. In Sections 1.2 through 1.4, we review some concepts from linear algebra. The general theory of vector spaces which are equipped with a non-degenerate symmetric inner product is discussed in Section 1.2. Jordan equivalence of linear transformations is introduced and it is shown that two nilpotent linear maps are Jordan equivalent if and only if they have the same rank. We prove that if a projective map is linearizable, then two linearizations differ only by a multiple; i.e. the linearization is unique modulo rescaling. In Section 1.3, we discuss self-adjoint maps in the positive definite setting. We give a generalized Rayleigh–Ritz minimax principle for computing the eigenvalues of a self-adjoint map and show that the eigenvalues of a 1 parameter family of self-adjoint maps vary continuously. In Section 1.4, we review work of Adams [1] giving possible splittings of the tangent bundle of the unit sphere. We also present a brief discussion of Clifford algebras.

Let (M, g) be a pseudo-Riemannian manifold. In Section 1.5, we define the Levi-Civita connection and the Riemann curvature tensor:

$$({}^g\nabla, {}^gR).$$

Algebraic curvature tensors are introduced to put things in an abstract setting. We present various natural operators which are associated to the curvature tensor; the Jacobi operator, the higher order Jacobi operator, and the Szabó operator are symmetric (self-adjoint) operators.

We also define the skew-symmetric curvature operator, a complex analogue, and a higher order analogue of the skew-symmetric curvature operator; these are skew-symmetric operators.

In Section 1.6, we study some of the elementary properties of algebraic curvature tensors. We show that an algebraic curvature tensor R is almost complex if and only if

$$R(x, y, z, w) = R(Jx, Jy, Jz, Jw) \text{ for all } x, y, z, w \in V.$$

We define the sectional curvature and establish some of its basic properties. We also review a result of Gray [91] concerning pseudo-Hermitian almost complex manifolds.

Let R be an algebraic curvature tensor and let

$$\mathcal{J}_R(x) : y \to R(y, x)x$$

be the Jacobi operator. We say that R is k-stein if there exist constants c_i so that

$$\text{(1.1.1.a)} \qquad \operatorname{Tr}\{\{\mathcal{J}_R(x)\}^i\} = c_i(x, x)^i \text{ for } 1 \le i \le k.$$

In Section 1.7, we study Einstein and k-stein algebraic curvature tensors. We prove that if R is k-stein for all k, then the Jacobi operator of a null vector is nilpotent. We use this observation to show that a 2-stein algebraic curvature tensor on a Lorentzian vector space ($p = 1$) has constant sectional curvature.

Let ϕ be a linear map of V with $\phi^* = \pm\phi$. Define:

$$\text{(1.1.1.b)} \quad R_\phi(x, y)z := \begin{cases} (\phi y, z)\phi x - (\phi x, z)\phi y & \text{if } \phi = \phi^*, \\ (\phi y, z)\phi x - (\phi x, z)\phi y - 2(\phi x, y)\phi z & \text{if } \phi = -\phi^*. \end{cases}$$

The tensor R_{Id} is the tensor of constant sectional curvature $+1$. More generally, if S is the *shape* or *Weingarten operator* of a non-degenerate hypersurface (M, g) in $\mathbb{R}^{(r,s)}$, then S is self-adjoint. We will show in Lemma 1.12.3 that ${}^gR = \varepsilon R_S$, where $\varepsilon = +1$ if the normal is spacelike and $\varepsilon = -1$ if the normal is timelike. If J is the almost complex structure on a complex projective space, then the curvature tensor of the Fubini-Study metric is $R_{\mathrm{Id}} + R_J$; see Lemma 3.6.4 for details. Thus, the tensors R_ϕ for $\phi = \pm\phi^*$ arise naturally in differential geometry; they will play a central role in our investigations.

In Section 1.8, we show that R_ϕ is an algebraic curvature tensor and establish some basic properties of this tensor. Let J be an almost complex structure on V. We prove that the tensor R_ϕ is an algebraic curvature tensor; R_ϕ is almost complex if $\phi J = \pm J\phi$. We then generalize a result of Fiedler [56] to show that the tensors R_ϕ for ϕ symmetric (resp anti-symmetric) generate the space of all algebraic curvature tensors. We show that if $\phi J = J\phi$, then the tensors R_ϕ can be realized geometrically by holomorphic manifolds.

In Section 1.9, we recall Weyl's theorem [158] for invariants of the orthogonal group and show that the higher order Jacobi operator and the higher order skew-symmetric curvature operator can be defined either in terms of averaging over the Grassmannian or in terms of contractions of indices.

We say that an algebraic curvature tensor R is *spacelike Osserman* (resp. *timelike Osserman*) if the Jacobi operator $\mathcal{J}_R(\cdot)$ has constant eigenvalues on the pseudo-spheres of unit spacelike (+) and timelike (-) vectors:

$$S^{\pm}V = \{v \in v : (v, v) = \pm 1\}.$$

In Section 1.10, we complexify and use the principle of analytic continuation to show that the notions of spacelike Osserman and timelike Osserman are the same if $p \geq 1$ and $q \geq 1$; thus the type of the vector x is irrelevant when considering the constancy of eigenvalues and we shall simply speak of Osserman algebraic curvature tensors. We shall also establish similar theorems for the higher order Jacobi operator, the skew-symmetric curvature operator, the higher order skew-symmetric curvature operator, the skew-symmetric curvature operator in the complex setting, and the Szabó operator.

In Sections 1.11, 1.12, 1.13, 1.14, and 1.15 we turn to geometric matters. In Section 1.11, we discuss exponential coordinates and establish a few facts concerning Jacobi vector fields which we shall need in Chapter 3. In Section 1.12, we deal with geometric realizations. We express the second derivatives of the metric in terms of the curvature tensor and the curvature tensor in terms of the second derivatives of the metric at the center of a geodesic polar coordinate system. We then show that every algebraic curvature tensor is geometrically realizable. If ϕ is self-adjoint, then we show that $\pm R_\phi$ is geometrically realizable as the curvature tensor of a hypersurface in a flat space. In Section 1.13, we present two Schur type results regarding the Einstein constants μ_1 and μ_2 which were defined in Equation (1.1.1.a).

In Section 1.14, we give some introductory results concerning space forms and in Section 1.15, we present similar results for complex space forms and for para-complex spaceforms; we complete our discussion of these manifolds in Section 3.6 after establishing some results concerning the Osserman conjecture.

1.2 Results from linear algebra

Let $(\cdot,\cdot)$ be a symmetric bilinear form on a finite dimensional real vector space V. We suppose that the form $(\cdot,\cdot)$ is non-degenerate, i.e. given a non-zero vector x in V, there is some vector y in V so that $(x,y)\neq 0$. We can then choose a basis $\{e_i\}$ for V so that

$$(e_i,e_j)=\begin{cases}0 & \text{if } i\neq j,\\ \pm 1 & \text{if } i=j.\end{cases}$$

Such a basis is called an *orthonormal basis.* We set $\varepsilon_i := (e_i,e_i)$. Let p be the number of indices i with $\varepsilon_i=-1$. Let $q=\dim V-p$ be the complementary index; q is the number of indices i with $\varepsilon_i=+1$. The inner product is then said to have *signature* (p,q); the integers p and q are independent of the particular orthonormal basis chosen.

If $\{e_1,...,e_{p+q}\}$ is the standard basis for Euclidean space $\mathbb{R}^m$, then we shall let $\mathbb{R}^{(p,q)}$ be $\mathbb{R}^m$ with the inner product given by:

$$(1.2.1.a)\qquad (e_i,e_j)=\begin{cases}0 & \text{if } i\neq j,\\ -1 & \text{if } i=j\leq p,\\ 1 & \text{if } i=j>p.\end{cases}$$

If V is a vector space of signature (p,q), then we can construct an isometry between V and $\mathbb{R}^{(p,q)}$ by choosing an orthonormal basis for V of the form given in Equation (1.2.1.a). However, it is often useful to work in a basis free setting with an abstract vector space V rather than with the concrete realization $\mathbb{R}^{(p,q)}$.

The following is a technical result that will be useful in several contexts. Let δ_{ij} be the Kronecker symbol:

$$\delta_{ij}:=\begin{cases}0 & \text{if } i\neq j,\\ 1 & \text{if } i=j.\end{cases}$$

1.2.2 Lemma. *Let $\mathcal{S}=\{v_1,...,v_k\}\subset V$, where V is a vector space of signature (p,q).*

(1) *If $\mathcal{S}$ is a set of linearly independent vectors, then there exist a set of vectors $\{w_1,...,w_k\}\subset V$ so that $(v_i,w_j)=\delta_{ij}$.*

(2) *Let $S(v):=(v_1,v)v_1+...+(v_k,v)v_k$. Then* $\operatorname{rank} S=k$ *if and only if $\mathcal{S}$ is a set of linearly independent vectors.*

(3) *Let $T(v):=(v_2,v)v_1-(v_1,v)v_2+...+(v_{2\ell},v)v_{2\ell-1}-(v_{2\ell-1},v)v_{2\ell}$, where $k=2\ell$. Then* $\operatorname{rank} T=2\ell$ *if and only if $\mathcal{S}$ is a set of linearly independent vectors.*

Proof. We use the inner product to define a linear map $\psi : V \to V^*$ by the identity $\psi(w)(v) = (v, w)$. If $w \neq 0$, then there exists v so $(w, v) \neq 0$ and thus ψ is injective. Since $\dim V = \dim V^*$, it follows that ψ is a linear isomorphism. Let $\{v_1, ..., v_k\}$ be a set of linearly independent elements of V. We extend this set to a basis $\{v_1, ..., v_{p+q}\}$ for V to assume without loss of generality that $k = p+q$ in the proof of Assertion (1). Let $\{v^1, ..., v^m\}$ be the associated *dual basis* for V^*; this means if $v \in V$, then $v = \sum_i v^i(v)v_i$. Since $v^i(v_j) = \delta_{ij}$, we may complete the proof of Assertion (1) by setting:

$$w_j = \psi^{-1}(v^j).$$

Assertions (2) and (3) now follow immediately from Assertion (1).

The following technical fact will be useful in our definition of the skew-symmetric curvature operator:

1.2.3 Lemma. *Let V be a vector space of signature (p, q). Let x_1, x_2, y_1, and y_2 belong to V, where $x_2 = ax_1 + by_1$ and $y_2 = cx_1 + dy_1$ for some constants a, b, c, and d. Then*

$$(x_2, x_2)(y_2, y_2) - (x_2, y_2)^2 = (ad - bc)^2\{(x_1, x_1)(y_1, y_1) - (x_1, y_1)^2\}.$$

Proof. Let $h_{11} := (x_1, x_1)$, $h_{12} := (x_1, y_1)$, and $h_{22} := (y_1, y_1)$. We compute:

$$\begin{aligned}
&(x_2, x_2)(y_2, y_2) - (x_2, y_2)^2 \\
=&(a^2h_{11} + 2abh_{12} + b^2h_{22})(c^2h_{11} + 2cdh_{12} + d^2h_{22}) \\
&-(ach_{11} + (ad + bc)h_{12} + bdh_{22})^2 \\
=&(a^2c^2 - a^2c^2)h_{11}^2 + (2a^2cd + 2abc^2 - 2ac(ad + bc))h_{11}h_{12} \\
&+(a^2d^2 + b^2c^2 - 2acbd)h_{11}h_{22} + (4abcd - (ad + bc)^2)h_{12}h_{12} \\
&+(2abd^2 + 2cdb^2 - 2(ad + bc)bd)h_{12}h_{22} + (b^2d^2 - b^2d^2)h_{22}h_{22} \\
=&(ad - bc)^2(h_{11}h_{22} - h_{12}^2). \quad \square
\end{aligned}$$

A vector $x \in V$ is said to be *spacelike* if $(x, x) > 0$, *timelike* if $(x, x) < 0$, and *degenerate* or *null* if $(x, x) = 0$. We emphasize that the zero vector is null; thus to say a vector is timelike or spacelike means that the vector is necessarily non-zero.

The following is a useful technical observation.

1.2.4 Lemma. *Let V be a vector space of signature (p,q).*

(1) *If $p>0$, then we can choose a basis $\mathcal{B}:=\{e_1,...,e_{p+q}\}$ for V so each element of $\mathcal{B}$ is timelike. In particular, V is spanned by the timelike vectors in V.*

(2) *If $q>0$, then we can choose a basis $\mathcal{B}:=\{e_1,...,e_{p+q}\}$ for V so each element of $\mathcal{B}$ is spacelike. In particular, V is spanned by the spacelike vectors in V.*

Proof. Suppose that $p>0$. Choose e_1 timelike. Since $0\neq e_1$, we can extend $\{e_1\}$ to a basis $\{e_1,\tilde{e}_2,...,\tilde{e}_{p+q}\}$ for V. Let ε be a real parameter. Then $\{e_1,\tilde{e}_2+\varepsilon e_1,...,\tilde{e}_{p+q}+\varepsilon e_1\}$ is a basis for V for any value of ε. For $2\leq i\leq p+q$, set:

$$F_i(\varepsilon):=(\tilde{e}_i+\varepsilon e_1,\tilde{e}_i+\varepsilon e_1)=\varepsilon^2(e_1,e_1)+2\varepsilon(e_1,\tilde{e}_i)+(\tilde{e}_i,\tilde{e}_i);$$

$F_i(\varepsilon)$ is a quadratic polynomial in ε with negative leading coefficient (e_1,e_1). Thus, there exists $\varepsilon_i>>0$ so that F_i is negative for $\varepsilon>\varepsilon_i$. Let ε be a real number which is greater than ε_i for $2\leq i\leq p+q$. Then $\tilde{e}_i+\varepsilon e_1$ will be timelike for $2\leq i\leq p+q$ and Assertion (1) now follows; the proof of Assertion (2) is similar. □

We say that an orthonormal basis

$$\{e_1^-,...,e_p^-,e_1^+,...,e_q^+\}$$

for V is a *normalized orthonormal basis* if the vectors $\{e_1^-,...,e_p^-\}$ are timelike and if the vectors $\{e_1^+,...,e_q^+\}$ are spacelike. Such a basis will play an important role in Sections 2.2, 3.2, and 3.3.

A bilinear map is determined by its restriction to the set of all spacelike 2 planes if $q\geq 2$. More precisely:

1.2.5 Lemma. *Let V be a vector space of signature (p,q), where $q\geq 2$. Let T be a bilinear map from $V\otimes V$ to some auxiliary vector space. Assume that $T(v_1,v_2)=0$ whenever v_1 and v_2 span a spacelike 2 plane. Then $T=0$.*

Proof. Let $\{e_1^-,...,e_p^-,e_1^+,...,e_q^+\}$ be a normalized orthonormal basis for V. We consider the collection of pairs:

$$\begin{aligned}C:=\{&(e_{j_1}^+,e_{j_1}^++e_{j_2}^+),\ (e_{j_1}^+,e_{j_2}^+),\ (e_{j_1}^++\tfrac{1}{2}e_{i_1}^-,e_{j_2}^+),\\&(e_{j_1}^+,e_{j_2}^++\tfrac{1}{2}e_{i_1}^-),\ (e_{j_1}^++\tfrac{1}{2}e_{i_1}^-,e_{j_2}^++\tfrac{1}{2}e_{i_2}^-)\},\end{aligned}$$

where

$$1\leq i_1\leq p,\ 1\leq i_2\leq p,$$
$$1\leq j_1\leq q,\ 1\leq j_2\leq q,\text{ and } j_1\neq j_2.$$

The elements of C all span spacelike 2 planes of V and generate $V \otimes V$ additively. Since T is bilinear and since, by assumption, T vanishes on these elements, we may conclude that $T = 0$. $\square$

A k dimensional subspace σ of V is said to be a *k plane*; let $\mathrm{Gr}_k(V)$ (resp. $\mathrm{Gr}_k^+(V)$) be the Grassmannian of all k planes (resp. oriented k planes). A k plane σ is said to be *non-degenerate* if the restriction of the metric to the plane is non-degenerate. A non-degenerate k plane is said to be *spacelike* or *timelike* if the restriction of the metric to σ is positive or negative definite. A non-degenerate 2 plane which is neither spacelike nor timelike is said to be *mixed* - the restriction of the metric has signature (1,1) in this case. A degenerate k plane is said to be *totally isotropic* if the restriction of the metric to σ vanishes identically.

We let $\mathrm{Gr}_{r,s}(V)$ (resp. $\mathrm{Gr}_{r,s}^+(V)$) be the *Grassmannian* consisting of all non-degenerate unoriented (resp. oriented) subspaces of V of signature (r, s). Similarly, let

$$\widetilde{\mathrm{Gr}}_k(V) = \cup_{r+s=k} \mathrm{Gr}_{r,s}(V) \text{ and } \widetilde{\mathrm{Gr}}_k^+(V) = \cup_{r+s=k} \mathrm{Gr}_{r,s}^+(V)$$

be the Grassmannians of all non-degenerate unoriented and oriented k planes in V, respectively. The maps which forget the orientation are normal coverings with deck group $\mathbb{Z}_2$:

$$\mathrm{Gr}_{r,s}^+(V) \to \mathrm{Gr}_{r,s}(V) \text{ and } \widetilde{\mathrm{Gr}}_k^+(V) \to \widetilde{\mathrm{Gr}}_k(V).$$

Let σ be a timelike (resp. spacelike) subspace. We can find a maximal timelike (resp. spacelike) subspace $\tilde{\sigma}$ of dimension p (resp. q) which contains σ. The orthogonal complement $\tilde{\sigma}^\perp$ will then be a maximal spacelike subspace (resp. timelike subspace) of the complementary dimension $q = m - p$ (resp. $p = m - q$) and we will have an orthogonal direct sum decomposition

$$V = \tilde{\sigma} \oplus \tilde{\sigma}^\perp.$$

A linear transformation ϕ of V is said to be *self-adjoint* (resp. *skew-adjoint*) if $\phi^* = \phi$ (resp. $\phi^* = -\phi$). We will also sometimes say that such a map is *symmetric* (resp. *anti-symmetric*). A linear transformation ϕ is said to be an *isometry* (resp. *para-isometry*) if $(\phi x, \phi x) = (x, x)$ (resp. $(\phi x, \phi x) = -(x, x)$) for all x; para-isometries exist if and only if $p = q$.

A linear transformation J of V is said to be a *pseudo-Hermitian almost complex structure* if:

$$J^2 = -\operatorname{Id}, \ J^* = -J, \text{ and } (Jx, Jy) = (x, y) \text{ for all } x, y \in V.$$

We use J to give a complex structure to V by defining $\sqrt{-1}x := Jx$. A real linear transformation of T is said to be *complex* if $JT = TJ$. A 2 plane π is said to be a *complex line* if $J\pi = \pi$. We let $\mathbb{CP}(V)$ be the projective space of complex lines. Any complex line is either spacelike, timelike, or totally isotropic; there are no mixed complex lines.

If $p = 0$, then $(\cdot,\cdot)$ is positive definite and we shall say that we are in the *Riemannian setting*; if $p = 1$, then there is one timelike direction and we shall say that we are in the *Lorentzian setting.* If $p = q$, then the inner product is said to be a *balanced innerproduct.* By changing the sign of $(\cdot,\cdot)$, i.e. by replacing $(\cdot,\cdot)$ by $-(\cdot,\cdot)$, we can interchange the roles of p and q and of timelike and spacelike vectors.

Let V be a vector space of dimension m. In studying the eigenvalues of a linear transformation, we shall always consider the complex eigenvalues counted with multiplicity. The following is a useful technical observation. We refer to [72, Lemma 2.1.6] for the proof.

1.2.6 Lemma. *Let ϕ_1 and ϕ_2 be two linear maps of a vector space V. Then the following assertions are equivalent:*

(1) *ϕ_1 and ϕ_2 have the same eigenvalues, counted with multiplicity.*
(2) $\det(\phi_1 - \lambda) = \det(\phi_2 - \lambda)$ *for all* $\lambda \in \mathbb{R}$.
(3) $\operatorname{Tr}(\phi_1^k) = \operatorname{Tr}(\phi_2^k)$ *for* $1 \le k \le \dim V$.

The eigenvalue structure does not determine a linear map ϕ up to conjugacy. We define a Jordan block of size k corresponding to a real eigenvalue $a \in \mathbb{R}$ by setting:

$$(1.2.6.a)\qquad \mathfrak{J}(k,a) := \begin{pmatrix} a & 1 & 0 & \dots & 0 & 0 \\ 0 & a & 1 & \dots & 0 & 0 \\ \dots & \dots & \dots & \dots & \dots & \dots \\ 0 & 0 & 0 & \dots & a & 1 \\ 0 & 0 & 0 & \dots & 0 & a \end{pmatrix} \text{ on } \mathbb{R}^k.$$

We define a Jordan block of size $2k \times 2k$ corresponding to the pair of complex conjugate eigenvalues $\{a + \sqrt{-1}b, a - \sqrt{-1}b\}$ by first setting

$$A := \begin{pmatrix} a & b \\ -b & a \end{pmatrix}, \text{ and } I_2 := \begin{pmatrix} 1 & 0 \\ 0 & 1 \end{pmatrix} \text{ on } \mathbb{R}^2,$$

and then setting

$$(1.2.6.b)\qquad \mathfrak{J}(k,a,b) := \begin{pmatrix} A & I_2 & 0 & \dots & 0 & 0 \\ 0 & A & I_2 & \dots & 0 & 0 \\ \dots & \dots & \dots & \dots & \dots & \dots \\ 0 & 0 & 0 & \dots & A & I_2 \\ 0 & 0 & 0 & \dots & 0 & A \end{pmatrix} \text{ on } \mathbb{R}^{2k}.$$

We refer to Adkins and Weintraub [2] for the proof of the following result.

1.2.7 Lemma. *Let ϕ be an linear transformation of a vector space V. Relative to a suitably chosen basis for V, ϕ decomposes as a direct sum of the Jordan blocks described in Equations* (1.2.6.a) *and* (1.2.6.b). *Furthermore, the unordered collection of Jordan blocks is determined by ϕ.*

The *Jordan normal form* of ϕ is the unordered collection of Jordan blocks described in Lemma 1.2.7. We say that ϕ and $\tilde{\phi}$ are *Jordan equivalent* if any of the following three equivalent conditions are satisfied:

(1) There exist bases $\mathcal{B} = \{e_1, ..., e_m\}$ and $\tilde{\mathcal{B}} = \{\tilde{e}_1, ..., \tilde{e}_m\}$ for V so that the matrix representation of ϕ with respect to the basis $\mathcal{B}$ is equal to the matrix representation of $\tilde{\phi}$ with respect to the basis $\tilde{\mathcal{B}}$.

(2) There exists an isomorphism ψ of V so $\phi = \psi\tilde{\phi}\psi^{-1}$, i.e. ϕ and $\tilde{\phi}$ are *conjugate*.

(3) The Jordan normal forms of ϕ and $\tilde{\phi}$ are equal.

The following is an immediate consequence of Lemma 1.2.7:

1.2.8 Lemma. *Let ϕ be a linear map of a vector space V. If 0 is the only eigenvalue of ϕ, then $\phi^{\dim V} = 0$.*

The following is a useful technical observation:

1.2.9 Lemma. *Let ϕ be a linear transformation of a vector space V. There exists a non-degenerate inner product $(\cdot, \cdot)$ on V so that ϕ is self-adjoint with respect to this inner product.*

Proof. Since, by Lemma 1.2.7, ϕ can be decomposed as the sum of Jordan blocks, it suffices to prove Lemma 1.2.9 for $\phi = \mathfrak{J}(k, a)$ or $\phi = \mathfrak{J}(k, a, b)$. Let $\{e_1, ..., e_k\}$ be the standard basis for $\mathbb{R}^k$. The Jordan block $\mathfrak{J}(k, a)$ defines the linear transformation:

$$\mathfrak{J}(k,a)e_i := \begin{cases} ae_i + e_{i-1} & \text{if } i > 1, \\ ae_i & \text{if } i = 1. \end{cases}$$

We define a non-degenerate innerproduct $(\cdot, \cdot)$ on $\mathbb{R}^k$ by setting:

$$(e_i, e_j) = \begin{cases} 1 & \text{if } i + j = k + 1, \\ 0 & \text{if } i + j \neq k + 1. \end{cases}$$

Since $\mathfrak{J}(k, a) = a \cdot \mathrm{Id} + \mathfrak{J}(k, 0)$, and since Id is self-adjoint with respect to any inner product, we may take $a = 0$. We have

$$(\mathfrak{J}(k,0)e_i, e_j) = \begin{cases} 1 & \text{if } i \geq 2 \text{ and } i - 1 + j = k + 1, \\ 0 & \text{if } i = 1 \quad \text{or } i - 1 + j \neq k + 1, \end{cases}$$

$$(e_i, \mathfrak{J}(k,0)e_j) = \begin{cases} 1 & \text{if } j \geq 2 \text{ and } i + j - 1 = k + 1, \\ 0 & \text{if } j = 1 \quad \text{or } i - 1 + j \neq k + 1. \end{cases}$$

Since $i-1+j=k+1$ implies $i\geq 2$ and $j\geq 2$, we have as desired:

$$(\mathfrak{J}(k,0)e_i,e_j)=(e_i,\mathfrak{J}(k,0)e_j)\text{ for all } i,j.$$

To study the Jordan block $\mathfrak{J}(k,a,b)$, let $\{e_1,f_1,...,e_k,f_k\}$ be the usual basis for $\mathbb{R}^{2k}$. Then

$$\mathfrak{J}(k,a,b)e_i=\begin{cases} ae_i-bf_i+e_{i-1} & \text{if } i>1,\\ ae_i-bf_i & \text{if } i=1,\end{cases}$$

$$\mathfrak{J}(k,a,b)f_i=\begin{cases} be_i+af_i+f_{i-1} & \text{if } i>1,\\ be_i+af_i & \text{if } i=1.\end{cases}$$

We define the inner product

$$(e_i,e_j)=\begin{cases} 1 & \text{if } i+j=k+1,\\ 0 & \text{if } i+j\neq k+1,\end{cases}$$

$$(e_i,f_j)=\ 0 \text{ for all } i,j,$$

$$(f_i,f_j)=\begin{cases} -1 & \text{if } i+j=k+1,\\ 0 & \text{if } i+j\neq k+1.\end{cases}$$

We may decompose $\mathfrak{J}(k,a,b)=\mathfrak{J}(k,0,0)+A$ where $Ae_i=ae_i-bf_i$ and $Af_i=be_i+af_j$. The proof that $\mathfrak{J}(k,0,0)$ is self-adjoint is the same as that given above to show $\mathfrak{J}(k,0)$ is self-adjoint and is therefore omitted; the bases $\{e_1,...,e_k\}$ and $\{f_1,...,f_k\}$ do not interact. Let δ_{ij} be the Kronecker symbol. To show that A is self-adjoint, we compute:

$$(Ae_i,e_j)=\delta_{ij}a,\quad (Af_i,f_j)=-\delta_{ij}a$$
$$(Ae_i,f_j)=\delta_{ij}b,\quad (e_i,Af_j)=\ \delta_{ij}b.\quad \square$$

In the positive definite context, the Jordan normal form of a symmetric or a skew-symmetric map is determined by the eigenvalue structure. This is *not* the case in the higher signature setting. Although two Jordan equivalent maps have the same eigenvalues, two maps with the same eigenvalues need not be Jordan equivalent. The following two families of maps will play an important role in the construction of higher order Osserman algebraic curvature tensors in Sections 3.2 and 3.3. Let $\{e_1^-,...,e_p^-,e_1^+,...,e_q^+\}$ be a normalized orthonormal basis for a vector space V of signature (p,q), where $p\geq 2$ and $q\geq 2$. Let $2\leq 2a\leq\min(p,q)$ and let $1\leq b\leq\min(p,q)$. Define:

$$\phi_a e_k^\pm=\begin{cases} \pm(e_{2i}^-+e_{2i}^+) & \text{if } k=2i-1\leq 2a,\\ \mp(e_{2i-1}^-+e_{2i-1}^+) & \text{if } k=2i\leq 2a,\\ 0 & \text{if } k>2a,\end{cases}$$

$$\Phi_b e_i^\pm=\begin{cases} \pm(e_i^+ + e_i^-) & \text{if } i\leq b,\\ 0 & \text{if } i>b.\end{cases}$$

The map ϕ_a is a skew-adjoint linear map of rank $2a$; the map Φ_b is a self-adjoint linear map of rank b. We have $\phi_a^2 = 0$ and $\Phi_b^2 = 0$. These maps are not diagonalizable, but 0 is the only eigenvalue. The following Lemma shows that the maps ϕ_a and Φ_{2a} are Jordan equivalent.

1.2.10 Lemma. *Let V be a vector space of signature (p, q).*

(1) *Let A be a linear map of V with $A^2 = 0$. Then 0 is the only eigenvalue of A.*

(2) *Let A_i be linear maps of V with $A_i^2 = 0$ for $i = 1, 2$. Then A_1 and A_2 are Jordan equivalent if and only if* $\operatorname{rank}(A_1) = \operatorname{rank}(A_2)$.

Proof. Let $m = \dim(V)$ and let A be a linear map of V. Assume that $A^2 = 0$. Let $p_A(\lambda) := \det(\lambda - A)$ be the characteristic polynomial of A. Then

$$\begin{aligned} p_A(\lambda)p_A(-\lambda) &= \det(\lambda - A)\det(-\lambda - A) = \det((\lambda - A)(-\lambda - A)) \\ &= \det(-\lambda^2 + A^2) = \det(-\lambda^2) = (-1)^m \lambda^{2m}. \end{aligned}$$

Thus, if $p_A(\lambda) = 0$, then $\lambda^{2m} = 0$ so $\lambda = 0$. Thus, $\lambda = 0$ is the only eigenvalue of A and Assertion (1) follows.

Suppose that $A^2 = 0$ and that A has rank k. Choose $v_i \in V$ so $\{\phi v_1, ..., \phi v_k\}$ is a basis for range(A). Since $A^2 = 0$, $w_i := \phi v_i \in \ker(A)$. Since $\operatorname{rank}(A) = k$, $\dim\ker(A) = m - k$. Thus, we can choose elements $\{x_1, ..., x_{m-2k}\}$ of V so that

$$\{w_1, ..., w_k, x_1, ..., x_{m-2k}\}$$

is a basis for $\ker(A)$. Suppose that we have a dependence relation:

$$a_1 v_1 + ... + a_k v_k + b_1 w_1 + ... + b_k w_k + c_1 x_1 + ... + c_{m-2k} x_{m-2k} = 0. \tag{1.2.10.a}$$

We apply A to Equation (1.2.10.a) to see $a_1 w_1 + ... + a_k w_k = 0$ so all the $a_i = 0$. Since $\{w_1, ..., w_k, x_1, ..., x_{m-2k}\}$ are linearly independent, we have that all the b_i and c_i vanish as well. Thus, the collection of vectors

$$\{w_1, v_1, ..., w_k, v_k, x_1, ..., x_{m-k}\}$$

forms a basis for V. Relative to this basis, A is represented by the matrix

$$A = \begin{pmatrix} 0 & 1 & 0 & 0 & ... & 0 & 0 & 0 & ... & 0 \\ 0 & 0 & 0 & 0 & ... & 0 & 0 & 0 & ... & 0 \\ 0 & 0 & 0 & 1 & ... & 0 & 0 & 0 & ... & 0 \\ 0 & 0 & 0 & 0 & ... & 0 & 0 & 0 & ... & 0 \\ ... & ... & ... & ... & ... & ... & ... & ... & ... & ... \\ 0 & 0 & 0 & 0 & ... & 0 & 1 & 0 & ... & 0 \\ 0 & 0 & 0 & 0 & ... & 0 & 0 & 0 & ... & 0 \\ ... & ... & ... & ... & ... & ... & ... & ... & ... & ... \\ 0 & 0 & 0 & 0 & ... & 0 & 0 & 0 & ... & 0 \end{pmatrix}.$$

This gives the Jordan normal form of A. Thus, $\mathrm{rank}(A_1) = \mathrm{rank}(A_2) = k$ implies A_1 and A_2 are Jordan equivalent. Since $\mathrm{rank}(\cdot)$ is preserved by Jordan equivalence, $\mathrm{rank}(A_1) \neq \mathrm{rank}(A_2)$ implies A_1 and A_2 are not Jordan equivalent. Assertion (2) now follows. □

Let A and B be vector spaces of signatures (p_A, q_A) and (p_B, q_B), respectively. Let $\mathbb{P}(A)$ and $\mathbb{P}(B)$ be the associated projective spaces of 1 dimensional linear subspaces of A and B. Let $\mathbb{P}(A^+) \subset \mathbb{P}(A)$ be the subset of spacelike lines. If a is spacelike, let $[a] := \mathbb{R} \cdot a$ be the associated spacelike line. Let $\Phi : \mathbb{P}(A^+) \to \mathbb{P}(B)$. A map $\phi : A \to B$ is said *linearize a projective map* Φ if ϕ is linear, if $\ker \phi$ contains no spacelike vectors, and if $\phi(a) \in \Phi([a])$ for any spacelike vector a.

The following technical lemma will be useful in our study of IP algebraic curvature tensors in Section 2.3.

1.2.11 Lemma. *Let A and B be vector spaces of signatures (p_A, q_A) and (p_B, q_B), respectively, where $q_A \geq 3$. Let $\Phi : \mathbb{P}(A^+) \to \mathbb{P}(B)$ be given. If ϕ_1 and ϕ_2 are two linearizations of Φ, then ϕ_1 is a multiple of ϕ_2.*

Proof. By Lemma 1.2.4, the spacelike vectors span A. Thus, to prove that ϕ_1 is a multiple of ϕ_2, it suffices to prove that there exists a constant c so that $\phi_1(a) = c\phi_2(a)$ for every spacelike a.

Let a_1 and a_2 be arbitrary spacelike vectors. Since $q_A \geq 3$, we can choose a_0 spacelike so that

$$a_0 \perp a_1 \text{ and } a_0 \perp a_2.$$

Since $[\phi_1 a_i] = [\phi_2 a_i]$, there are non-zero real numbers c_i so that

$$\phi_1(a_i) = c_i \phi_2(a_i) \text{ for } 0 \leq i \leq 2.$$

We prove Lemma 1.2.11 by showing that $c_1 = c_0 = c_2$. Since the roles of a_1 and a_2 are symmetric, we must only show $c_0 = c_1$. We have that $\ker \phi_2$ contains no spacelike vectors. Thus, as ϕ_2 is injective on the spacelike 2 plane $\mathrm{span}\{a_0, a_1\}$, $\phi_2(a_0)$ and $\phi_2(a_1)$ are linearly independent vectors. Choose c_{01} so that

$$\phi_1(a_0 + a_1) = c_{01}\phi_2(a_0 + a_1).$$

We show $c_0 = c_{01} = c_1$ and complete the proof by computing:

$$\begin{aligned} c_{01}\phi_2(a_0) + c_{01}\phi_2(a_1) &= c_{01}\phi_2(a_0 + a_1) = \phi_1(a_0 + a_1) \\ =&\phi_1(a_0) + \phi_1(a_1) = c_0\phi_2(a_0) + c_1\phi_2(a_1). \quad \square \end{aligned}$$

The following Lemma reduces the question of linearizations of projective maps to finding linearizations of the restrictions to hyperplanes. More precisely:

1.2.12 Lemma. *Let A and B be Riemannian vector spaces, where we have $\dim(A) \geq 5$. Suppose given a map $\Phi : \mathbb{P}(A) \to \mathbb{P}(B)$. Let $\mathcal{F}$ be a family of codimension 1 subspaces of A. Suppose that Φ is linearizable on every $F \in \mathcal{F}$ and that every 2 plane in A is contained in some element of $\mathcal{F}$. Then Φ is linearizable.*

Proof. For each $F \in \mathcal{F}$, let ϕ_F be a linearization of Φ restricted to F. Fix a hyperplane $\bar{F} \in \mathcal{F}$ to serve as a normalization. Both ϕ_F and $\phi_{\bar{F}}$ are linearizations of Φ on $F \cap \bar{F} \neq \{0\}$. Since

$$\dim(F \cap \bar{F}) \geq \dim A - 2 \geq 3,$$

we can use Lemma 1.2.11 to see that there exists a non-zero constant $c(F)$ so that

$$\phi_F = c(F)\phi_{\bar{F}} \text{ on } F \cap \bar{F}.$$

We replace ϕ_F by $c(F)^{-1}\phi_F$ to assume without loss of generality that

$$\phi_F = \phi_{\bar{F}} \text{ on } F \cap \bar{F} \text{ for all } F \in \mathcal{F}.$$

Let $F_i \in \mathcal{F}$. Since $\phi_{F_1} = \phi_{\bar{F}} = \phi_{F_2}$ on $F_1 \cap F_2 \cap \bar{F} \neq \{0\}$ and since $\dim(F_1 \cap F_2) \geq 3$, we use Lemma 1.2.11 to see $\phi_{F_1} = \phi_{F_2}$ on the (possibly) larger intersection $F_1 \cap F_2$. Thus, the map

$$\phi(a) := \phi_F(a)$$

is well defined on $\cup_{F \in \mathcal{F}} F = A$. Since every 2 plane is contained in some element of $\mathcal{F}$, ϕ is linear and provides the required linearization of Φ. □

Let $O(V)$ be the *orthogonal group* of linear transformations of V which preserve the inner product $(\cdot,\cdot)$; $O(V)$ is the isometry group of the pseudo-spheres. The associated Lie algebra $\mathfrak{so}(V)$ is the subspace of skew-symmetric linear transformations.

The next Lemma will be used to replace an indefinite metric on V by a positive definite inner product.

1.2.13 Lemma. *Let $(\cdot,\cdot)$ be an inner product of signature (p,q) on a vector space V. There exists a self-adjoint linear map ψ from V to V with $\psi^2 = \mathrm{Id}$ so that if we define $(v_1, v_2)_+ := (\psi v_1, v_2)$, then:*

(1) *$(\cdot,\cdot)_+$ is a positive definite inner product on V.*

(2) *$(v_1, v_2) = (\psi v_1, v_2)_+ = (v_1, \psi v_2)_+$ for all $v_1, v_2 \in V$.*

(3) *$(v_1, v_2)_+ = (\psi v_1, v_2) = (v_1, \psi v_2)$ for all $v_1, v_2 \in V$.*

(4) *The map $\mathcal{T} \to \psi\mathcal{T}$ defines a linear isomorphism from $\mathfrak{so}(V)$ to $\mathfrak{so}(V, (\cdot,\cdot)_+)$.*

Proof. Let $\{e_1^-, ..., e_p^-, e_1^+, ..., e_q^+\}$ be a normalized orthonormal basis for V. We define $\psi(e_i^-) = -e_i^-$ for $1 \le i \le p$ and $\psi(e_j^+) = e_j^+$ for $1 \le j \le q$. Assertions (1)-(3) are now immediate. We complete the proof of Lemma 1.2.13 by establishing the following chain of equivalent statements:

(1) $\mathcal{T} \in \mathfrak{so}(V)$.
(2) $(\mathcal{T}v_1, v_2) + (v_1, \mathcal{T}v_2) = 0$ for all $v_1, v_2 \in V$.
(3) $(\psi\mathcal{T}v_1, v_2)_+ + (v_1, \psi\mathcal{T}v_2)_+ = 0$ for all $v_1, v_2 \in V$.
(4) $\psi\mathcal{T} \in \mathfrak{so}(V, (\cdot,\cdot)_+)$. □

Let $\mathcal{T} \in \mathfrak{so}(V)$. We use the metric on V to define a correspondence $\mathcal{T} \to \omega(\mathcal{T})$, which identifies $\mathfrak{so}(V)$ with $\Lambda^2(V^*)$, by setting:

$$\omega(\mathcal{T})(v_1, v_2) := (\mathcal{T}v_1, v_2).$$

This correspondence will be crucial in studying the rank of $\mathcal{T}$.

1.2.14 Lemma. *Let V be a vector space of signature (p, q). Let $\mathcal{T}$ be a non-zero element of $\mathfrak{so}(V)$.*

(1) *We have* $\operatorname{rank}(\mathcal{T})$ *is even.*
(2) *We have* $\operatorname{rank}(\mathcal{T}) = 2$ *if and only if* $\omega(\mathcal{T}) \wedge \omega(\mathcal{T}) = 0$.

Proof. We use Lemma 1.2.13 to suppose without loss of generality that the metric on V is positive definite since

$$\operatorname{rank}(\mathcal{T}) = \operatorname{rank}(\psi\mathcal{T}) \text{ and } \omega_{(\cdot,\cdot)}(\mathcal{T}) = \omega_{(\cdot,\cdot)_+}(\psi\mathcal{T}).$$

Since $\mathcal{T}$ is skew-adjoint and since the metric is positive definite, the eigenvalues of $\mathcal{T}$ are purely imaginary and we can diagonalize $\mathcal{T}$ over $\mathbb{C}$. Consequently, we may find an orthonormal basis $\{e_i\}$ for V and positive real numbers λ_μ so that:

$$\begin{array}{llll} \mathcal{T}e_{2\mu} & = -\lambda_\mu e_{2\mu-1} & \text{for} & \mu \le \ell, \\ \mathcal{T}e_{2\mu-1} & = \lambda_\mu e_{2\mu} & \text{for} & \mu \le \ell, \text{ and} \\ \mathcal{T}e_j & = 0 & \text{for} & j > 2\ell. \end{array}$$

This shows that $\operatorname{rank}(\mathcal{T}) = 2\ell$ is even. Let $\{e_i^*\}$ be the associated dual basis for the dual vector space V^*. We complete the proof of Lemma 1.2.14 by computing:

$$\begin{array}{ll} \omega(\mathcal{T}) = \sum_{1\le\mu\le\ell} \lambda_\mu e_{2\mu-1}^* \wedge e_{2\mu}^* & \\ \omega(\mathcal{T})^r \ne 0 \text{ for } r \le \ell, & \text{and} \\ \omega(\mathcal{T})^r = 0 \text{ for } r > \ell. & \square \end{array}$$

We shall need the following result for elements of $\mathfrak{so}(V)$ of rank 2:

1.2.15 Lemma. *Let V be a vector space of signature (p,q). Let $\mathcal{T}_1$ and $\mathcal{T}_2$ be elements of $\mathfrak{so}(V)$ which have rank 2. We have:*

(1) *$\mathcal{T}_1$ is a multiple of $\mathcal{T}_2$ if and only if* $\text{range}(\mathcal{T}_1) = \text{range}(\mathcal{T}_2)$.

(2) *Let $0 \neq v \in \text{range}(\mathcal{T}_1)$. If $p = 0$, then $\{v, \mathcal{T}_1 v\}$ is an orthogonal basis for* $\text{range}(\mathcal{T}_1)$.

Proof. We use Lemma 1.2.13 to replace $\mathcal{T}_i$ by $\psi\mathcal{T}_i$ and thereby assume without loss of generality that the metric is positive definite in the proof of Assertion (1). Since $\text{rank}(\mathcal{T}_i) = 2$, $\mathcal{T}_i$ is a multiple of a 90 degree rotation in the 2 plane $\text{range}(\mathcal{T}_i)$ and vanishes on $\text{range}(\mathcal{T}_i)^\perp$. Assertions (1) and (2) now follow. □

We conclude this subsection with two final definitions we shall need presently. We can *polarize* any quadratic polynomial f defined on a vector space V to define an associated symmetric bilinear form on V by defining:

$$F(U,V) := \tfrac{1}{2}\partial_\epsilon|_{\varepsilon=0} f(U + \epsilon V).$$

We then have $f(W) = F(W,W)$. Similarly, we can polarize a cubic or quartic polynomial to define an associated totally symmetric trilinear or quadrilinear form.

Let $\psi : V \to W$ be a linear map. If $\Xi \in \otimes^k W^*$, then we define the *pull-back* $\psi^*\Xi \in \otimes^k V^*$ by setting

$$\psi^*\Xi(v_1, ..., v_k) := \Xi(\psi v_1, ..., \psi v_k).$$

In particular, ψ is an isometric embedding if and only if

$$\psi^*(\cdot,\cdot)_W = (\cdot,\cdot)_V.$$

1.3 Self-adjoint maps of a spacelike vector space

Throughout this section, let V be a vector space of dimension m with a positive definite inner product. In Lemmas 1.3.2 and 1.3.3, we give a variational principle which characterizes the eigenvalues of a self-adjoint map of V. In Lemma 1.3.4, we show the eigenvalues of a continuous 1 parameter family of self-adjoint maps vary continuously. In Lemma 1.3.5, we discuss a continuity result for the eigenvalues of a non-self adjoint map.

Let $\mathcal{B} := \{v_1, ..., v_m\}$ be an orthonormal basis for V. If A is a linear map of V, then let $Ae_i = \sum_j A_{ij} v_j$ define the matrix A_{ij} of A relative to $\mathcal{B}$. Since $\{v_i\}$ is an orthonormal basis, $A_{ij} = (Av_i, v_j)$ so

$$\text{Tr}(A) = \textstyle\sum_{1 \leq i \leq m} A_{ii} = \sum_{1 \leq i \leq m} (Av_i, v_i). \tag{1.3.1.a}$$

If $\tau \in \text{Gr}_k(V)$ is a k dimensional subspace of V, then let ρ_τ be orthogonal projection on τ; ρ_τ is a self-adjoint linear map of V such that $\rho_\tau^2 = \rho_\tau$. If $\{v_1, ..., v_k\}$ is an orthonormal basis for τ, then

$$\rho_\tau(v) = \textstyle\sum_{1 \le i \le k} (v_i, v) v_i \text{ for any } v \in V.$$

We extend the collection $\{v_1, ..., v_k\}$ to an orthonormal basis $\{v_1, ..., v_m\}$ for V. Then

$$(\rho_\tau A v_i, v_j) = (A v_i, \rho_\tau v_j) = \begin{cases} (A v_i, v_j) & \text{if } j \le k, \\ 0 & \text{if } j > k. \end{cases}$$

Thus, Equation (1.3.1.a) could be generalized to yield:

$$\text{Tr}(\rho_\tau A) = \textstyle\sum_{1 \le i \le k} (A v_i, v_i).$$

Let A be a self-adjoint map of V. We say that an orthonormal basis $\{v_1, ..., v_m\}$ for V *diagonalizes* A if

$$A v_i = \lambda_i v_i \text{ for } 1 \le i \le m.$$

Such bases always exist. We choose the notation so that $\lambda_1 \le ... \le \lambda_m$. We recall the method of *Rayleigh–Ritz* for determining the minimal eigenvalue λ_1 of a self-adjoint map.

1.3.2 Lemma. *Let V be a vector space of dimension m with a positive definite inner product. Let λ_1 be the minimal eigenvalue of a self-adjoint map A on V. Then $\lambda_1 = \min_{v \in S(V)} (Av, v)$. Furthermore, $(Av, v) = \lambda_1 |v|^2$ if and only if $Av = \lambda_1 v$.*

Proof. Let $\{v_1, ..., v_m\}$ be an orthonormal basis for V which diagonalizes A. If $v \in V$, then let $c_i := (Av, v_i)$ be the Fourier coefficients. We then have:

$$v = \textstyle\sum_i c_i v_i,\ (v, v) = \sum_i c_i^2, \text{ and } Av = \sum_i \lambda_i c_i v_i.$$

If $v \in S(V)$, then $\sum_i c_i^2 = 1$. Consequently, we have that:

$$\text{(1.3.2.a)} \qquad \lambda_1 = \textstyle\sum_i \lambda_1 c_i^2 \le \sum_i \lambda_i c_i^2 = \sum_{i,j} \lambda_i c_i c_j (v_i, v_j) = (Av, v).$$

The inequality in Display (1.3.2.a) is an equality if and only if $c_i = 0$ for $\lambda_i \ne \lambda_1$ or, equivalently, if and only if $Av = \lambda_1 v$. □

We remark that there is a similar characterization of the maximal eigenvalue. We can generalize Lemma 1.3.2 to obtain a variational principle which determines the other eigenvalues of a self-adjoint linear map.

1.3.3 Lemma. *Let V be a vector space of dimension m with a positive definite inner product. Let A be a self-adjoint linear map of V with eigenvalues $\lambda_1 \le \cdots \le \lambda_m$. Set $\Lambda_k := \lambda_1 + \cdots + \lambda_k$. Then $\Lambda_k = \min_{\tau \in Gr_k(V)} \operatorname{Tr}\{\rho_\tau A\}$. Furthermore, $\Lambda_k = \operatorname{Tr}\{\rho_\tau A\}$ if and only if A preserves τ and if the eigenvalues of A restricted to τ are $\lambda_1 \le \cdots \le \lambda_k$.*

Proof. We give a proof due to JH Park [128]. Let $\{v_1, ..., v_m\}$ be an orthonormal basis for V which diagonalizes A. Suppose first that $k = 1$. If $v \in S(V)$, then let $\tau(v) := \operatorname{span}\{v\}$. Then $\operatorname{Tr}(\rho_\tau A) = (Av, v)$ so Lemma 1.3.3 follows from Lemma 1.3.2 in this special case. We therefore suppose $k \ge 2$ and proceed by induction on k. Let

$$\sigma_k := \operatorname{span}\{v_k, ..., v_m\} \in Gr_{m+1-k}(V).$$

The subspace σ_k has codimension $k-1$ in V. Thus, for any $\tau \in \operatorname{Gr}_k(m)$, we have $\dim \sigma_k \cap \tau \ge 1$. Choose $w_k \in S(\sigma_k \cap \tau)$ and let $\tau_1 := \operatorname{span}\{w_k\}$. Since A preserves σ_k, and since the smallest eigenvalue of A on σ_k is λ_k, we apply Lemma 1.3.2 to $A|_{\sigma_k}$ to see that:

$$\lambda_k \le \operatorname{Tr}(\rho_{\tau_1} A). \tag{1.3.3.a}$$

Let $\tau_2 := \tau \cap \tau_1^\perp$. Since $\dim(\tau_2) < k$, we can use the induction hypothesis to see that:

$$\Lambda_{k-1} \le \operatorname{Tr}(\rho_{\tau_2} A). \tag{1.3.3.b}$$

Since τ is the orthogonal direct sum of τ_1 and τ_2, we have $\rho_\tau = \rho_{\tau_1} + \rho_{\tau_2}$. Thus, we may use Equations (1.3.3.a) and (1.3.3.b) to see:

$$\Lambda_k = \Lambda_{k-1} + \lambda_k \le \operatorname{Tr}(\rho_{\tau_2} A) + \operatorname{Tr}(\rho_{\tau_1} A) = \operatorname{Tr}(\rho_\tau A). \tag{1.3.3.c}$$

We let $\tau := \operatorname{span}\{v_1, ..., v_k\}$ to see that equality is possible in Equation (1.3.3.c). Conversely, suppose that we have equality in Equation (1.3.3.c). Then we must have equality in both Equations (1.3.3.a) and (1.3.3.b). Thus, A must preserve the subspaces τ_1 and τ_2 and hence it must preserve τ. Furthermore, the eigenvalue of $A|_{\tau_1}$ must be λ_k and the eigenvalues of $A|_{\tau_2}$ must be $\lambda_1 \le \cdots \le \lambda_{k-1}$. □

We can use Lemma 1.3.3 to give a proof of the following continuity result:

1.3.4 Lemma. *Let V be a vector space of dimension m with a positive definite inner product. Let $A(\varepsilon)$ be a continuous 1 parameter family of self-adjoint linear maps of V for $\varepsilon \in [0,1]$ with eigenvalues $\lambda_1(\varepsilon) \le ... \le \lambda_m(\varepsilon)$. Then the functions $\lambda_i(\varepsilon)$ are continuous functions of ε.*

Proof. We define $\Lambda_k(\varepsilon) := \lambda_1(\varepsilon) + ... + \lambda_k(\varepsilon)$ for $1 \le k \le m$. We set $\Lambda_0(\varepsilon) := 0$. To prove the Lemma, it suffices to prove the functions $\Lambda_i(\varepsilon)$ are continuous since $\lambda_i(\varepsilon) = \Lambda_i(\varepsilon) - \Lambda_{i-1}(\varepsilon)$.

Let $||$ denote the usual operator norm on the space of linear maps of V:

$$||T|| = \max_{v \in S(V)} |Tv|.$$

We then have

$$||T_1 T_2|| \le ||T_1|| \cdot ||T_2|| \text{ and } |\operatorname{Tr}(T)| = |\textstyle\sum_i (Te_i, e_i)| \le m||T||.$$

Let $\epsilon > 0$ be given. Since the family $A(\varepsilon)$ is continuous, we choose $\delta > 0$ so that if $|\varepsilon_1 - \varepsilon_2| < \delta$, then $||A(\varepsilon_1) - A(\varepsilon_2)|| < \epsilon$. We suppose $|\varepsilon_1 - \varepsilon_2| < \delta$ henceforth. If $\tau \in \operatorname{Gr}_k(V)$, then $||\rho_\tau|| = 1$. Choose $\tau \in \operatorname{Gr}_k(V)$ so that

$$\Lambda_k(\varepsilon_2) = \operatorname{Tr}\{\rho_\tau A(\varepsilon_2)\}.$$

We may then use Lemma 1.3.3 to estimate:

$$\begin{aligned}
&\Lambda_k(\varepsilon_1) - \Lambda_k(\varepsilon_2) \le \operatorname{Tr}\{\rho_\tau A(\varepsilon_1)\} - \Lambda_k(\varepsilon_2) \\
&= \operatorname{Tr}\{\rho_\tau A(\varepsilon_1)\} - \operatorname{Tr}\{\rho_\tau A(\varepsilon_2)\} \le |\operatorname{Tr}\{\rho_\tau (A(\varepsilon_1) - A(\varepsilon_2))\}| \\
&\le m||\rho_\tau|| \cdot ||A(\varepsilon_1) - A(\varepsilon_2)|| \le m\epsilon.
\end{aligned}$$

Since the roles of ε_1 and ε_2 are symmetric, we have similarly

$$\Lambda_k(\varepsilon_2) - \Lambda_k(\varepsilon_1) \le m\epsilon. \quad \square$$

Let A be a linear map of a vector space V; we no longer assume that A is self-adjoint. Let $p_A(\lambda) := \det(A - \lambda I)$ be the associated characteristic polynomial. The multiplicity μ of an eigenvalue can then be computed using Cauchy's integral formula:

$$\mu(A, \lambda) = \tfrac{1}{2\pi i} \int_{|\sigma - \lambda| = \epsilon} \tfrac{p_A'}{p_A}(\sigma) d\sigma$$

over any sufficiently small circle about the eigenvalue λ in question. The following Lemma is now immediate:

1.3.5 Lemma. *Let $A(t)$ be a continuous 1 parameter of maps of a vector space V. Fix t_0. There exists $\epsilon > 0$ and $\delta > 0$ so that if $|t - t_0| < \delta$, then*

$$\sum_{\{\lambda : |\lambda - \lambda_0| < \epsilon\}} \mu(A(t), \lambda) = \mu(A(t_0), \lambda_0).$$

1.4 Clifford algebras and matrices

1.4.1 Definition. We define the *Adams number* $\nu(m)$ by setting:

$$\begin{aligned} &\nu(1) = 0,\ \nu(2) = 1,\ \nu(4) = 3,\ \nu(8) = 7, \\ &\nu(16m) = \nu(m) + 8,\ \text{and} \\ &\nu(a2^s) = \nu(2^s)\ \text{for } a \text{ odd.} \end{aligned} \tag{1.4.1.a}$$

Let S^{m-1} be the unit sphere in $\mathbb{R}^m$, where $\mathbb{R}^m$ is equipped with the usual positive definite inner product. We refer to Adams [1] for the proof of the following result:

1.4.2 Theorem. *Suppose given a non-trivial decomposition of the tangent bundle* $T(S^{m-1}) = E_0 \oplus ... \oplus E_\ell$ *as an orthogonal direct sum of vector bundles of dimension* $\mu_i := \dim(E_i)$*, where* $\mu_0 \geq ... \geq \mu_\ell$*. Then* $\mu_1 + ... + \mu_\ell \leq \nu(m)$.

We use Clifford algebras to show that the estimate given in Theorem 1.4.2 is sharp.

1.4.3 Lemma. *We have that:*

(1) *There exist skew-adjoint linear maps* $\{c_1, ..., c_{\nu(m)}\}$ *on* $\mathbb{R}^m$ *so that* $c_i c_j + c_j c_i = -2\delta_{ij} \operatorname{Id}$ *for* $1 \leq i, j \leq \nu(m)$.

(2) *There exists a decomposition of* $T(S^{m-1}) = E_0 \oplus E_1 \oplus ... \oplus E_{\nu(m)}$*, where* $\dim E_i = 1$ *for* $1 \leq i \leq \nu(m)$.

Proof. If m is odd, then $\nu(m) = 0$ and there is nothing to prove. We therefore suppose m even so $\nu(m) \geq 1$.

Let $\{e_1, ..., e_k\}$ be the standard orthonormal basis for $\mathbb{R}^k$. The *Clifford algebra* $\operatorname{Cliff}(k)$ is the universal unital algebra generated by $\mathbb{R}^k$ subject to the *Clifford commutation relations*

$$v * w + w * v = -2(v, w) \operatorname{Id} \ \text{for}\ v, w \in \mathbb{R}^k.$$

Let $\mathbb{H}$ be the quaternions. Let $M_k(\mathbb{A})$ be the set of $k \times k$ matrices over a unital algebra $\mathbb{A}$. We have the following structure theorem due to Atiyah, Bott, and Shapiro [10]:

1.4.4 Theorem.

$$\begin{aligned} &\operatorname{Cliff}(0) = \mathbb{R}, && \operatorname{Cliff}(1) = \mathbb{C}, \\ &\operatorname{Cliff}(2) = \mathbb{H}, && \operatorname{Cliff}(3) = \mathbb{H} \oplus \mathbb{H}, \\ &\operatorname{Cliff}(4) = \mathbb{M}_2(\mathbb{H}), && \operatorname{Cliff}(5) = \mathbb{M}_4(\mathbb{C}), \\ &\operatorname{Cliff}(6) = \mathbb{M}_8(\mathbb{R}), && \operatorname{Cliff}(7) = \mathbb{M}_8(\mathbb{R}) \oplus \mathbb{M}_8(\mathbb{R}), \\ &\operatorname{Cliff}(8) = \mathbb{M}_{16}(\mathbb{R}), && \operatorname{Cliff}(k+8) = \mathbb{M}_{16}(\operatorname{Cliff}(k)). \end{aligned}$$

We use Theorem 1.4.4 to see that there exist unital representations

$$\begin{aligned}c &: \mathrm{Cliff}(1) = \mathbb{C} \to \mathbb{M}_2(\mathbb{R}),\\ c &: \mathrm{Cliff}(3) = \mathbb{H} \oplus \mathbb{H} \to \mathbb{M}_4(\mathbb{R}),\\ c &: \mathrm{Cliff}(7) = \mathbb{M}_8(\mathbb{R}) \oplus \mathbb{M}_8(\mathbb{R}) \to \mathbb{M}_8(\mathbb{R}),\\ c &: \mathrm{Cliff}(8) = \mathbb{M}_{16}(\mathbb{R}) \to \mathbb{M}_{16}(\mathbb{R}).\end{aligned}$$

Thus, if $m = 2^s a$ for $s = 1, 2, 3, 4$ and a odd, then there exists a unital representation of $\mathrm{Cliff}(\nu(m))$ on $\mathbb{R}^m$. As $M_{16}(\mathrm{Cliff}(k)) = \mathrm{Cliff}(k+8)$ and as $\nu(16m) = \nu(m) + 8$, there exists a unital representation of $\mathrm{Cliff}(\nu(m))$ on $\mathbb{R}^m$ for any m.

Set $c_i = c(e_i)$, where $\{e_1, ..., e_{\nu(m)}\}$ is the usual orthonormal basis for $\mathbb{R}^{\nu(m)}$. We then have $c_i c_j + c_j c_i = -2\delta_{ij}$. Let $\mathbb{G}$ be the finite subgroup of $\mathbb{GL}_m(\mathbb{R})$ generated by the elements c_i for $1 \le i \le \nu(m)$. By averaging the standard positive definite metric on $\mathbb{R}^m$ over this group, we can find a new positive definite metric on $\mathbb{R}^m$ so that the c_i act by isometries; any two positive definite metrics on $\mathbb{R}^m$ are equivalent so we replace the original metric by this new metric. Since $c_i^2 = -\mathrm{Id}$, this implies $c_i^* = -c_i$ and completes the proof of Assertion (1).

To prove Assertion (2), we let the maps $\{c_i\}$ be as in Assertion (1). Let $x \in S^{m-1}$. We show that $\{x, c_1x, ..., c_{\nu(m)}x\}$ is an orthonormal set of vectors by computing:

$$\begin{aligned}(c_ix, c_ix) &= -(c_ic_ix, x) = (x, x) = 1, \ (x, c_ix) = -(c_ix, x) = 0,\\ (c_ix, c_jx) &= -(c_ic_jx, x) = (c_jc_ix, x) = -(c_ix, c_jx) = 0 \text{ for } i \ne j.\end{aligned}$$

We complete the proof of Assertion (2) by defining:

$$\begin{aligned}E_i(x) &:= \mathrm{span}\{c_ix\} \text{ for } 1 \le i \le \nu(m) \text{ and}\\ E_0 &:= \{E_1 \oplus ... \oplus E_{\nu(m)}\}^\perp. \quad \square\end{aligned}$$

We now consider vector spaces of arbitrary signature.

1.4.5 Lemma.

(1) *Let $\{\phi_1, ..., \phi_\nu\}$ be skew-symmetric linear maps of a vector space V of signature (p, q) so that $\phi_i^2 = \pm\mathrm{Id}$ and so that $\phi_i\phi_j + \phi_j\phi_i = 0$ for $i \ne j$. If $x \in S^\pm(V)$, then $\{x, \phi_1x, ..., \phi_\nu x\}$ is an orthonormal set.*

(2) *There exist skew-adjoint linear maps of $\mathbb{R}^{(2,2)}$ so that $\phi_1^2 = \mathrm{Id}$, so that $\phi_2^2 = \mathrm{Id}$, so that $\phi_3^2 = -\mathrm{Id}$, and so that $\phi_i\phi_j + \phi_j\phi_i = 0$ for $i \ne j$.*

(3) *Let $\ell \equiv 0$ mod 2^{p+1}. Suppose given constants $\varepsilon_i = \pm 1$ for $0 \le i \le 2p$. There exist skew-symmetric linear maps $\{\phi_0, \phi_1, ..., \phi_{2p}\}$ of $\mathbb{R}^{(\ell,\ell)}$ so that $\phi_i^2 = -\varepsilon_i\,\mathrm{Id}$ and so that $\phi_i\phi_j + \phi_j\phi_i = 0$ for $i \ne j$.*

Proof. Let $\{\phi_1, ..., \phi_\nu\}$ satisfy the assumptions of Assertion (1). We prove Assertion (1) by computing:

$$(x, \phi_i x) = -(\phi_i x, x) = 0, \ (\phi_i x, \phi_i x) = -(\phi_i^2 x, x) = \pm 1, \text{ and}$$
$$(\phi_i x, \phi_j x) = -(\phi_j \phi_i x, x) = (\phi_i \phi_j x, x) = -(\phi_j x, \phi_i x) = 0 \text{ for } i \neq j.$$

We introduce some additional notation before beginning the proof of Assertions (2) and (3). Let

$$\alpha_0 := \begin{pmatrix} 1 & 0 \\ 0 & -1 \end{pmatrix}, \ \alpha_1 := \begin{pmatrix} 0 & 1 \\ 1 & 0 \end{pmatrix}, \text{ and } \alpha_2 := \alpha_0 \alpha_1 = \begin{pmatrix} 0 & 1 \\ -1 & 0 \end{pmatrix}.$$

We then have that

$$\alpha_i \alpha_j + \alpha_j \alpha_i = 0 \text{ for } i \neq j,$$
$$\alpha_0^2 = \mathrm{Id}, \ \alpha_1^2 = \mathrm{Id}, \text{ and } \alpha_2^2 = -\mathrm{Id}.$$

The adjoints are given by:

$$\alpha_0^* = \alpha_0, \ \alpha_1^* = -\alpha_1, \ \alpha_2^* = \alpha_2 \text{ on } \mathbb{R}^{(1,1)}$$
$$\alpha_0^* = \alpha_0, \ \alpha_1^* = \alpha_1, \ \alpha_2^* = -\alpha_2 \text{ on } \mathbb{R}^{(0,2)}.$$

We identify $\mathbb{R}^{(2,2)} = \mathbb{R}^{(1,1)} \otimes \mathbb{R}^{(0,2)}$ and prove Assertion (2) by taking

$$\phi_1 := \alpha_2 \otimes \alpha_2, \ \phi_2 := \alpha_1 \otimes \mathrm{Id}, \text{ and } \phi_3 := \alpha_0 \otimes \alpha_2.$$

If $k = 1$, then we set $\alpha_{i,1} = \alpha_i$ for $0 \leq i \leq 2$. For $k > 1$, we use induction to define the family $\{\alpha_{0,k}, ..., \alpha_{2k,k}\}$ of $2^k \times 2^k$ matrices by setting:

$$\alpha_{i,k} := \alpha_{i,k-1} \otimes \mathrm{Id}, \text{ for } i < 2k-2 \text{ and}$$
$$\alpha_{2k-2+j,k} := \alpha_{2k-2,k-1} \otimes \alpha_j \text{ for } 0 \leq j \leq 2.$$

We then have the Clifford commutation relations:

$$\alpha_{i,k}\alpha_{j,k} + \alpha_{j,k}\alpha_{i,k} = 0 \text{ for } 0 \leq i < j \leq 2k.$$

We set $k = p$ to ensure we have $2p + 1$ Clifford matrices which act on $\mathbb{R}^{(2^{p-1}, 2^{p-1})}$. Let $\ell = 2^{p+1} \cdot \bar{\ell}$. Let

$$\phi_i := \alpha_{i,p} \otimes \beta_i \otimes \gamma_i \otimes \mathrm{Id} \text{ on}$$
$$\mathbb{R}^{(\ell,\ell)} = \mathbb{R}^{(2^{p-1}, 2^{p-1})} \otimes \mathbb{R}^{(1,1)} \otimes \mathbb{R}^{(1,1)} \otimes \mathbb{R}^{\bar{\ell}}.$$

We set $\beta_i = \text{Id}$ if $\alpha_{i,p}$ is skew-adjoint and $\beta_i = \alpha_1$ if $\alpha_{i,p}$ is self-adjoint to ensure that ϕ_i is skew-adjoint. We set $\gamma_i = \text{Id}$ or $\gamma_i = \alpha_2$ to ensure that $\phi_i^2 = -\varepsilon_i \text{Id}$. □

If V is a vector space of arbitrary signature (r, s), then the *Clifford algebra* $\text{Cliff}(V)$ is the universal unital algebra generated by V subject to the relations

$$v * w + w * v = -2(v, w)\,\text{Id} \text{ for all } v, w \in V.$$

Let $2p+1 \geq r+s$ and let $\ell \equiv 0 \bmod 2^{p+1}$. Lemma 1.4.5 shows that $\text{Cliff}(V)$ has a unital representation on $\mathbb{R}^{(\ell,\ell)}$. This bound for the powers of 2 which appear need not be optimal; the structure of these Clifford algebras is given in Karoubi [109] and can often be used to provide better bounds.

Let A and B be Riemannian vector spaces. A bilinear map T from $A \otimes A$ to $\mathfrak{so}(B)$ is said to be *alternating* if $T(a_1, a_2) = -T(a_2, a_1)$. Such a map is said to be a *bilinear Clifford module structure* if $T(a_1, a_2)^2 = -\text{Id}$ for any orthonormal set $\{a_1, a_2\}$. Such structures will play an important role the discussion of four and eight dimensional geometry which will be given in Sections 2.8 and 2.10. If we fix a_1, then the map $T : a_2 \to T(a_1, a_2)$ gives B a $\text{Cliff}(a_1^\perp)$ module structure.

We note that if $c : A \to \mathfrak{so}(B)$ gives B a Clifford module structure, then we may define a bilinear Clifford module structure by defining

$$c(\xi, \eta) := \tfrac{1}{2}\{c(\xi)c(\eta) - c(\eta)c(\xi)\}.$$

However, not every bilinear Clifford module structure arises in this way from a Clifford module structure.

1.5 Natural operators

Let M be a smooth connected manifold of dimension m. We say that g is a *pseudo-Riemannian metric* of signature (p, q) on M if g is a smooth non-degenerate inner product of signature (p, q) on the tangent bundle TM; (M, g) is then said to be a *pseudo-Riemannian manifold* of signature (p, q). We are interested in local theory so we can take M to be a small neighborhood of a point P in Euclidean space and let g be the germ of a metric defined near P. Let ${}^g\nabla$ be the Levi-Civita connection of the metric g; ${}^g\nabla$ is characterized by the identities:

(1.5.1.a) $${}^g\nabla_x y - {}^g\nabla_y x - [x, y] = 0, \text{ and}$$
(1.5.1.b) $$x g(y, z) = g({}^g\nabla_x y, z) + g(y, {}^g\nabla_x z)$$

for any smooth vector fields x, y, and z. Equation (1.5.1.a) means that ${}^g\nabla$ is *torsion free* and Equation (1.5.1.b) means that ${}^g\nabla$ is *pseudo-Riemannian.* The *associated curvature operator* and *curvature tensor* are then given by:

(1.5.1.c) $\quad {}^gR(x,y) := {}^g\nabla_x{}^g\nabla_y - {}^g\nabla_y{}^g\nabla_x - {}^g\nabla_{[x,y]}$, and

(1.5.1.d) $\quad {}^gR(x,y,z,w) := g({}^gR(x,y)z,w)$.

This operator is *tensorial* i.e. if f is a smooth function on M, then:

$$\begin{aligned} f\cdot{}^gR(x,y,z,w) &= {}^gR(fx,y,z,w) = {}^gR(x,fy,z,w)\\ &= {}^gR(x,y,fz,w) = {}^gR(x,y,z,fw). \end{aligned}$$

This means that the value of ${}^gR(x,y,z,w)$ at a point $P \in M$ is determined by the values $x(P)$, $y(P)$, $z(P)$, and $w(P)$. Thus, we may regard ${}^gR$ as defining an element of $\otimes^4 T^*M$.

We have the following symmetries of the curvature tensor (see Lemma 1.12.1 for details):

(1.5.1.e) $\quad {}^gR(x,y,z,w) = -{}^gR(y,x,z,w) = -{}^gR(x,y,w,z)$,

(1.5.1.f) $\quad {}^gR(x,y,z,w) = {}^gR(z,w,x,y)$, and

(1.5.1.g) $\quad {}^gR(x,y,z,w) + {}^gR(y,z,x,w) + {}^gR(z,x,y,w) = 0$.

Equation (1.5.1.e) means that ${}^gR$ is skew-symmetric in the variables (x,y) and (z,w) and hence that ${}^gR(x,y) \in \mathfrak{so}(TM)$. Equation (1.5.1.f) means that ${}^gR$ is symmetric in the pairs (x,y) and (z,w); Equation (1.5.1.g) is the *first Bianchi identity.*

It is convenient to work in a purely algebraic setting. Let V be a vector space with an inner product of signature (p,q). A tensor $R \in \otimes^4 V^*$ is said to be an *algebraic curvature tensor* if R satisfies the symmetries given in Equations (1.5.1.e), (1.5.1.f), and (1.5.1.g). We use Equation (1.5.1.d) to define the associated curvature operator $R(v_1,v_2) \in \mathfrak{so}(V)$.

A pseudo-Riemannian manifold (M,g) is said to be a *geometric realization* of an algebraic curvature tensor R at a point P of M if there exists an isometry Ψ from the tangent space T_PM to M at P to the vector space V so that $\Psi^*R = {}^gR_P$ on T_PM:

$$\begin{aligned} &(\Psi x, \Psi y) = (x,y) \text{ for all } x,y \in T_PM, \text{ and}\\ &R(\Psi x,\Psi y,\Psi z,\Psi w) = {}^gR(x,y,z,w) \text{ for all } x,y,z,w \in T_PM. \end{aligned}$$

Every algebraic curvature tensor is geometrically realizable by the germ of a pseudo-Riemannian metric - see Theorem 1.12.2 for details. Thus, the algebraic curvature tensors are important in differential geometry.

Since the curvature tensor encodes much of the geometry of the manifold, it can be a very complicated object to investigate. There are natural operators associated to R which are useful to study. One wants to know what are the geometric consequences if the eigenvalues or more generally the Jordan form (i.e. conjugacy class) of such an operator are constant on the natural domain of definition.

Let R be an algebraic curvature tensor on a vector space V of signature (p, q). Let ${}^gR$ be the curvature tensor of the Levi-Civita connection on a pseudo-Riemannian manifold (M, g).

1.5.2 Definition. The *Jacobi operator*

$$\mathcal{J}_R(x) : y \to R(y, x)x \tag{1.5.2.a}$$

is a self-adjoint operator which is important in the study of geodesic variations; see Section 1.11 for details. The natural domains of the Jacobi operator are the pseudo-spheres $S^{\pm}(V)$ and pseudo-sphere bundles $S^{\pm}(M, g)$ of unit spacelike (+) and timelike (−) vectors.

Suppose $p = 0$ so we are in the Riemannian setting. If (M, g) is a rank 1 symmetric space or is flat, then the local isometries of (M, g) act transitively on the unit sphere bundle $S(M, g)$ and thus the eigenvalues of $\mathcal{J}_R$ are constant on $S(M, g)$. Osserman [126] wondered if the converse was true - i.e. if the eigenvalues of $\mathcal{J}_R$ are constant on $S(M, g)$, then is (M, g) locally a rank 1 symmetric space or is flat.

This question has been called *the Osserman conjecture* by later authors and has been generalized to the higher signature setting. We refer to Section 3.1 for further details.

We say that an algebraic curvature tensor R is *spacelike or timelike Osserman* if the eigenvalues of $\mathcal{J}_R$ are constant on the pseudospheres $S^{\pm}(V)$. In Lemma 1.10.1, we will show that these two notions coincide and we will simply speak of Osserman algebraic curvature tensors. Similarly, we will say that a pseudo-Riemannian manifold is *Osserman* if the eigenvalues of $\mathcal{J}_R$ are constant on $S^{\pm}(M, g)$. We do not permit the eigenvalues to vary with the point of the manifold.

In the Riemannian setting ($p = 0$), $\mathcal{J}_R$ is diagonalizable so the eigenvalues determine the operator up to conjugacy. In the higher signature setting, this need not be the case. We say that an algebraic curvature tensor R or that a pseudo-Riemannian manifold (M, g) is *spacelike Jordan Osserman* (resp. *timelike Jordan Osserman*) if the Jordan normal form of $\mathcal{J}_R$ is constant on $S^+(V)$ or $S^+(M, g)$ (resp. on $S^-(V)$ or $S^-(M, g)$). Equivalently, this means $\mathcal{J}_R(x_1)$ and $\mathcal{J}_R(x_2)$ are conjugate linear transformations for every pair of unit spacelike (resp. timelike) vectors x_1 and x_2. In Theorem 3.2.2, we will show that these are inequivalent notions. We will also show that there exist Osserman algebraic curvature tensors which are neither spacelike Jordan Osserman nor timelike Jordan Osserman.

1.5.3 Definition. Stanilov and Videv [144] have defined the *higher order Jacobi operator*. Let $\mathcal{B} := \{e_1, ..., e_k\}$ be a basis for a non-degenerate k plane σ in V. Let $h_{ij} := (e_i, e_j)$ for $1 \leq i, j \leq k$ give the components (with respect to the basis $\mathcal{B}$) of the metric restricted to V. Since σ is non-degenerate, $\det(h_{ij}) \neq 0$ and we can let h^{ij} be the inverse matrix. Let

$$\text{(1.5.3.a)} \qquad \mathcal{J}_R(\sigma)y := \textstyle\sum_{i,j} h^{ij} R(y, e_i)e_j.$$

We will show in Theorem 1.9.4 that $\mathcal{J}_R(\sigma)$ is independent of the particular basis $\mathcal{B}$ which was chosen and furthermore, that if σ is spacelike, then $\mathcal{J}_R(\sigma)$ can be regarded as the average Jacobi operator:

$$\text{(1.5.3.b)} \qquad \mathcal{J}_R(\sigma) = \frac{k}{\mathrm{vol}(S^{k-1})} \int_{x\in\sigma,\ |x|=1} \mathcal{J}_R(x)dx.$$

We say that (r, s) is an *admissible pair* if $\mathrm{Gr}_{r,s}(V)$ is non empty and does not consist of a single point. Equivalently, this means that

$$0 \leq r \leq p,\ 0 \leq s \leq q, \text{ and } 1 \leq r + s \leq p + q - 1.$$

We say that R is *Osserman of type* (r, s) if the eigenvalues of $\mathcal{J}_R$ are constant on $\mathrm{Gr}_{r,s}(V)$. Let (r, s) and $(\tilde{r}, \tilde{s})$ be admissible pairs such that we have $r + s = \tilde{r} + \tilde{s} = k$. In Lemma 1.10.2, we will show that if R is Osserman of type (r, s), then R is Osserman of type $(\tilde{r}, \tilde{s})$. Thus, only the value k is relevant so we shall simply say that an algebraic curvature tensor is *k Osserman*. We say that R is *Jordan Osserman of type (r,s)* if the Jordan normal form of $\mathcal{J}_R$ is constant on the unoriented Grassmannian of planes of type (r, s). In Sections 3.2 and 3.3, we shall present examples that show that these are inequivalent notions; we will also construct examples of algebraic curvature tensors which are k-Osserman but which are not Jordan Osserman of type (r, s).

1.5.4 Definition. The *Szabó operator*, in contrast to the other operators we shall be discussing, is defined by the covariant derivative of the curvature tensor, $\nabla^g R \in \otimes^5 TM$. Let:

$$\text{(1.5.4.a)} \qquad \mathcal{S}_{^g\nabla R}(x) : y \to (\nabla_x{}^g R)(y, x)x.$$

In the algebraic setting, we shall say that a 5 tensor $\nabla R \in \otimes^5 V^*$ is an *algebraic covariant derivative* curvature tensor if it satisfies the symmetries of the covariant derivative of the Riemann curvature tensor (see Lemma 1.12.1 for details):

$$(1.5.4.b)\quad \nabla R(a,b,c,d;e) = -\nabla R(b,a,c,d;e) = \nabla R(c,d,a,b;e)$$
$$(1.5.4.c)\quad \nabla R(a,b,c,d;e) + \nabla R(a,c,d,b;e) + \nabla R(a,d,b,c;e) = 0$$
$$(1.5.4.d)\quad \nabla R(a,b,c,d;e) + \nabla R(a,b,d,e;c) + \nabla R(a,b,e,c;d) = 0.$$

Equation (1.5.4.c) is the covariant derivative of the first Bianchi identity and Equation (1.5.4.d) is the second Bianchi identity. Here, the notation $\nabla R \in \otimes^5 V^*$ is to be understood in a purely formal sense as there is, after all, no covariant derivative in the algebraic setting.

The corresponding curvature operator $\nabla_e R(a,b)c$ and Szabó operator $\mathcal{S}_{\nabla R}(x)$ are defined by the identities:

$$(\nabla_e R(a,b)c, d) = \nabla R(a,b,c,d;e) \text{ and } \mathcal{S}_{\nabla R}(x)y = \nabla_x R(y,x)x.$$

We say that an algebraic covariant derivative curvature tensor ∇R or that a pseudo-Riemannian manifold (M,g) is *spacelike Szabó* (resp. *timelike Szabó*) if the eigenvalues of $\mathcal{S}_{\nabla R}$ are constant on $S^+(V)$ or $S^+(M,g)$ (resp. $S^-(V)$ or $S^-(M,g)$). In Lemma 1.10.9 we will show that these two notions are equivalent and hence we shall simply talk of Szabó algebraic curvature tensors and metrics.

Similarly, we say that ∇R or that (M,g) is *spacelike Jordan Szabó* (resp. *timelike Jordan Szabó*) if the Jordan normal form of $\mathcal{S}_{\nabla R}$ is constant on the set of unit spacelike (resp. timelike) vectors.

Let (M,g) be a Riemannian manifold. We say that (M,g) is a 2 *point homogeneous space* if the local isometries of (M,g) act transitively on the unit sphere bundle $S(M,g)$. This implies that the eigenvalues of $\mathcal{S}_{\nabla R}$ are constant on $S(M,g)$. Szabó [150] showed this implied $\nabla R = 0$ and used this observation to give an elementary proof of the fact that any 2 point homogeneous Riemannian manifold either is flat or is a rank 1 symmetric space. For this reason, the operator has become known in this context as the Szabó operator.

In Theorem 3.1.9, we will show that a Szabó Lorentzian algebraic curvature tensor is trivial. This result fails in the higher signature setting as we shall show in Theorem 3.1.10.

1.5.5 Definition. If $\{x,y\}$ is an oriented basis for a non-degenerate 2 plane π, then the *skew-symmetric curvature operator*

$$(1.5.5.a)\qquad R(\pi) := |(x,x)(y,y) - (x,y)^2|^{-1/2} R(x,y)$$

is a skew-adjoint operator which is independent of the particular oriented basis chosen for π; see Lemma 1.9.1 for details. The natural domains of definition for the skew-symmetric curvature operator $R(\pi)$ are the oriented Grassmannians $\mathrm{Gr}^+_{0,2}(V)$, $\mathrm{Gr}^+_{1,1}(V)$, and $\mathrm{Gr}^+_{0,2}(V)$ of oriented spacelike, mixed, or timelike 2 planes, respectively. We say that R is *spacelike, mixed, or timelike IP* if the eigenvalues of $R(\pi)$ are constant on $\mathrm{Gr}^+_{0,2}(V)$, $\mathrm{Gr}^+_{1,1}(V)$, or $\mathrm{Gr}^+_{0,2}(V)$, respectively. In Lemma 1.10.5, we will use analytic continuation to show that these are equivalent notions and hence we shall simply speak of IP algebraic curvature tensors. The notation 'IP' has been used by many authors since Ivanov and Petrova [97] classified these tensors and metrics in signature $(0,4)$.

Again, in the higher signature setting, the eigenvalues do not determine the operator up to conjugacy so we say that an algebraic curvature tensor R is *spacelike, mixed, or timelike Jordan IP* if the Jordan normal form of the skew-symmetric curvature operator is constant on the appropriate Grassmannian. In Section 2.2, we will present various examples to show that these are different notions; we will also show that there exist IP algebraic curvature tensors which are not Jordan IP.

We say that a pseudo-Riemannian manifold (M,g) is *spacelike, mixed, or timelike (Jordan) IP* if the associated algebraic curvature tensor ${}^gR$ is spacelike, mixed, or timelike (Jordan) IP at each point of M. In contrast to the situation with the Jacobi operator, we permit the eigenvalues or Jordan form of the skew-symmetric curvature operator to vary with the point of the manifold in question.

1.5.6 Definition. Let J be a pseudo-Hermitian almost complex structure on V. An algebraic curvature tensor R is said to be an *almost complex algebraic curvature tensor* if

$$JR(\pi) = R(\pi)J$$

for every non-degenerate complex line π. We refer to Lemma 1.6.2 for equivalent formulations of this property.

The natural domains of definition for the skew-symmetric curvature operator $R(\pi)$ in the complex setting are the projective spaces of spacelike and timelike complex lines; there are no complex lines of mixed type. Let R be almost complex. We say that an algebraic curvature tensor R is *almost complex spacelike IP* or *almost complex timelike IP* if the eigenvalues of the skew-symmetric curvature operator defined by R are constant on the appropriate projective spaces. We will show in Lemma 1.10.7 that the type of the complex line plays no role; almost complex spacelike IP and almost complex timelike IP are equivalent notions. Thus, we will simply speak of *almost complex IP* algebraic curvature tensors.

Again, the Jordan form plays a crucial role. We say that R is *almost complex spacelike Jordan IP* or *almost complex timelike Jordan IP* if the Jordan normal form of an almost complex algebraic curvature tensor R is constant on the projective space of spacelike or timelike complex lines. In Section 2.11, we will show that these notions are distinct. We will also construct examples of algebraic curvature tensors which are almost complex IP but which are neither almost complex spacelike Jordan IP nor almost complex timelike Jordan IP.

If (M, g, J) is a pseudo-Hermitian almost complex manifold, then we say that (M, g, J) is *almost complex IP*, *almost complex spacelike Jordan IP*, or *almost complex timelike Jordan IP* if the associated curvature tensor ${}^gR$ is almost complex IP, almost complex spacelike Jordan IP, or almost complex timelike Jordan IP at each point P of the manifold; the eigenvalues or Jordan normal form are allowed to vary with the point in question.

1.5.7 Definition. Stanilov [141] has also defined a higher order skew-symmetric curvature operator analogously to the higher order Jacobi operator discussed in Definition 1.5.3. Let R be an algebraic curvature tensor. Let $\{e_1, ..., e_k\}$ be a basis for a non-degenerate k dimensional subspace $\sigma \subset V$. Let $h_{ij} := (e_i, e_j)$ describe the metric on σ and let h^{ij} be the inverse matrix. Define:

$$\text{(1.5.7.a)} \qquad \mathfrak{S}_R(\sigma) := \textstyle\sum_{i,j,k,l} h^{ik}h^{jl}R(e_i, e_j)R(e_k, e_l);$$

we will show in Theorem 1.9.4 that this operator is independent of the particular basis chosen for σ; furthermore, if σ is positive definite, then we will also show that

$$\text{(1.5.7.b)} \qquad \mathfrak{S}_R(\sigma) := \frac{k(k-1)}{\text{vol}(Gr_2(k))} \textstyle\int_{\pi\subset\sigma} R(\pi)^2.$$

Consequently, $\mathfrak{S}_R(\sigma)$ can be regarded as the average value of $R(\cdot)^2$ on the unoriented Grassmannian of 2 planes in σ. We remark that it is necessary to square the skew-symmetric curvature operator to get a non-trivial endomorphism.

We say that R is *IP of type* (r, s) if the eigenvalues of $\mathfrak{S}_R$ are constant on the unoriented Grassmannian $\text{Gr}_{r,s}(V)$. In Lemma 1.10.8 we will show that these are equivalent notions for $0 \le r \le p, 0 \le s \le q$ and $r + s = k$; only the sum $k = r + s$ plays a crucial role and hence we shall simply say that an algebraic curvature tensor is k *IP*. We say that R is *Jordan IP of type* (r, s) if the Jordan normal form of $\mathfrak{S}_R$ is constant on $\text{Gr}_{r,s}(V)$. In Section 2.12, we shall present two families of such algebraic curvature tensors.

1.6 Algebraic curvature tensors

In this section, we establish some of the basic properties of the algebraic curvature tensors. Let R be an algebraic curvature tensor on a vector space V of signature (p, q).

1.6.1 Definition. We define $R_\Lambda : \Lambda^2(V) \to \Lambda^2(V)$ by the identity:

$$(R_\Lambda(x \wedge y), z \wedge w) = R(x, y, z, w). \tag{1.6.1.a}$$

Since $R(x, y, z, w) = R(z, w, x, y)$, R_Λ is self-adjoint.

Let J be a pseudo-Hermitian almost complex structure on V. We extend J to $\Lambda^2(V)$ by linearity. The following Lemma provides several useful characterizations of the almost complex algebraic curvature tensors.

1.6.2 Lemma. *Let R be an algebraic curvature tensor on a vector space V of signature (p, q). Let J be pseudo-Hermitian almost complex structure on V. Then the following assertions are equivalent and all define the notion of an almost complex algebraic curvature tensor:*

(1) *If $q \geq 2$, then $JR(\pi) = R(\pi)J$ for every spacelike complex line π.*
(2) *If $p \geq 2$, then $JR(\pi) = R(\pi)J$ for every timelike complex line π.*
(3) *$JR(x, Jx) = R(x, Jx)J$ for all x in V.*
(4) *$J^*R = R$, i.e. $R(x, y, z, w) = R(Jx, Jy, Jz, Jw)$ $\forall\ x, y, z, w \in V$.*
(5) *$JR_\Lambda = R_\Lambda J$ on $\Lambda^2(V)$.*

Proof. Suppose that Assertion (1) holds. Let x be a spacelike vector. We define the associated spacelike complex line $\pi := \operatorname{span}\{x, Jx\}$. Then

$$R(\pi) = |x|^{-2} R(x, Jx),$$

so $JR(x, Jx) = R(x, Jx)J$ for any spacelike x. Let y be arbitrary and let x be spacelike. Then for large values of ε, $y + \varepsilon x$ is spacelike and hence

$$\begin{aligned}
&J\{R(y, Jy) + \varepsilon R(y, Jx) + \varepsilon R(x, Jy) + \varepsilon^2 R(x, Jx)\} \\
=&JR(y + \varepsilon x, J(y + \varepsilon x)) = R(y + \varepsilon x, J(y + \varepsilon x))J \\
=&\{R(y, Jy) + \varepsilon R(y, Jx) + \varepsilon R(x, Jy) + \varepsilon^2 R(x, Jx)\}J.
\end{aligned} \tag{1.6.2.a}$$

Since Equation (1.6.2.a) is a quadratic identity in ε which holds for an infinite number of values of the parameter ε, it holds for all values of ε and in particular for $\varepsilon = 0$. Thus, $JR(y, Jy) = R(y, Jy)J$ for all y and Assertion

(3) holds. It is immediate that (3) implies (1). Similarly, Assertions (2) and (3) are equivalent.

Assertion (5) holds if and only if $R(x,y,Jz,Jw)=R(Jx,Jy,z,w)$ for all x,y,z,w. We replace (x,y) by (Jx,Jy) to see that Assertions (4) and (5) are equivalent.

We follow an argument shown to us by Salamon to complete the proof by establishing the equivalence of Assertions (3) and (5). We have the following series of equivalences and implications:

a) $JR(x,Jx)=R(x,Jx)J\ \forall x$,

$\Leftrightarrow$ b) $(JR(x,Jx)z,w)-(R(x,Jx)Jz,w)=0$ for all x,z,w,

$\Leftrightarrow$ c) $R(x,Jx,z,Jw)+R(x,Jx,Jz,w)=0$ for all x,z,w.

We polarize Assertion (c), which is a quadratic identity in x, to obtain an equivalent multilinear identity:

$\Leftrightarrow$ d) $R(y,Jx,z,Jw)+R(x,Jy,z,Jw)+R(y,Jx,Jz,w)$
$+R(x,Jy,Jz,w)=0$ for all x,y,z,w.

We use symmetry (1.5.1.e) to interchange the first two arguments in the first and third terms to see:

$\Leftrightarrow$ e) $-R(Jx,y,z,Jw)+R(x,Jy,z,Jw)-R(Jx,y,Jz,w)$
$+R(x,Jy,Jz,w)=0$ for all x,y,z,w.

We replace (x,w) by (Jx,Jw) to show:

$\Leftrightarrow$ f) $-R(x,y,z,w)-R(Jx,Jy,z,w)+R(x,y,Jz,Jw)$
$+R(Jx,Jy,Jz,Jw)=0$ for all x,y,z,w.

We use symmetry (1.5.1.f) to interchange the first two arguments with the final two arguments and demonstrate:

$\Leftrightarrow$ g) $-R(z,w,x,y)-R(z,w,Jx,Jy)+R(Jz,Jw,x,y)$
$+R(Jz,Jw,Jx,Jy)=0$ for all x,y,z,w.

We now change notation to interchange x and z and y and w to see:

$\Leftrightarrow$ h) $-R(x,y,z,w)-R(x,y,Jz,Jw)+R(Jx,Jy,z,w)$
$+R(Jx,Jy,Jz,Jw)=0$ for all x,y,z,w.

We add f) and h) to derive the one sided implication:

$\Rightarrow$ i) $-R(x,y,z,w)+R(Jx,Jy,Jz,Jw)=0$ for all x,y,z,w

$\Leftrightarrow$ j) $J^*R=R$.

We replace (x,y) by (Jx,Jy) in (i) to see

$\Leftrightarrow$ k) $-R(Jx,Jy,z,w)+R(x,y,Jz,Jw)=0.$ for all x,y,z,w.

We use i) and k) to derive f) which is equivalent to a). □

We shall need two elementary results from the theory of several complex variables. As they are well known, we shall omit the proof in the interests of brevity. Assertion (1) is known as the *identity theorem.*

1.6.3 Lemma.

(1) *Let $f(z_1, ..., z_k)$ be a holomorphic function of k complex variables which is defined on a connected open subset $\mathcal{O} \subset \mathbb{C}^k$. Let $\mathcal{O}_r$ be a non-empty open subset of $\mathcal{O} \cap \mathbb{R}^k$. If $f(z) = 0$ on $\mathcal{O}_r$, then $f(z) = 0$ on $\mathcal{O}$.*

(2) *Let $p(z_1, ..., z_k)$ be a non-constant polynomial function of k complex variables. Then the set $\mathcal{O} := \{z \in \mathbb{C}^k : p(z) \neq 0\}$ is a connected non-empty dense open subset of $\mathbb{C}^k$.*

Let R be an algebraic curvature tensor on a vector space V of signature (p, q). Let $\{e_1, e_2\}$ be a basis for a non-degenerate 2 plane π. The *sectional curvature* $\kappa(\pi)$ is defined to be

$$\kappa(\pi) = \frac{R(e_1, e_2, e_2, e_1)}{(e_1, e_1)(e_2, e_2) - (e_1, e_2)^2}.$$

We say that R has *constant sectional curvature* c if $\kappa(\pi) = c$ for every non-degenerate 2 plane π in V. We use Equation (1.1.1.b) to define:

$$R_{\mathrm{Id}}(x, y)z := (y, z)x - (x, z)y.$$

1.6.4 Lemma. *Let V be a vector space of signature (p, q). Let R be an algebraic curvature tensor on V.*

(1) *If π is a non-degenerate 2 plane, then the sectional curvature $\kappa(\pi)$ is independent of the particular basis chosen for π.*

(2) *Let $(r, s) = (2, 0)$, let $(r, s) = (1, 1)$, or let $(r, s) = (0, 2)$. Suppose that $p \geq r$, that $q \geq s$, and that $\kappa(\pi) = c$ for every $\pi \in Gr_{r,s}(V)$. Then $R = cR_{\mathrm{Id}}$ and R has constant sectional curvature c.*

(3) *If R has constant sectional curvature 0, then $R = 0$.*

Proof. Let $\{e_1, e_2\}$ and $\{\tilde{e}_1, \tilde{e}_2\}$ be bases for a non-degenerate 2 plane π. We expand $\tilde{e}_1 = ae_1 + be_2$ and $\tilde{e}_2 = ce_1 + de_2$ and then compute:

$$\begin{aligned} R(\tilde{e}_1, \tilde{e}_2, \tilde{e}_2, \tilde{e}_1) &= R(ae_1 + be_2, ce_1 + de_2, ce_1 + de_2, ae_1 + be_2) \\ &= (ad - bc)R(e_1, e_2, ce_1 + de_2, ae_1 + be_2) \\ &= (ad - bc)^2 R(e_1, e_2, e_2, e_1). \end{aligned} \tag{1.6.4.a}$$

We use Lemma 1.2.3 to see that:

$$\text{(1.6.4.b)}\qquad \begin{aligned}&(\tilde{e}_1,\tilde{e}_1)(\tilde{e}_2,\tilde{e}_2)-(\tilde{e}_1,\tilde{e}_2)^2\\ &=(ad-bc)^2\{(e_1,e_1)(e_2,e_2)-(e_1,e_2)^2\}.\end{aligned}$$

Assertion (1) now follows from Equations (1.6.4.a) and (1.6.4.b).

Let $m = p + q = \dim V$. Assertion (2) is immediate if $m = 2$, so we suppose $m \geq 3$. Suppose that $\kappa(\pi) = c$ for all $\pi \in Gr_{r,s}(V)$. We complexify and set $V_c := V \otimes \mathbb{C}$. We extend the inner product and the algebraic curvature tensor R to be complex multi-linear. Let $\mathcal{O}$ be the set of non-degenerate complex 2 frames:

$$\mathcal{O} := \{(e_1,e_2) \in V_c \times V_c : (e_1,e_1)(e_2,e_2) - (e_1,e_2)^2 \neq 0\}.$$

By choosing a basis for V, we may identify $V_c = \mathbb{C}^m$ and $V_c \times V_c = \mathbb{C}^{2m}$. Since the inner product is polynomial in the components of the vectors e_1 and e_2, we may use Lemma 1.6.3 (2) to see that $\mathcal{O}$ is a non-empty connected open subset of $V_c \times V_c$. We define:

$$\kappa(e_1,e_2) := \frac{R(e_1,e_2,e_2,e_1)}{(e_1,e_1)(e_2,e_2)-(e_1,e_2)^2} \text{ for } (e_1,e_2) \in \mathcal{O}.$$

Let

$$\mathcal{O}_{r,s} := \{(e_1,e_2) \in V \times V : \operatorname{span}\{e_1,e_2\} \in \operatorname{Gr}_{r,s}(V)\}$$

be a non-empty open subset of $\mathcal{O} \cap (V \times V)$. If $(e_1,e_2) \in \mathcal{O}_{r,s}$, then $\kappa(e_1,e_2)$ is the sectional curvature of $\operatorname{span}\{e_1,e_2\}$ so $\kappa(e_1,e_2) - c = 0$. Thus, by Lemma 1.6.3 (1), we have:

$$\text{(1.6.4.c)}\qquad \frac{R(e_1,e_2,e_2,e_1)}{(e_1,e_1)(e_2,e_2)-(e_1,e_2)^2} = c \text{ on } \mathcal{O}.$$

Let $m = p + q = \dim V$. Let $\mathcal{B} = \{e_1, ..., e_m\}$ be a complex orthonormal basis for V_c so that $(e_i,e_j) = \delta_{ij}$ for $1 \leq i,j \leq m$. Let

$$R^{\mathcal{B}}_{ijkl} := R(e_i,e_j,e_k,e_l)$$

be the components of the curvature tensor relative to this basis. To prove Assertion (2), we must show

$$\text{(1.6.4.d)}\qquad R^{\mathcal{B}}_{ijkl} = c\{\delta_{jk}\delta_{il} - \delta_{ik}\delta_{jl}\}.$$

Equation (1.6.4.d) is equivalent to the three assertions:

(1.6.4.e) $R^{\mathcal{B}}_{ijji} = c$ if $\{i, j\}$ are distinct

(1.6.4.f) $R^{\mathcal{B}}_{ijki} = 0$ if $\{i, j, k\}$ are distinct, and

(1.6.4.g) $R^{\mathcal{B}}_{ijkl} = 0$ if $\{i, j, k, l\}$ are distinct.

Equation (1.6.4.e) follows immediately from Equation (1.6.4.c). Let i, j, and k be distinct indices. Let

$$e^\theta := \cos\theta e_j + \sin\theta e_k.$$

Then $\{e_i, e^\theta\}$ is an orthonormal set so we may use Equation (1.6.4.c) to see:

$$\begin{aligned} c =& R(e_i, e^\theta, e^\theta, e_i) \\ =& \cos^2\theta R^{\mathcal{B}}_{ijji} + \sin^2\theta R^{\mathcal{B}}_{ikki} + 2\sin\theta\cos\theta R^{\mathcal{B}}_{ijki} \\ =& c + 2\sin\theta\cos\theta R^{\mathcal{B}}_{ijki}. \end{aligned}$$

Consequently, $R^{\mathcal{B}}_{ijki} = 0$. Assertion (2) now follows if $m = 3$. We therefore suppose that $m \geq 3$.

Since the orthonormal basis $\mathcal{B}$ was arbitrary, we may conclude that $R(x, y, z, x) = 0$ if $\{x, y, z\}$ is an orthonormal set. Let $\{i, j, k, l\}$ be distinct indices. Let $e^\theta := \cos\theta e_i + \sin\theta e_l$. We then have:

$$\begin{aligned} 0 =& R(e^\theta, e_j, e_k, e^\theta) \\ =& \cos^2\theta R^{\mathcal{B}}_{ijki} + \sin^2\theta R^{\mathcal{B}}_{ljkl} + \cos\theta\sin\theta\{R^{\mathcal{B}}_{ijkl} + R^{\mathcal{B}}_{ljki}\}. \end{aligned}$$

Therefore, $R^{\mathcal{B}}_{ijkl} + R^{\mathcal{B}}_{ljki} = 0$ or, equivalently, $R^{\mathcal{B}}_{ijkl} = R^{\mathcal{B}}_{iklj}$. Similarly, we have that $R^{\mathcal{B}}_{iklj} = R^{\mathcal{B}}_{iljk}$. Thus, the first Bianchi identity

$$R^{\mathcal{B}}_{ijkl} + R^{\mathcal{B}}_{iklj} + R^{\mathcal{B}}_{ilkj} = 0$$

implies that $R^{\mathcal{B}}_{ijkl} = 0$ if all 4 indices are distinct. This completes the proof of Assertion (2); Assertion (3) is now immediate. □

Let (M, g) be a pseudo-Riemannian manifold of signature (p, q), where $p + q \geq 3$. In Lemma 1.14.1, we will show that if the associated algebraic curvature tensor ${}^gR_P$ has constant sectional curvature κ_P at each point $P \in M$, then κ_P is a constant which is independent of P; thus the sectional curvature really is constant. We will also show that any two such manifolds are locally isometric; see Lemma 1.14.2 for details. Thus, the constancy of the sectional curvature is a geometrically rigid property. In Section 2.6, we will show that the pseudo-spheres provide examples of pseudo-Riemannian manifolds of arbitrary signatures and arbitrary constant sectional curvature.

Let (M,g) be a pseudo-Riemannian manifold. Let J be a skew-symmetric map of TM so that $J^2 = -\operatorname{Id}$. Then J is said to be a *Hermitian almost complex structure* on M and (M,g,J) is said to be an *almost complex pseudo-Hermitian* manifold. We complexify and define distributions

$$\mathcal{F}^{\pm} := \{X \in TM \otimes \mathbb{C} : JX = \pm\sqrt{-1}X\}.$$

We say that a triple (M,g,J) is *holomorphic* if there exist local complex coordinates $z_i = x_i + \sqrt{-1}y_i$ such that

$$J\partial_i^x = \partial_i^y \text{ and } J\partial_i^y = -\partial_i^x, \text{ i.e. } \mathcal{F}^{\pm} = \operatorname{span}\{\partial_i^x \mp \sqrt{-1}\partial_i^y\}.$$

This means that the distributions $\mathcal{F}^{\pm}$ are integrable, i.e. if X and Y are sections to $\mathcal{F}^{\pm}$, then $[X,Y]$ is also a section to $\mathcal{F}^{\pm}$. Not every almost complex structure defines a holomorphic manifold; the *Nirenberg–Newlander* theorem [122] shows that (M,g,J) defines a holomorphic structure if and only if the distributions $\mathcal{F}^{\pm}$ are integrable. Gray [91, Corollary 3.2] showed that if an algebraic curvature tensor is geometrically realizable by a holomorphic Hermitian manifold, then there is an additional curvature symmetry:

$$\begin{aligned} &R(x,y,z,w) + R(Jx,Jy,Jz,Jw) \\ &\quad = R(Jx,Jy,z,w) + R(Jx,y,Jz,w) + R(Jx,y,z,Jw) \\ &\quad + R(x,Jy,Jz,w) + R(x,Jy,z,Jw) + R(x,y,Jz,Jw). \end{aligned} \tag{1.6.4.h}$$

1.7 Einstein and k-stein algebraic curvature tensors

Throughout this section, we shall let R be an algebraic curvature tensor on a vector V space of signature (p,q) and dimension $m = p+q$. If $\{e_1,...,e_m\}$ is an orthonormal basis for V, we shall let $\varepsilon_i := (e_i,e_i) = \pm 1$. We may expand any vector v in V in the form $v = \sum_i \varepsilon_i(v,e_i)e_i$. We begin by defining the Ricci tensor.

1.7.1 Definition. Let R be an algebraic curvature tensor on V. Let

$$\mathcal{J}_R(x) : y \to R(y,x)x$$

be the *Jacobi operator* defined in Section 1.5.2. The *Ricci tensor* is defined by:

$$\rho(x,y) := \operatorname{Tr}\{z \to R(z,x)y\}. \tag{1.7.1.a}$$

Let $\{e_1, ..., e_m\}$ be an orthonormal basis for V. Then we have that:

$$R(e_i, x)y = \textstyle\sum_i \varepsilon_j R(e_i, x, y, e_j)e_j \text{ so} \tag{1.7.1.b}$$
$$\rho(x, y) = \textstyle\sum_i \varepsilon_i R(e_i, x, y, e_i).$$

Since $R(e_i, x, y, e_i) = R(y, e_i, e_i, x) = R(e_i, y, x, e_i)$, $\rho(x, y) = \rho(y, x)$. Consequently, the Ricci tensor is a symmetric bilinear form.

The following Lemma provides several different characterizations of Einstein algebraic curvature tensors.

1.7.2 Lemma. *Let R be an algebraic curvature tensor on a vector space V of signature (p, q). The following conditions are equivalent and all define the notion of an Einstein algebraic curvature tensor:*

(1) *If $p \geq 1$, then $\operatorname{Tr}\{\mathcal{J}_R(\cdot)\}$ is constant on $S^-(V)$.*
(2) *If $q \geq 1$, then $\operatorname{Tr}\{\mathcal{J}_R(\cdot)\}$ is constant on $S^+(V)$.*
(3) *There exists a constant c_1 so $\rho(x, y) = c_1(x, y)$.*
(4) *We have $\sum_i \varepsilon_i \mathcal{J}_R(e_i) = c_1 \operatorname{Id}$ for any orthonormal basis $\{e_1, ..., e_m\}$.*

Proof. Since $\rho(x, x) = \operatorname{Tr} \mathcal{J}_R(x)$, Assertion (3) implies Assertions (1) and (2). Suppose that Assertion (1) holds. Choose c_1 so $\operatorname{Tr} \mathcal{J}_R(x) = -c_1$ for all $x \in S^-(V)$. Since $\mathcal{J}_R(\lambda x) = \lambda^2 \mathcal{J}_R(x)$,

$$\operatorname{Tr}\{\mathcal{J}_R(x)\} = c_1(x, x)$$

for any timelike vector x. Since $p \geq 1$, there exists a timelike vector x_0. Let v be an arbitrary vector in V and let $\varepsilon \in \mathbb{R}$ be a real parameter. Let

$$\begin{aligned} f(\varepsilon) :=& \rho(\varepsilon x_0 + v, \varepsilon x_0 + v) - c_1(\varepsilon x_0 + v, \varepsilon x_0 + v) \\ =& \operatorname{Tr}\{\mathcal{J}_R(\varepsilon x_0 + v)\} - c_1(\varepsilon x_0 + v, \varepsilon x_0 + v). \end{aligned}$$

Since $\varepsilon x_0 + v$ is timelike for large values of ε, f vanishes for an infinite number of values of ε. Since the function f is a quadratic polynomial in ε, this implies f vanishes identically. We set $\varepsilon = 0$ to see

$$\rho(v, v) = c_1(v, v) \text{ for all } v \in V. \tag{1.7.2.a}$$

We polarize Equation (1.7.2.a) to derive Assertion (3). The proof that Assertion (2) implies Assertion (3) is similar and is therefore omitted.

We have shown that Assertions (1), (2), and (3) are all equivalent. Suppose that Assertion (4) holds. We show that Assertion (3) holds by using Equation (1.7.1.b) to compute:

$$\rho(x, y) = \textstyle\sum_i \varepsilon_i R(x, e_i, e_i, y) = \sum_i \varepsilon_i (\mathcal{J}_R(e_i)x, y) = c_1(x, y).$$

Conversely, suppose that Assertion (3) holds. Then $\operatorname{Tr}\{\mathcal{J}_R(x)\} = c_1(x,x)$ for all $x \in V$. Let $\mathcal{J} := \sum_i \varepsilon_i \mathcal{J}_R(e_i)$. We use Equation (1.7.1.b) to see:

$$\begin{aligned} c_1(x,x) &= \operatorname{Tr}\{\mathcal{J}_R(x)\} = \textstyle\sum_i \varepsilon_i R(e_i,x,x,e_i) \\ &= \textstyle\sum_i \varepsilon_i R(x,e_i,e_i,x) = (\sum_i \varepsilon_i \mathcal{J}_R(e_i)x,x) \\ &= (\mathcal{J}x,x). \end{aligned}$$

Since $\mathcal{J}$ is self-adjoint, we can polarize this identity to see $c_1(x,y) = (\mathcal{J}x,y)$ for all $x,y \in V$. It now follows that $\mathcal{J} = c_1 \operatorname{Id}$ and Assertion (4) follows. □

We follow the discussion in Carpenter, Gray, and Willmore [40] and say that R is *k-stein* if there exist constants c_i for $1 \le i \le k$ so that

$$\operatorname{Tr}(\mathcal{J}_R(x)^i) = c_i(x,x)^i \text{ for all } x \in V.$$

Note that 1-stein and Einstein are equivalent notions. The proof of the equivalence of the first 3 assertions of Lemma 1.7.2 generalizes to this setting immediately to yield the following result; we omit details in the interests of brevity:

1.7.3 Lemma. *Let R be an algebraic curvature tensor on a vector space V of signature (p,q). The following conditions are equivalent and all define the notion of a k-stein algebraic curvature tensor:*

(1) *If $p \ge 1$, then $\operatorname{Tr}\{\mathcal{J}_R(x)^i\}$ is constant on $S^-(V)$ for $1 \le i \le k$.*
(2) *If $q \ge 1$, then $\operatorname{Tr}\{\mathcal{J}_R(x)^i\}$ is constant on $S^+(V)$ for $1 \le i \le k$.*
(3) *There exist constants c_i so that $\operatorname{Tr}\{\mathcal{J}_R(x)^i\} = c_i(x,x)^i$ for all $x \in V$ and for $1 \le i \le k$.*

We say that a linear map A is *nilpotent* if $A^k = 0$ for some k. There is a close relationship between the m-stein condition and nilpotency of the Jacobi operator on null vectors.

1.7.4 Lemma. *Let R be an algebraic curvature tensor on a vector space V of signature (p,q), where $q \ge 2$. Let v be a null vector of V. If R is k-stein, then $\operatorname{Tr}\{\mathcal{J}_R(v)^i\} = 0$ for all $i \le k$. If R is m-stein, then $\mathcal{J}_R(v)$ is nilpotent.*

Proof. Suppose that R is k-stein. Let v be a null vector. Since $q \ge 2$, we can choose a unit spacelike vector w so that $w \perp v$. Let $v(t) := t^{-1}v + w$. Since $v(t) \in S^+(V)$ and since R is k-stein, there are constants c_i so that

$$c_i = \operatorname{Tr}\{\mathcal{J}_R(t^{-1}v + w)^i\} \text{ for all } 1 \le i \le k \text{ and } t > 0.$$

Since $\mathcal{J}_R(\cdot)$ is quadratic, we can rewrite this identity in the form:

$$t^{2i}c_i = t^{2i}\operatorname{Tr}\{\mathcal{J}_R(t^{-1}v+w)^i\} = \operatorname{Tr}\{\mathcal{J}_R(v+tw)^i\}.$$

Taking the limit as $t \downarrow 0$ then implies $\operatorname{Tr}\{\mathcal{J}_R(v)^i\} = 0$ for $1 \le i \le k$.

If R is m-stein, then $\operatorname{Tr}\{\mathcal{J}_R(v)^i\} = 0$ for $1 \le i \le m$. This implies that the characteristic polynomial of $\mathcal{J}_R(v)$ is t^m and hence, by Lemma 1.2.6, $\mathcal{J}_R(v)$ has only the zero eigenvalue. Thus, $\mathcal{J}_R(v)^m = 0$. □

Recall that a vector space V is said to be *Lorentzian* if V has signature $(1, q)$.

1.7.5 Lemma. *Let R be an algebraic curvature tensor on a Lorentzian vector space V. If $\operatorname{Tr}\{\mathcal{J}_R(v)^2\} = 0$ for every null vector v of V, then R has constant sectional curvature.*

Proof. Let V have signature $(1, q)$ and dimension $m = q + 1$. We suppose without loss of generality that $q \ge 2$ and follow an argument due to Stavrov [145]. We polarize to define the associated quadratic form

$$\mathcal{J}_R(u,v) : y \to \tfrac{1}{2}\{R(y,u)v + R(y,v)u\}.$$

Let $\mathcal{B} = \{e_0, ..., e_m\}$ be an orthonormal basis for V, where e_0 is timelike. Then $\varepsilon_0 = -1$ and $\varepsilon_i = +1$ for $i \ge 1$. By assumption, $\operatorname{Tr}\{J_R(e_0 \pm e_1)^2\} = 0$. We may express

$$\begin{aligned}
0 =& \tfrac{1}{2}\operatorname{Tr}\{\mathcal{J}_R(e_0+e_1)^2 + \mathcal{J}_R(e_0-e_1)^2\}\\
=& \tfrac{1}{2}\operatorname{Tr}\{\{\mathcal{J}_R(e_0,e_0) + \mathcal{J}_R(e_1,e_1) + 2\mathcal{J}_R(e_0,e_1)\}^2\}\\
&+\tfrac{1}{2}\operatorname{Tr}\{\{\mathcal{J}_R(e_0,e_0) + \mathcal{J}_R(e_1,e_1) - 2\mathcal{J}_R(e_0,e_1)\}^2\}\\
=& \operatorname{Tr}\{\{\mathcal{J}_R(e_0,e_0) + \mathcal{J}_R(e_1,e_1)\}^2 + 4\mathcal{J}_R(e_0,e_1)^2\}.
\end{aligned} \tag{1.7.5.a}$$

Let $R^{\mathcal{B}}_{ijkl} := R(e_i, e_j, e_k, e_l)$. We use Equation (1.7.5.a) to compute:

$$0 = \textstyle\sum_{ij} \varepsilon_i\varepsilon_j\{(R^{\mathcal{B}}_{i00j} + R^{\mathcal{B}}_{i11j})^2 + (R^{\mathcal{B}}_{i01j} + R^{\mathcal{B}}_{i10j})^2\}. \tag{1.7.5.b}$$

Let j be an arbitrary index. The two terms in Equation (1.7.5.b) with $i = 0$ and $i = 1$ are given by:

$$\begin{aligned}
&\varepsilon_0\varepsilon_j\{(R^{\mathcal{B}}_{000j} + R^{\mathcal{B}}_{011j})^2 + (R^{\mathcal{B}}_{001j} + R^{\mathcal{B}}_{010j})^2\}\\
&+\varepsilon_1\varepsilon_j\{(R^{\mathcal{B}}_{100j} + R^{\mathcal{B}}_{111j})^2 + (R^{\mathcal{B}}_{101j} + R^{\mathcal{B}}_{110j})^2\}\\
&= -\varepsilon_j(R^{\mathcal{B}}_{011j})^2 - \varepsilon_j(R^{\mathcal{B}}_{010j})^2 + \varepsilon_j(R^{\mathcal{B}}_{100j})^2 + \varepsilon_j(R^{\mathcal{B}}_{101j})^2\\
&=0.
\end{aligned}$$

Thus, the terms in Equation (1.7.5.b) with $i = 1$ and $i = 0$ cancel. We may therefore restrict of the index i in this equation to the range $2 \leq i \leq q$. A similar argument shows that we may restrict j to the range $2 \leq j \leq q$. Since $\varepsilon_i = \varepsilon_j = +1$ in this range, Equation (1.7.5.b) shows that a sum of squares is zero. Consequently, we have that:

$$R^{\mathcal{B}}_{i11i} = -R^{\mathcal{B}}_{i00i} \text{ for } 2 \leq i \leq q.$$

By permuting the elements of the basis, we see that there exists a constant $\kappa_{\mathcal{B}}$ so

$$\text{(1.7.5.c)} \qquad R^{\mathcal{B}}_{ijji} = -R^{\mathcal{B}}_{i00i} = \kappa_{\mathcal{B}} \text{ for } 1 \leq i, j \leq q, i \neq j.$$

To complete the proof, we must show $R^{\mathcal{B}}_{abcd} = 0$ if at least 3 of the indices are different. This is automatic if $m = 2$, so we suppose $m \geq 3$. The timelike index plays a distinguished role and must be treated separately.

Let $\{0, i, j\}$ be distinct indices. We polarize Equation (1.7.5.c) by considering the 1 parameter family

$$e_0^\theta := \cosh\theta e_0 + \sinh\theta e_i \text{ and } e_i^\theta := \sinh\theta e_0 + \cosh\theta e_i.$$

We have $\{e_0^\theta, e_i^\theta, e_j\}$ is an orthonormal set. Note that $\pi := \operatorname{span}\{e_0^\theta, e_i^\theta\}$ is independent of θ. By Lemma 1.6.4, the sectional curvature of π is independent of the orthonormal basis. Thus, by Equation (1.7.5.c), we have:

$$\begin{aligned}
-\kappa_{\mathcal{B}} &= - R(e_i^\theta, e_0^\theta, e_0^\theta, e_i^\theta) = R(e_i^\theta, e_j, e_j, e_i^\theta) \\
&= \cosh^2\theta R^{\mathcal{B}}_{ijji} + \sinh^2\theta R^{\mathcal{B}}_{0jj0} + 2\cosh\theta\sinh\theta R^{\mathcal{B}}_{ijj0} \\
&= -\cosh^2\theta R^{\mathcal{B}}_{i00i} + \sinh^2\theta R^{\mathcal{B}}_{0ii0} + 2\cosh\theta\sinh\theta R^{\mathcal{B}}_{ijj0} \\
&= -\kappa_{\mathcal{B}} + 2\cosh\theta\sinh\theta R^{\mathcal{B}}_{ijj0}.
\end{aligned}$$

This shows $R^{\mathcal{B}}_{ijj0} = 0$. Next, we consider the 1 parameter family:

$$e_i^\theta := \cos\theta e_i + \sin\theta e_j \text{ and } e_j^\theta := \cos\theta e_j - \sin\theta e_i.$$

We have $\{e_0, e_i^\theta, e_j^\theta\}$ is an orthonormal set. Since $\pi := \operatorname{span}\{e_i^\theta, e_j^\theta\}$ is independent of θ, we may use Equation (1.7.5.c) to compute that:

$$\begin{aligned}
\kappa_{\mathcal{B}} &= R(e_i^\theta, e_j^\theta, e_j^\theta, e_i^\theta) = -R(e_i^\theta, e_0, e_0, e_i^\theta) \\
&= -\cos^2\theta R^{\mathcal{B}}_{i00i} - \sin^2\theta R^{\mathcal{B}}_{j00j} - 2\sin\theta\cos\theta R^{\mathcal{B}}_{i00j} \\
&= \kappa_{\mathcal{B}} - 2\sin\theta\cos\theta R^{\mathcal{B}}_{i00j} \text{ so} \\
R^{\mathcal{B}}_{i00j} &= 0.
\end{aligned}$$

We summarize our results so far. If $\{0, i, j\}$ are distinct indices, then:

$$R^{\mathcal{B}}_{ijji} = -R^{\mathcal{B}}_{i00i} = \kappa_{\mathcal{B}} \text{ and } R^{\mathcal{B}}_{ijj0} = R^{\mathcal{B}}_{0ij0} = 0. \tag{1.7.5.d}$$

This completes the proof if $m = 3$, so we assume $m \geq 4$.

Let $\{0, i, j, k\}$ be distinct indices. We can polarize Display (1.7.5.d) to see that $R^{\mathcal{B}}_{ijk0} + R^{\mathcal{B}}_{ikj0} = 0$. This shows $R^{\mathcal{B}}_{0jki} = R^{\mathcal{B}}_{0kij}$. Similarly, we have that $R^{\mathcal{B}}_{0kij} = R^{\mathcal{B}}_{0ijk}$. The first Bianchi identity then implies $R^{\mathcal{B}}_{0ijk} = 0$. We polarize the identity $R^{\mathcal{B}}_{ijji} = -R^{\mathcal{B}}_{i00i}$ to see $R^{\mathcal{B}}_{ijki} = 0$. This completes the proof in case $m = 4$ since

$$R^{\mathcal{B}}_{ijji} = -R^{\mathcal{B}}_{i00i} = \kappa_{\mathcal{B}} \text{ and } R^{\mathcal{B}}_{ijj0} = R^{\mathcal{B}}_{0ij0} = R^{\mathcal{B}}_{ijki} = R^{\mathcal{B}}_{0ijk} = 0.$$

If $m \geq 5$, then let $\{0, i, j, k, l\}$ be distinct indices. We polarize the identity $R^{\mathcal{B}}_{ijki} = 0$ to see $R^{\mathcal{B}}_{ijkl} + R^{\mathcal{B}}_{ljki} = 0$. Again, we use the first Bianchi identity to show $R^{\mathcal{B}}_{ijkl} = 0$. □

The condition that an algebraic curvature tensor is 2-stein is very restrictive in Lorentzian geometry. The following result, due to Blažić, Bokan and Gilkey [19] is now an immediate consequence of Lemmas 1.7.4 and 1.7.5; see Gilkey and Stavrov [86] for details:

1.7.6 Corollary. *Let R be a 2-stein algebraic curvature tensor on a vector space V of signature $(1, q)$. Then R has constant sectional curvature.*

1.8 Properties of the curvature tensors R_ϕ

The following tensors, defined previously in Equation (1.1.1.b), will play a fundamental role in our development. Let ϕ be a linear map of a vector space V of signature (p, q). If $\phi^* = \pm\phi$, we define:

$$R_\phi(x,y)z := \begin{cases} (\phi y, z)\phi x - (\phi x, z)\phi y & \text{if } \phi = \phi^*, \\ (\phi y, z)\phi x - (\phi x, z)\phi y - 2(\phi x, y)\phi z & \text{if } \phi = -\phi^*. \end{cases}$$

1.8.1 Lemma. *Let V be a vector space of signature (p, q). Let ϕ be a linear transformation of V. Assume that $\phi^* = \pm\phi$.*

1. *The tensor R_ϕ is an algebraic curvature tensor.*
2. *If J is a pseudo-Hermitian almost complex structure on V and if $J\phi = \pm\phi J$, then R_ϕ is almost complex.*
3. *If $\phi^2 = 0$, then $R_\phi(x, y)R_\phi(\tilde{x}, \tilde{y}) = 0$ for all $x, y, \tilde{x}, \tilde{y} \in V$.*

Proof. Let $\phi^* = \pm\phi$. We must verify that Equations (1.5.1.e), (1.5.1.f), and (1.5.1.g) are satisfied to prove Assertion (1).

Suppose first that ϕ is self-adjoint. We establish the curvature symmetry $R(x,y,z,w) = -R(y,x,z,w) = -R(x,y,w,z)$ of Equation (1.5.1.e) by computing:

$$\begin{aligned} R_\phi(x,y,z,w) =&(\phi y,z)(\phi x,w) - (\phi x,z)(\phi y,w) \\ =& - R_\phi(y,x,z,w) = -R_\phi(x,y,w,z). \end{aligned}$$

If ϕ is skew-adjoint, then $(\phi x, y) = -(x, \phi y)$ and thus we can show that Equation (1.5.1.e) holds by computing:

$$\begin{aligned} R_\phi(x,y,z,w) =&(\phi y,z)(\phi x,w) - (\phi x,z)(\phi y,w) - 2(\phi x,y)(\phi z,w) \\ =& - R_\phi(x,y,z,w) = -R_\phi(x,y,w,z). \end{aligned}$$

As $\phi^* = \pm\phi$, we have that:

$$\begin{aligned} (\phi y,z)(\phi x,w) &= (y,\phi z)(x,\phi w), \\ (\phi x,z)(\phi y,w) &= (x,\phi z)(y,\phi w), \text{ and} \\ (\phi x,y)(\phi z,w) &= (\phi z,w)(\phi x,y). \end{aligned}$$

As R_ϕ is a linear combination of these three terms, and as the pairs (x, y) and (z, w) play symmetric roles in these three terms, we see that the curvature symmetry $R(x,y,z,w) = R(z,w,x,y)$ of Equation (1.5.1.f) is satisfied.

Let $\phi = \phi^*$. We verify that R_ϕ satisfies the first Bianchi identity:

$$\begin{aligned} &R_\phi(x,y)z + R_\phi(y,z)x + R_\phi(z,x)y \\ =&(\phi y,z)\phi x - (\phi x,z)\phi y \\ &+(\phi z,x)\phi y - (\phi y,x)\phi z \\ &+(\phi x,y)\phi z - (\phi z,y)\phi x = 0. \end{aligned}$$

If $\phi^* = -\phi$, then we complete the proof that R_ϕ is an algebraic curvature tensor by computing:

$$\begin{aligned} &R_\phi(x,y)z + R_\phi(y,z)x + R_\phi(z,x)y \\ =&(\phi y,z)\phi x - (\phi x,z)\phi y - 2(\phi x,y)\phi z \\ &+(\phi z,x)\phi y - (\phi y,x)\phi z - 2(\phi y,z)\phi x \\ &+(\phi x,y)\phi z - (\phi z,y)\phi x - 2(\phi z,x)\phi y = 0. \end{aligned}$$

Let $\phi^* = \pm\phi$ and let $\phi J = \pm J\phi$. To prove Assertion (2), we compute:

$$\begin{aligned}
(\phi Jy, Jz)(\phi Jx, Jw) &= (J\phi y, Jz)(J\phi x, Jw) = (\phi y, z)(\phi x, w),\\
(\phi Jx, Jz)(\phi Jy, Jw) &= (J\phi x, Jz)(J\phi y, Jw) = (\phi x, z)(\phi y, w),\\
(\phi Jx, Jy)(\phi Jz, Jw) &= (J\phi x, Jy)(J\phi z, Jw) = (\phi x, y)(\phi z, w).
\end{aligned}$$

Since R_ϕ is a linear combination of these terms, we may use Lemma 1.6.2 to see that R_ϕ is almost complex by checking that we have the identity:

$$R(Jx, Jy, Jz, Jw) = R(x, y, z, w).$$

Suppose $\phi^2 = 0$. It is immediate that $\text{range}\{R_\phi(x,y)\} \subset \text{range}\,\phi$. Since $(\phi u, \phi v) = \pm(\phi^2 u, v) = 0$, we have $R_\phi(x,y)\phi z = 0$ for any $x, y, z \in V$. Assertion (3) now follows. □

Let V be a vector space of signature (p,q). Let $\mathcal{C}(V)$ be the space of algebraic curvature tensors on V. Let

$$\begin{aligned}
\mathcal{A}(V) &:= \text{span}\{R_\phi : \phi^* = -\phi\} \subset \mathcal{C}(V) \text{ and}\\
\mathcal{S}(V) &:= \text{span}\{R_\phi : \phi^* = \phi\} \subset \mathcal{C}(V).
\end{aligned}$$

Fiedler [56] used representation theory to show that

$$\mathcal{C}(V) = \mathcal{A}(V) + \mathcal{S}(V).$$

We have also been informed that Semmelman has derived this result independently. In fact, a slightly more general result is true.

1.8.2 Theorem. *We have $\mathcal{C}(V) = \mathcal{A}(V) = \mathcal{S}(V)$.*

Proof. We use the metric to define a bijective linear correspondence $\phi \to \tilde\phi$ between the set of linear transformations of V and the set of bilinear maps of V by the identity:

$$\tilde\phi(v, w) := (\phi v, w).$$

The map ϕ is self-adjoint if and only if the bilinear form $\tilde\phi$ is symmetric; similarly, the map ϕ is skew-adjoint if and only if the bilinear form $\tilde\phi$ is skew-symmetric. It is convenient at this point to work with bilinear forms rather than with the associated linear transformations as the signature of V then plays no role in our arguments. We then have, depending on whether $\phi = \phi^*$ or $\phi = -\phi^*$, that

$$R_{\tilde\phi}(x,y,z,w) := \begin{cases} \tilde\phi(y,z)\tilde\phi(x,w) - \tilde\phi(y,w)\tilde\phi(x,z) \text{ or}\\ \tilde\phi(y,z)\tilde\phi(x,w) - \tilde\phi(y,w)\tilde\phi(x,z) - 2\tilde\phi(x,y)\tilde\phi(z,w). \end{cases}$$

Let $\{e_i\}$ be a basis for V and let $\{e^i\}$ be the corresponding dual basis for V^*. Then $\{e^i \otimes e^j \otimes e^k \otimes e^l\}$ is a basis for $\otimes^4 V^*$. We define a tensor which satisfies the first two curvature symmetries given in Equations (1.5.1.e) and (1.5.1.f) by setting:

$$\begin{aligned} T_{ijkl} :=& e^i \otimes e^j \otimes e^k \otimes e^l + e^k \otimes e^l \otimes e^i \otimes e^j \\ & - e^j \otimes e^i \otimes e^k \otimes e^l - e^k \otimes e^l \otimes e^j \otimes e^i \\ & - e^i \otimes e^j \otimes e^l \otimes e^k - e^l \otimes e^k \otimes e^i \otimes e^j \\ & + e^j \otimes e^i \otimes e^l \otimes e^k + e^l \otimes e^k \otimes e^j \otimes e^i. \end{aligned}$$

Let R be an algebraic curvature tensor. Since R is a linear combination of the tensors T_{ijkl}, we may decompose R in the form:

(1.8.2.a) $\quad R = \sum_{i,j \text{ distinct}} c_{ijji} T_{ijji}$

(1.8.2.b) $\quad + \sum_{i,j,k \text{ distinct}} c_{ijki} T_{ijki}$

(1.8.2.c) $\quad + \sum_{i,j,k,l \text{ distinct}} c_{ijkl} T_{ijkl},$

where the coefficients c_{ijkl} may be chosen so $c_{ijkl} = c_{klij} = -c_{jikl}$. Since R satisfies the first Bianchi identity, we have that:

$$c_{ijkl} + c_{iklj} + c_{iljk} = 0.$$

We shall prove Theorem 1.8.2 by studying the tensors which appear in Displays (1.8.2.a), (1.8.2.b), and (1.8.2.c) separately.

We first study the tensors in Display (1.8.2.a). Let i and j be distinct indices. Define $\tilde{\phi} \in S^2(V^*)$ by requiring that

$$\tilde{\phi}(e_i, e_j) = \tilde{\phi}(e_j, e_i) = 1 \text{ and } \tilde{\phi}(e_a, e_b) = 0 \text{ otherwise.}$$

The only non-trivial term is

$$R_\phi(e_i, e_j, e_j, e_i) = -1 \text{ so } T_{ijji} = -2R_\phi \in \mathcal{S}.$$

Similarly, define $\tilde{\phi} \in \Lambda^2(V^*)$ so that

$$\tilde{\phi}(e_i, e_j) = -\tilde{\phi}(e_j, e_i) = 1 \text{ and } \tilde{\phi}(e_a, e_b) = 0 \text{ otherwise.}$$

The only non-trivial term is

$$R_\phi(e_i, e_j, e_j, e_i) = 3 \text{ so } T_{ijji} = \tfrac{2}{3} R_\phi \in \mathcal{A}.$$

Next, we study the tensors in Display (1.8.2.b). Let $\{i, j, k\}$ be distinct indices. Define $\tilde{\phi} \in S^2(V^*)$ by requiring that

$$\begin{aligned}&\tilde{\phi}(e_i, e_j) = \tilde{\phi}(e_j, e_i) = \tilde{\phi}(e_i, e_k) = \tilde{\phi}(e_k, e_i) = 1 \text{ and}\\&\tilde{\phi}(e_a, e_b) = 0 \text{ otherwise.}\end{aligned}$$

The only non-trivial terms are

$$\begin{aligned}&R_\phi(e_i, e_j, e_j, e_i) = R_\phi(e_i, e_k, e_k, e_i) = R_\phi(e_i, e_j, e_k, e_i) = -1 \text{ so that}\\&R_\phi = -\tfrac{1}{2}T_{ijji} - \tfrac{1}{2}T_{ikki} - T_{ijki}.\end{aligned}$$

Since $T_{ijji} \in \mathcal{S}$ and $T_{ikki} \in \mathcal{S}$, we may conclude that

$$T_{ijki} \in \mathcal{S}.$$

Similarly, define $\tilde{\phi} \in \Lambda^2(V^*)$ by requiring that

$$\begin{aligned}&\tilde{\phi}(e_i, e_j) = -\tilde{\phi}(e_j, e_i) = \tilde{\phi}(e_i, e_k) = -\tilde{\phi}(e_k, e_i) = 1 \text{ and}\\&\tilde{\phi}(e_a, e_b) = 0 \text{ otherwise.}\end{aligned}$$

The only non-trivial terms are

$$\begin{aligned}&R_\phi(e_i, e_j, e_j, e_i) = R_\phi(e_i, e_k, e_k, e_i) = R_\phi(e_i, e_j, e_k, e_i) = 3 \text{ so that}\\&T_{ijki} \in \mathcal{A}.\end{aligned}$$

Finally, let $\{i, j, k, l\}$ be distinct indices. We define $\tilde{\phi} \in S^2(V^*)$ by requiring that

$$\begin{aligned}&\tilde{\phi}(e_i, e_k) = \tilde{\phi}(e_k, e_i) = \tilde{\phi}(e_j, e_l) = \tilde{\phi}(e_l, e_j) = 1 \text{ and}\\&\tilde{\phi}(e_a, e_b) = 0 \text{ otherwise.}\end{aligned}$$

The only non trivial terms are:

$$\begin{aligned}&R_{\tilde{\phi}}(e_i, e_j, e_k, e_l) = -1,\ R_{\tilde{\phi}}(e_i, e_k, e_l, e_j) = 0,\ R_{\tilde{\phi}}(e_i, e_l, e_j, e_k) = +1\\&R_{\tilde{\phi}}(e_i, e_k, e_k, e_i) = R_{\tilde{\phi}}(e_j, e_l, e_l, e_j) = -1 \text{ so}\\&T_{ijkl} - T_{iljk} \in \mathcal{S}.\end{aligned}$$

We cyclically permute the indices $\{j, k, l\}$ to see

$$T_{iklj} - T_{ijkl} \in \mathcal{S}.$$

We define $\tilde{\phi} \in \mathcal{A}$ by requiring

$$\begin{aligned}&\tilde{\phi}(e_i,e_k) = -\tilde{\phi}(e_k,e_i) = \tilde{\phi}(e_j,e_l) = -\tilde{\phi}(e_l,e_j) = 1\\&\tilde{\phi}(e_a,e_b) = 0 \text{ otherwise.}\end{aligned}$$

The only non-trivial terms are

$$\begin{aligned}&R_\phi(e_i,e_j,e_k,e_l) = -1,\ R_\phi(e_i,e_k,e_l,e_j) = 2,\ R_\phi(e_i,e_l,e_j,e_k) = -1\\&R_\phi(e_i,e_k,e_k,e_i) = R_\phi(e_j,e_l,e_l,e_j) = 3, \text{ so}\\&-T_{ijkl} + 2T_{iklj} - T_{iljk} \in \mathcal{A}.\end{aligned}$$

We cyclically permute the indices $\{j,k,l\}$ to see that

$$-T_{iklj} + 2T_{iljk} - T_{ijkl} \in \mathcal{A}.$$

We subtract to see $T_{iklj} - T_{iljk} \in \mathcal{A}$. We cyclically permute the indices $\{j,k,l\}$ to see

$$T_{ijkl} - T_{iklj} \in \mathcal{A} \text{ and } T_{iljk} - T_{ijkl} \in \mathcal{A}.$$

Since R satisfies the first Bianchi identity, we have $c_{ijkl} + c_{iklj} + c_{iljk} = 0$ and thus the terms in Display (1.8.2.c) are expressible in terms of the tensors $T_{ijkl} - T_{iljk}$ and $T_{ijkl} - T_{iklj}$ which belong to $\mathcal{A}$ and to $\mathcal{S}$. □

We now extend Theorem 1.8.2 to the complex category. Let J be a pseudo-Hermitian almost complex structure on V. Let

$$\mathcal{C}_J(V) := \{R \in \mathcal{C}(V) : J^*R = R\} \subset \mathcal{C}(V)$$

be the set of almost complex algebraic curvature tensors on V. Let

$$\begin{aligned}&\mathcal{A}_J^\pm(V) := \operatorname{span}\{R_\phi : \phi^* = -\phi \text{ and } J^*\phi = \pm\phi\} \text{ and}\\&\mathcal{S}_J^\pm(V) := \operatorname{span}\{R_\phi : \phi^* = \phi \text{ and } J^*\phi = \pm\phi\}.\end{aligned}$$

By Lemma 1.8.1,

$$\mathcal{A}_J^\pm(V) \subset \mathcal{C}_J(V) \text{ and } \mathcal{S}_J^\pm(V) \subset \mathcal{C}_J(V).$$

1.8.3 Theorem. *We have* $\mathcal{C}_J(V) = \mathcal{A}_J^+(V) + \mathcal{A}_J^-(V) = \mathcal{S}_J^+(V) + \mathcal{S}_J^-(V)$.

Proof. Let $\varepsilon = \pm 1$. Let ϕ_1 and ϕ_2 be maps of V with $\phi_1^* = \varepsilon\phi_1$ and $\phi_2^* = \varepsilon\phi_2$. Since the mixed terms cancel, we have that:

$$R_{(\phi_1+\phi_2)} + R_{(\phi_1-\phi_2)} = 2R_{\phi_1} + 2R_{\phi_2} \tag{1.8.3.a}$$

Suppose that ϕ is a linear map of V with $\phi^* = \pm\phi$. We then have that:

$$\begin{aligned}
&J^*R_\phi(x,y,z,w) = R_\phi(Jx,Jy,Jz,Jw)\\
&=\begin{cases}(\phi Jy,Jz)(\phi Jx,Jw)-(\phi Jy,Jw)(\phi Jx,Jz) \text{ or}\\ (\phi Jy,Jz)(\phi Jx,Jw)-(\phi Jy,Jw)(\phi Jx,Jz)-2(\phi Jx,Jy)(\phi Jz,Jz)\end{cases}\\
&=\begin{cases}(J\phi Jy,z)(J\phi Jx,w)-(J\phi Jy,w)(J\phi Jx,z) \text{ or}\\ (J\phi Jy,z)(J\phi Jx,w)-(J\phi Jy,w)(J\phi Jx,z)-2(J\phi Jx,y)(J\phi Jz,w)\end{cases}\\
&=R_{J\phi J}(x,y,z,w).
\end{aligned}$$

Let R be an algebraic curvature tensor. We use Theorem 1.8.2 to decompose $R = \sum_i c_i R_{\phi_i}$, where the ϕ_i are either self-adjoint or skew-adjoint linear transformations. If R is an almost complex algebraic curvature tensor, then $R = J^*R$. Consequently, we may use Equation (1.8.3.a) to decompose

$$\begin{aligned}
R =&\tfrac{1}{2}\textstyle\sum_i c_i(R_{\phi_i} + J^*R_{\phi_i}) = \tfrac{1}{2}\sum_i c_i(R_{\phi_i} + R_{J\phi_i J})\\
=&\tfrac{1}{4}\textstyle\sum_i c_i\{R_{(\phi_i+J\phi_i J)} + R_{(\phi_i - J\phi_i J)}\}.
\end{aligned}$$

The desired result follows since

$$J(\phi_i \pm J\phi_i J) = \mp(\phi_i \pm J\phi_i J)J. \quad \square$$

We can use the proof of Theorem 1.8.2 to compute the dimension of $\mathcal{C}(V)$:

1.8.4 Corollary. *Let V be a vector space of dimension n. Then we have that* $\dim \mathcal{C}(V) = \frac{1}{12}n^2(n^2-1)$.

Proof. In the proof of Theorem 1.8.2, we showed that the tensors T_{ijji}, T_{ijki}, and $T_{ijkl} - T_{iklj}$ spanned the space of algebraic curvature tensors $\mathcal{C}(V)$. By taking the redundancies in this spanning set into account, we see that:

$$\begin{aligned}
\dim \mathcal{C}(V) =&\tfrac{1}{2}n(n-1) + \tfrac{1}{2}n(n-1)(n-2) + \tfrac{2}{24}n(n-1)(n-2)(n-3)\\
=&\tfrac{1}{12}n(n-1)\{6 + 6(n-2) + (n^2-5n+6)\}\\
=&\tfrac{1}{12}n(n-1)(n^2+n) = \tfrac{1}{12}n^2(n^2-1). \quad \square
\end{aligned}$$

As noted in Section 1.8, there is an additional curvature symmetry

$$\begin{aligned}
&R(x,y,z,w) + R(Jx,Jy,Jz,Jw)\\
&= R(Jx,Jy,z,w) + R(Jx,y,Jz,w) + R(Jx,y,z,Jw)\\
&+ R(x,Jy,Jz,w) + R(x,Jy,z,Jw) + R(x,y,Jz,Jw)
\end{aligned} \tag{1.8.4.a}$$

that is satisfied if an algebraic curvature tensor can be geometrically realized by a holomorphic Hermitian manifold.

1.8.5 Lemma. *Let V be a vector space of signature (p,q). Let ϕ be a linear transformation of V with $\phi^* = \pm\phi$. Let $R = R_\phi$ be the associated algebraic curvature tensor.*

(1) *If $J\phi = \phi J$, then Equation* (1.8.4.a) *holds.*
(2) *If $J\phi = -\phi J$ and if* $\operatorname{rank}\phi \le 2$, *then Equation* (1.8.4.a) *holds.*
(3) *If $J\phi = -\phi J$ and if* $\operatorname{rank}\phi > 2$, *then Equation* (1.8.4.a) *fails.*

Proof. Let $\phi^* = \pm\phi$ and suppose $J\phi = \pm\phi J$. By Lemma 1.8.1, R_ϕ is almost complex. Since $J^*R = R$, we have that:

$$\begin{aligned} R(x,y,z,w) &= R(Jx,Jy,Jz,Jw),\\ R(Jx,Jy,z,w) &= R(x,y,Jz,Jw),\\ R(Jx,y,Jz,w) &= R(x,Jy,z,Jw), \text{ and}\\ R(Jx,y,z,Jw) &= R(x,Jy,Jz,w). \end{aligned}$$

Thus, Equation (1.8.4.a) holds if and only if

$$R(x,y,z,w) = R(Jx,Jy,z,w) + R(Jx,y,Jz,w) + R(Jx,y,z,Jw). \tag{1.8.5.a}$$

To prove Assertion (1), we suppose that $J\phi = \phi J$. We then have:

$$\begin{aligned} &(\phi Jy,z)(\phi Jx,w) + (\phi y,Jz)(\phi Jx,w) + (\phi y,z)(\phi Jx,Jw) = (\phi y,z)(\phi x,w),\\ &(\phi Jx,Jz)(\phi y,w) + (\phi Jx,z)(\phi Jy,w) + (\phi Jx,z)(\phi y,Jw) = (\phi x,z)(\phi y,w),\\ &(\phi Jx,Jy)(\phi z,w) + (\phi Jx,y)(\phi Jz,w) + (\phi Jx,y)(\phi z,Jw) = (\phi x,y)(\phi z,w). \end{aligned}$$

Since R_ϕ is a linear combination of these tensors, Equation (1.8.4.a) holds; this completes the proof of Assertion (1).

To prove Assertions (2) and (3), we suppose that $J\phi = -\phi J$. We also suppose ϕ does not vanish identically and choose a vector x_0 in V so that $\phi x_0 \ne 0$. We then choose a vector z_0 so that

$$(\phi x_0, z_0) \ne 0.$$

We have the following identities for any x,y,z,w in V:

$$\begin{aligned} &(\phi Jy,z)(\phi Jx,w) + (\phi y,Jz)(\phi Jx,w) + (\phi y,z)(\phi Jx,Jw)\\ &\qquad = 2(J\phi y,z)(J\phi x,w) - (\phi y,z)(\phi x,w)\\ &(\phi Jx,Jz)(\phi y,w) + (\phi Jx,z)(\phi Jy,w) + (\phi Jx,z)(\phi y,Jw),\\ &\qquad = 2(J\phi x,z)(J\phi y,w) - (\phi x,z)(\phi y,w), \text{ and}\\ &(\phi Jx,Jy)(\phi z,w) + (\phi Jx,y)(\phi Jz,w) + (\phi Jx,y)(\phi z,Jw)\\ &\qquad = 2(J\phi x,y)(J\phi z,w) - (\phi x,y)(\phi z,w) \end{aligned} \tag{1.8.5.b}$$

Suppose first that we have $\phi = \phi^*$. We use Display (1.8.5.b) to see that Equation (1.8.4.a) is satisfied if and only if for all x, y, z, w in V we have:

$$0 =(J\phi y, z)J\phi x - (\phi y, z)\phi x - (J\phi x, z)J\phi y + (\phi x, z)\phi y. \tag{1.8.5.c}$$

We have chosen $\{x_0, z_0\}$ so that $(\phi x_0, z_0) \neq 0$. Let

$$\pi_y := \operatorname{span}\{J\phi x_0, \phi x_0, J\phi y\}.$$

Suppose Equation (1.8.5.c) holds. We use Equation (1.8.5.c) with $x = x_0$ and $z = z_0$ to see that $\phi y \in \pi_y$. Thus, π_y is J invariant. Since we have $1 \leq \dim \pi_y \leq 3$, we must have $\dim \pi_y = 2$. Thus, $\pi_y = \operatorname{span}\{J\phi x_0, \phi x_0\}$ is independent of y. Since $\phi y \in \pi$, we may conclude that ϕ has rank 2.

On the other hand, suppose that ϕ has rank 2. Since

$$V = \operatorname{span}\{x_0, Jx_0\} \oplus \ker \phi,$$

it suffices to check that Equation (1.8.5.c) holds for $x, y \in \operatorname{span}\{x_0, Jx_0\}$. Since Equation (1.8.5.c) is skew in (x, y), we need only study the case $x = x_0$ and $y = Jx_0$. Thus, we may show that Equation (1.8.5.c) holds if $\operatorname{rank}\phi \leq 2$ by computing:

$$\begin{aligned}
&(J\phi Jx_0, z)J\phi x_0 - (\phi Jx_0, z)\phi x_0 - (J\phi x_0, z)J\phi Jx_0 + (\phi x_0, z)\phi Jx \\
=&(\phi x_0, z)J\phi x_0 + (J\phi x_0, z)\phi x_0 - (J\phi x_0, z)\phi x_0 - (\phi x_0, z)J\phi x_0 \\
=&0.
\end{aligned}$$

Suppose next that $\phi = -\phi^*$. We use Equation (1.8.5.b) to see that R satisfies Equation (1.8.4.a) if and only if

$$\begin{aligned}
0 =&(J\phi y, z)J\phi x - (\phi y, z)\phi x - (J\phi x, z)J\phi y + (\phi x, z)\phi y \\
&-2(J\phi x, y)J\phi z + 2(\phi x, y)\phi z.
\end{aligned} \tag{1.8.5.d}$$

Suppose that Equation (1.8.5.d) holds. Recall that we chose $\{x_0, z_0\}$ so that $(\phi x_0, z_0) \neq 0$. Let

$$\pi := \operatorname{span}\{\phi x_0, J\phi x_0\}.$$

We set $x = x_0$, $y = z_0$, and $z = z_0$ and use Equation (1.8.5.d) to see that

$$\phi z_0 \in \operatorname{span}\{\phi x_0, J\phi x_0, J\phi z_0\}.$$

This shows that $\phi z_0 \in \pi$ and $J\phi z_0 \in \pi$. Let y be arbitrary. We use Equation (1.8.5.d) with $x = x_0$ and $z = z_0$ to see that $\phi y_0 \in \operatorname{span}\{\phi x_0, J\phi x_0, J\phi y\}$ and hence $\phi y \in \pi$ so $\operatorname{rank}\phi = 2$.

Conversely, suppose that $\operatorname{rank}\phi = 2$. Again, we may suppose without loss of generality that x, y, and z belong to $\operatorname{span}\{x_0, Jx_0\}$ since Equation (1.8.5.d) holds trivially if x, y, or z belongs to $\ker\phi$. Since Equation (1.8.5.d) is alternating in $\{x, y\}$, we may suppose $x = x_0$ and $y = Jx_0$. The same argument, as given in the self adjoint case, shows

$$0 = (J\phi y, z)J\phi x - (\phi y, z)\phi x - (J\phi x, z)J\phi y + (\phi x, z)\phi y.$$

Since ϕ is a skew-adjoint linear map and since J is an isometry, we also have $(J\phi x_0, Jx_0) = 0$. □

The following technical Lemma will play an important role in our study of IP algebraic curvature tensors.

1.8.6 Lemma. *Let ϕ be a linear map from a vector space V of signature (p, q) to an auxiliary vector space W of signature (r, s). Assume that $q \geq 3$. Let $x, y \in V$. Define a transformation $T_\phi(x, y)$ from W to W by the identity:*

$$T_\phi(x, y)z := (\phi y, z)\phi x - (\phi x, z)\phi y.$$

Assume that $\ker\phi$ has no spacelike vectors. Then

(1) *If x and y span a spacelike 2 plane,* $\operatorname{range}(T_\phi(x, y)) = \operatorname{span}\{\phi x, \phi y\}$.

(2) *If $T_\phi(x, y) = T_{\tilde\phi}(x, y)$ for all $x, y \in V$, then $\phi = \pm\tilde\phi$.*

(3) *If $V = W$, then T_ϕ is an algebraic curvature tensor if and only if $\phi = \phi^*$.*

Proof. Assertion (1) is an immediate consequence of Lemma 1.2.2.

We prove Assertion (2) as follows. Let x be a spacelike vector. Let

$$\Phi([x]) := \cap_{y \in S^+(x^\perp)} \operatorname{range} T_\phi(x, y).$$

Let $\{x, y, z\}$ be an orthonormal spacelike set. Then $\{\phi x, \phi y, \phi z\}$ is a linearly independent set. We use Assertion (1) to see that:

$$\begin{aligned}\phi x \in \Phi(x) &\subset \operatorname{range}(T_\phi(x, y)) \cap \operatorname{range}(T_\phi(x, z))\\ &= \operatorname{span}\{\phi x, \phi y\} \cap \operatorname{span}\{\phi x, \phi z\} = \phi x \cdot \mathbb{R}.\end{aligned}$$

This implies that $\Phi(x)$ is a line. Thus, Φ induces a map from $S^+(V)$ to $\mathbb{P}(W)$. As ϕx is a non-zero point of $\Phi([x])$, ϕ linearizes Φ. As $T_\phi = T_{\tilde\phi}$,

$$\Phi = \tilde\Phi.$$

Thus, both ϕ and $\tilde{\phi}$ are linearizations of the same projective map. As $q \geq 3$, we may apply Lemma 1.2.11 to see that $\phi = c\tilde{\phi}$. Since $T_\phi = c^2 T_{\tilde{\phi}}$, $c^2 = 1$ so $c = \pm 1$. Assertion (2) now follows.

To prove Assertion (3), we set $V = W$ and study the associated 4 tensor:

$$\begin{aligned} T_\phi(z, w, x, y) &= (\phi w, x)(\phi z, y) - (\phi z, x)(\phi w, y) \\ &= (\phi^* x, w)(\phi^* y, z) - (\phi^* x, z)(\phi^* y, w) \\ &= T_{\phi^*}(x, y, z, w). \end{aligned}$$

Thus, the second curvature symmetry $R(x, y, z, w) = R(z, w, x, y)$, which was described in Equation (1.5.1.f), is satisfied if and only if $T_\phi = T_{\phi^*}$. Thus, by Assertion (2), $\phi = \pm\phi^*$. We suppose that $\phi = -\phi^*$ and argue for a contradiction. We use the first Bianchi identity to see:

$$\begin{aligned} 0 =& T_\phi(x, y)z + T_\phi(y, z)x + T_\phi(z, x)y \\ =& (\phi y, z)\phi x - (\phi x, z)\phi y \\ &+ (\phi z, x)\phi y - (\phi y, x)\phi z \\ &+ (\phi x, y)\phi z - (\phi z, y)\phi x \\ =& 2(\phi y, z)\phi x + 2(\phi z, x)\phi y + 2(\phi x, y)\phi z. \end{aligned} \tag{1.8.6.a}$$

Let $x \in S^+(V)$. Then $\phi x \neq 0$. Since $q \geq 3$, we may choose y so $y \perp \phi x$ and so $\{x, y\}$ is a spacelike orthonormal set. Suppose that $z \perp \phi x$. As $\phi x \neq 0$, we use Equation (1.8.6.a) to see that $z \perp \phi y$. Thus, ϕy is a multiple of ϕx. This is not possible as ϕ is injective on spacelike subspaces. Consequently, we have $\phi = \phi^*$. Conversely by Lemma 1.8.1, T is an algebraic curvature tensor if ϕ is self adjoint. □

1.9 Invariants of the orthogonal group

Throughout this section, let $(\cdot, \cdot)$ be an inner product of signature (p, q) on a vector space V. Recall that a map T from $V \times V$ to an auxiliary vector space W is said to be an *alternating bilinear map* if T is bilinear and if $T(x, y) = -T(y, x)$ for all $x, y \in V$.

If $\mathcal{B} = \{x, y\}$ is a basis for an oriented 2 plane π, then we have defined the *skew-symmetric curvature operator* by:

$$R(\pi) := \frac{R(x, y)}{|(x, x)(y, y) - (x, y)^2|^{1/2}}.$$

The fact that $R(\pi)$ is independent of $\mathcal{B}$ follows from the following more general result:

1.9.1 Lemma. *Let V be a vector space of signature (p,q). Let T be an alternating bilinear map from $V \times V$ to an auxiliary vector space W. Let $\{x_1, y_1\}$ and $\{x_2, y_2\}$ be bases for a non-degenerate 2 plane π which induce the same orientation on π. Then*

$$\frac{T(x_1,y_1)}{|(x_1,x_1)(y_1,y_1)-(x_1,y_1)^2|^{1/2}} = \frac{T(x_2,y_2)}{|(x_2,x_2)(y_2,y_2)-(x_2,y_2)^2|^{1/2}}.$$

Proof. The proof is similar to the proof given for Lemma 1.6.4. Let $\{x_1, y_1\}$ and $\{x_2, y_2\}$ be two bases which induce the same orientation on a non-degenerate 2 plane π. Expand $x_2 = ax_1 + by_1$ and $y_2 = cx_1 + dy_1$, where a, b, c, and d are suitably chosen constants. By Lemma 1.2.3,

$$(1.9.1.a) \quad (x_2,x_2)(y_2,y_2)-(x_2,y_2)^2 = (ad-bc)^2\{(x_1,x_1)(y_1,y_1)-(x_1,y_1)^2\}.$$

Since the orientation of π is preserved, $ad - bc > 0$. As T is alternating, we have that:

$$T(ax_1 + by_1, cx_1 + dy_1) = (ad - bc)T(x_1, y_1) = |ad - bc|T(x_1, y_1). \quad \square$$

Let $O(V)$ be the group of length preserving linear transformations:

$$O(V) := \{g \in \mathrm{Gl}(V) : (gv, gv) = (v, v) \text{ for all } v \in V\}.$$

We can polarize this identity to see that if $g \in O(V)$, then

$$(gv, gw) = (v, w) \text{ for all } v, w \in V.$$

If $g \in O(V)$, then we may extend g to act orthogonally on tensors of all types. We say that a map $\psi : \otimes^k V \to \mathbb{R}$ is *invariant* if

$$\psi(g \cdot w) = \psi(w) \text{ for all } g \in O(V) \text{ and for all } w \in \otimes^k V.$$

We can construct invariant mappings as follows. Let $k = 2\ell$ and let σ be a permutation of the integers from 1 to 2ℓ. We define

$$(1.9.1.b) \qquad \psi_\sigma(v_1, ..., v_{2\ell}) := (v_{\sigma(1)}, v_{\sigma(2)}) \cdot ... \cdot (v_{\sigma(2\ell-1)}, v_{\sigma(2\ell)}).$$

Since ψ_σ is multi-linear, we may extend ψ_σ to a map from $\otimes^{2\ell} V$ to $\mathbb{R}$. If $g \in O(V)$, then

$$\begin{aligned}\psi_\sigma(gv_1, ..., gv_{2\ell}) =& (gv_{\sigma(1)}, gv_{\sigma(2)}) \cdot ... \cdot (gv_{\sigma(2\ell-1)}, gv_{\sigma(2\ell)})\\ =& (v_{\sigma(1)}, v_{\sigma(2)}) \cdot ... \cdot (v_{\sigma(2\ell-1)}, v_{\sigma(2\ell)})\\ =& \psi_\sigma(v_1, ..., v_{2\ell}).\end{aligned}$$

Thus, ψ_σ is invariant. Weyl's theorem on orthogonal invariants can be stated in this language as follows; we refer to Weyl [158] for the proof.

1.9.2 Theorem. *Let V be a vector space of signature (p,q). If k is odd, then there are no invariant maps from $\otimes^k V$ to $\mathbb{R}$. If k is even, then the space of invariant maps from $\otimes^k V$ to $\mathbb{R}$ is spanned by the maps ψ_σ described in Equation* (1.9.1.b).

Let $\mathcal{B} := \{v_1, ..., v_{\dim V}\}$ be a basis for a vector space V of signature (p,q) and let $\mathcal{B}^* := \{v^1, ..., v^{\dim V}\}$ be the dual basis for the dual vector space V^*. If $w \in \otimes^k V^*$, then we may expand

$$w = \sum_{i_1,\dots,i_k} w(v_{i_1}, ..., v_{i_k}) v^{i_1} \otimes ... \otimes v^{i_k}.$$

Let $h_{ij} := (v_i, v_j)$ be the components of the inner product relative to the basis $\mathcal{B}$; the inverse matrix $h^{ij} = (v^i, v^j)$ describes the dual inner product on $\mathcal{B}^*$. We define

(1.9.2.a) $\quad \psi_1(w) := \sum_{ij} h^{ij} w(v_i, v_j)$ if $k = 2$,

(1.9.2.b) $\quad \psi_2(w) := \sum_{ijkl} h^{ij} h^{kl} w(v_i, v_j, v_k, v_l)$ if $k = 4$, and

(1.9.2.c) $\quad \psi_3(w) := \sum_{ijkl} h^{ik} h^{jl} w(v_i, v_k, v_j, v_l)$ if $k = 4$.

We replace V by V^* to derive the following Corollary from Theorem 1.9.2:

1.9.3 Corollary. *Let V be a vector space of signature (p,q).*

(1) *The map ψ_1 given in Equation* (1.9.2.a) *generates the space of invariant maps from $\otimes^2 V^*$ to $\mathbb{R}$.*

(2) *The maps ψ_2 and ψ_3 given in Equations* (1.9.2.b) *and* (1.9.2.c) *generate the space of invariant maps from $\otimes^4 V^*$ to $\mathbb{R}$.*

Let R be an algebraic curvature tensor on V. Let σ be a non-degenerate k dimensional subspace of V. In Section 1.5, we defined the higher order Jacobi operator $\mathcal{J}_R(\sigma)$ and the higher order skew-symmetric curvature operator $\mathfrak{S}_R(\sigma)$. We now show that the definitions of the higher order Jacobi operator and the higher order skew-symmetric curvature operator are independent of the bases chosen. If σ is spacelike, then $S(\sigma)$ and $\mathrm{Gr}_2(\sigma)$ are compact and inherit natural Riemannian measures.

1.9.4 Theorem. *Let R be an algebraic curvature tensor on a vector space V of signature (p,q). Let $\mathcal{B} := \{e_1, ..., e_k\}$ be a basis for a non-degenerate subspace σ of V. Let $h_{ij} := (e_i, e_j)$ and let h^{ij} be the inverse matrix.*

(1) *Let $\mathcal{J}_R(\sigma)y := \sum_{ij} h^{ij} R(y, e_i)e_j$ be the higher order Jacobi operator.*

1a) *$\mathcal{J}_R(\sigma)$ is independent of the choice of $\mathcal{B}$.*

1b) *If σ is spacelike, then $\mathcal{J}_R(\sigma) = \frac{k}{\mathrm{vol}(S(\sigma))} \int_{x \in S(\sigma)} \mathcal{J}_R(x)$.*

(2) *Let* $\mathfrak{S}_R(\sigma)y := \sum_{i,j,k,l} h^{ij}h^{kl}R(e_i,e_k)R(e_j,e_l)y$ *be the higher order skew-symmetric curvature operator.*

2a) $\mathfrak{S}_R(\sigma)$ *is independent of the basis* $\mathcal{B}$.

2b) *If* σ *is spacelike, then* $\mathfrak{S}_R(\sigma) = \frac{k(k-1)}{\mathrm{vol}(Gr_2^+(\sigma))} \int_{\pi\in Gr_2^+(\sigma)} R(\pi)^2$.

Proof. To prove Assertions (1a) and (2a), we apply Corollary 1.9.3 to the components of the maps involved to see that $\mathcal{J}_R(\sigma)$ and $\mathfrak{S}_R(\sigma)$ are independent of the basis $\mathcal{B}$. We suppose that σ is spacelike for the proof of Assertions (1b) and (2b). Let $\{e^1, ..., e^k\}$ be the corresponding dual basis for σ^*.

We first study the Jacobi operator. Let $T \in \otimes^2\sigma^*$. Then we may regard T as a linear map from $\sigma\otimes\sigma$ to $\mathbb{R}$. Define

$$J[T] := \int_{x\in S(\sigma)} T(x,x).$$

The map $T \to J[T]$ defines a linear map J from $\sigma^*\otimes\sigma^*$ to $\mathbb{R}$. Since the construction is basis free, the map is invariant. Let $T_{ij} = T(e_i, e_j)$. We apply Corollary 1.9.3 to see that there exists a real number c so that:

$$J[T] = c \cdot \sum_{ij} h^{ij}T_{ij}$$

for some constant c. We evaluate the constant c by considering the special case $T := \sum_{ij} h_{ij}e^i\otimes e^j$. Then $T(e_i,e_j) = h_{ij}$ so T represents the induced inner product on σ. We compute:

$$J[T] = \int_{x\in S(\sigma)} T(x,x) = \mathrm{vol}(S(\sigma)) = c\cdot\sum_{ij} h^{ij}h_{ij} = k\cdot c.$$

Consequently, $c^{-1} = \frac{k}{\mathrm{vol}(S(\sigma))}$ and we have the identity:

$$\text{(1.9.4.a)} \quad \sum_{ij} h^{ij}T(e_i,e_j) = \frac{k}{\mathrm{vol}\, S(\sigma)} \int_{x\in S(\sigma)} T(x,x) \text{ for any } T\in\sigma^*\otimes\sigma^*.$$

Let W be an auxiliary vector space and let $T : \sigma\otimes\sigma \to \mathrm{End}(W)$. We may then define:

$$J[T] = \int_{x\in S(\sigma)} T(x,x) \in \mathrm{End}(W).$$

By taking a basis for W and by studying the components of the matrix T, we apply the same argument to extend Equation (1.9.4.a) to this more general setting. Now let R be an algebraic curvature tensor and let W be the space of endomorphisms of V. Define $T : \sigma\otimes\sigma \to W$ by

$$T(x_1,x_2)y := R(y,x_1)x_2.$$

Assertion (1b) now follows from Equation (1.9.4.a).

The higher order skew-symmetric curvature operator can be discussed similarly. Let $T \in \otimes^4\sigma^*$. We anti-symmetrize to define

$$\begin{aligned}\tilde{T}(x,y,z,w) :=&T(x,y,z,w) - T(y,x,z,w)\\ &-T(x,y,w,z) + T(y,x,w,z).\end{aligned}$$

If $\pi := \operatorname{span}\{x,y\}$ is a 2 plane in σ, we may then define

$$T(\pi) := \{(x,x)(y,y) - (x,y)^2\}^{-1}\tilde{T}(x,y,x,y).$$

Since $\tilde{T}$ is alternating in the first and second pair of indices, we may use Equation (1.9.1.a) to see that $T(\pi)$ is independent of the particular basis chosen for π; the orientation plays no role here. We define

$$S[T] = \textstyle\int_{\pi\in\mathrm{Gr}_2(\sigma)} \tilde{T}(\pi).$$

The map $T \to S[T]$ is an invariant function. Thus, by Corollary 1.9.3, there exist universal constants c_1 and c_2 so that

(1.9.4.b) $S[T] = \sum_{ijkl}\{c_1h^{ij}h^{kl}T(e_i,e_j,e_k,e_l) + c_2h^{ij}h^{kl}T(e_i,e_k,e_j,e_l)\}$.

This discussion extends without change to vector valued maps T from $\otimes^4\sigma$ to an auxiliary vector space W.

Let R be an algebraic curvature tensor. For $x,y,z,w \in \sigma$, we define

$$T(x,y,z,w) := R(x,y)R(z,w)$$

as a map from $\otimes^4\sigma$ to the set of linear maps of V. We apply Equation (1.9.4.b); since $T(x,y,z,w) = -T(y,x,z,w)$,

$$\textstyle\sum_{ijkl} h^{ij}h^{kl}Te_i,e_j,e_k,e_l) = 0.$$

We therefore have:

$$\textstyle\int_{\pi\in\mathrm{Gr}_2(\sigma)} R(\pi)^2 = c_2\sum_{ijkl} h^{ij}h^{kl}R(e_i,e_k)R(e_j,e_l).$$

We complete the proof by evaluating the normalizing constant c_2. Let $R = R_{\mathrm{Id}}$ have constant sectional curvature $+1$. Then $-R(\pi)^2$ is orthogonal projection on π so $\operatorname{Tr}\{R(\pi)^2\} = -2$. Let $\{e_i\}$ be an orthonormal basis for σ. We take the trace to see:

$$\begin{aligned}&\operatorname{Tr}\{\textstyle\int_{\pi\in\mathrm{Gr}_2(\sigma)} R(\pi)^2\} = \int_{\pi\in\mathrm{Gr}_2(\sigma)} \operatorname{Tr}\{R(\pi)^2\} = -2\operatorname{vol}(\mathrm{Gr}_2(\sigma))\\ &= \operatorname{Tr}\{c_2\textstyle\sum_{ijkl} h^{ij}h^{kl}R(e_i,e_k)R(e_j,e_l)\} = \sum_{i,j}\operatorname{Tr}\{c_2R(e_i,e_j)^2\}\\ &= -2c_2k(k-1) \text{ so}\\ &c_2^{-1} = \tfrac{k(k-1)}{\operatorname{vol}(\mathrm{Gr}_2(\sigma))}. \quad \square\end{aligned}$$

1.10 Natural operators with constant eigenvalues

Let R be an algebraic curvature tensor which is defined on a vector space V of signature (p,q). Let $\mathcal{E}_R$ be a natural operator which is associated to R and whose natural domains of definition are the oriented or unoriented Grassmannians of non-degenerate planes of type (r,s) for $r+s=k$. For example, if $\mathcal{E}$ is the Jacobi operator, then the domains of definition are the oriented Grassmannians $\text{Gr}^+_{1,0}(V)=S^-(V)$ and $\text{Gr}^+_{0,1}(V)=S^+(V)$; here $(r,s)\in\{(0,1),(1,0)\}$, and $k=1$.

In this section, we show that if the eigenvalues of $\mathcal{E}_R$ are constant on one non-trivial Grassmannian of type (r,s), then they are constant on any other non-trivial Grassmannian of type $(\tilde{r},\tilde{s})$, where $\tilde{r}+\tilde{s}=r+s$. Lemma 1.10.1 deals with the Jacobi operator, Lemma 1.10.2 deals with the higher order Jacobi operator, Lemmas 1.10.5 and 1.10.7 deal with the skew-symmetric operator in the real and in the complex settings, Lemma 1.10.8 deals with the higher order skew-symmetric curvature operator, and Lemma 1.10.9 deals with the Szabó operator.

Throughout this section, let $\mathcal{B}=\{e_1,...,e_{\dim V}\}$ be a basis for V, let $h_{ij}:=(e_i,e_j)$ be the components of the inner product relative to the basis $\mathcal{B}$, and let h^{ij} be the inverse matrix.

We begin by discussing the *Jacobi operator* $\mathcal{J}_R(x):y\rightarrow R(y,x)x$. Recall that the *Ricci tensor* ρ_R is defined by:

$$\rho_R(x,y):=\text{Tr}(z\rightarrow R(z,x)y)=\textstyle\sum_{1\le i,j\le\dim(V)}h^{ij}R(e_i,x,y,e_j).$$

We say that R is *spacelike Osserman* (resp. *timelike Osserman*) if the eigenvalues of $\mathcal{J}_R$ are constant on the pseudosphere $S^+(V)$ (resp. $S^-(V)$). We say that R is *k-stein* if there exist constants c_i so that

$$\text{Tr}(\mathcal{J}_R(x)^k)=c_i(x,x)^k\text{ for }1\le i\le k.$$

1.10.1 Lemma. *Let R be an algebraic curvature tensor on a vector space V of signature (p,q) and dimension $m=p+q$. The following assertions are equivalent and all define the notion of an Osserman algebraic curvature tensor:*

(1) *If $p\ge1$, then R is timelike Osserman.*
(2) *If $q\ge1$, then R is spacelike Osserman.*
(3) *The tensor R is m-stein.*

Proof. Let R be an algebraic curvature tensor on a vector space V of signature (p,q) and dimension $m=p+q$. Let

$$f_{i,R}(x):=\text{Tr}(\mathcal{J}_R(x)^i).$$

We use Lemma 1.2.6 to see that the eigenvalues of $\mathcal{J}_R$ are constant on $S^\pm(V)$ if and only if the functions $f_{i,R}$ are constant on $S^\pm(V)$.

If R is m-stein, then there exist constants c_i so $f_{i,R}(x) = c_i(x,x)^i$ for $1 \leq i \leq m$ and for any $x \in V$. This implies that the functions $f_{i,R}$ are constant on the pseudo-spheres $S^\pm(V)$ and hence R is both timelike and spacelike Osserman. Thus, Assertion (3) implies Assertions (1) and (2).

To prove that Assertions (1) or (2) imply Assertion (3), we use the same argument as that used in the proof of Lemmas 1.6.2 and 1.7.2. We suppose that $q \geq 1$ and that R is timelike Osserman; the spacelike case is similar and is therefore omitted. We then have constants c_i for $1 \leq i \leq m$ so $f_{i,R}(\cdot) = (-1)^i c_i$ on $S^-(V)$. Since $f_{i,R}$ is homogeneous of degree $2i$, we may then conclude that

$$f_{i,R}(x) - c_i(x,x)^i = 0 \text{ for any timelike vector } x. \tag{1.10.1.a}$$

Since $q \geq 1$, there is at least one timelike vector x_0. Let y be an arbitrary element of V and let ε be a real parameter. For $i \leq m$, define:

$$F_i(\varepsilon) := \operatorname{Tr}((\mathcal{J}_R(y+\varepsilon x_0))^i) - c_i(y+\varepsilon x_0, y+\varepsilon x_0)^i.$$

For large values of ε, $y + \varepsilon x_0$ is timelike and hence, by Equation (1.10.1.a), $F_i(\varepsilon) = 0$. Since F_i is polynomial in the parameter ε, we may now conclude that F_i vanishes identically. Setting $\varepsilon = 0$, we see that $f_{i,R}(y) - c_i(y,y)^i = 0$ for all $y \in V$ and for $1 \leq i \leq m$. This implies that R is m-stein. □

Let $\mathcal{B} := \{e_1, ..., e_k\}$ be a basis for a non-degenerate k plane σ. In Section 1.5, we defined the *higher order Jacobi operator* by setting

$$\mathcal{J}_R(\sigma) : y \to \textstyle\sum_{1 \leq i,j \leq k} h^{ij} R(y, e_i) e_j.$$

In Theorem 1.9.4 we showed $\mathcal{J}_R(\sigma)$ was independent of $\mathcal{B}$. A pair (r,s) is an *admissible pair* if the Grassmannian $\operatorname{Gr}_{r,s}(V)$ is non-trivial, i.e.

$$0 \leq r \leq p,\ 0 \leq s \leq q, \text{ and } 1 \leq r+s \leq p+q-1.$$

An algebraic curvature tensor R is said to be *Osserman of type* (r,s) if the eigenvalues of $\mathcal{J}_R(\sigma)$ are constant on $\operatorname{Gr}_{r,s}(V)$. We can now show that only the value $r+s$ is relevant. Recall $\widetilde{\operatorname{Gr}}_k(V)$ is the Grassmannian of all non-degenerate k planes.

1.10.2 Lemma. *Let R be an algebraic curvature tensor on a vector space V of signature (p,q) and dimension $m = p+q$. Let $1 \leq k \leq m-1$. The* *following assertions are equivalent and all define the notion of a k Osserman algebraic curvature tensor:*

(1) *There exists an admissible pair* (r,s) *with* $r+s=k$ *so that the eigenvalues of* $\mathcal{J}_R$ *are constant on* $\mathrm{Gr}_{r,s}(V)$.

(2) *The eigenvalues of* $\mathcal{J}_R$ *are constant on* $\widetilde{\mathrm{Gr}}_k(V)$.

(3) *If* $1 \le i \le m$, *then* $\mathrm{Tr}\{\mathcal{J}_R(\cdot)^i\}$ *is constant on* $\widetilde{\mathrm{Gr}}_k(V)$.

Proof. It is immediate that Assertion (2) implies Assertion (1). Furthermore, by Lemma 1.2.6, Assertions (2) and (3) are equivalent. We complete the proof by establishing the reverse implication.

Let $V^k := V \times ... \times V$ be the space of all k tuples and let

$$\mathcal{O} := \{(v_1, ..., v_k) \in V^k : \det((v_i, v_j)) \neq 0\}$$

be the open subset of non-degenerate k frames. If $\vec{v} \in \mathcal{O}$, then we may set $\pi(\vec{v}) := \mathrm{span}\{v_1, ..., v_k\}$; this is a non-degenerate k plane. We may decompose $\mathcal{O}$ as the disjoint union

$$\mathcal{O} = \cup_{r+s=k}\mathcal{O}_{(r,s)},$$

where $\mathcal{O}_{(r,s)}$ is the open subset of k tuples $\vec{v}$ so that $\pi(\vec{v})$ has type (r,s). We define

$$f_{i,R}(\vec{v}) := \mathrm{Tr}\{\mathcal{J}_R(\pi(\vec{v}))^i\}.$$

Since, by Lemma 1.2.6, the functions $f_{i,R}(\vec{v})$ determine the characteristic polynomial of the higher order Jacobi operator $\mathcal{J}_R(\pi(\vec{v}))$, we see that the eigenvalues of $\mathcal{J}_R$ are constant on $\mathrm{Gr}_{r,s}(V)$ if and only if the functions $f_{i,R}$ are constant on $\mathcal{O}_{(r,s)}$.

We complexify and let $V_{\mathbb{C}} := V \otimes_{\mathbb{R}} \mathbb{C}$. We extend the metric $(\cdot,\cdot)$ and the algebraic curvature tensor R to be complex multi-linear. The functions $f_{i,R}$ then extend to holomorphic functions on

$$\mathcal{O}_{\mathbb{C}} := \{(v_1, ..., v_k) \in V_{\mathbb{C}}^k : \det((v_i, v_j)) \neq 0\}.$$

Let $m = \dim V$. By choosing a basis for V_C, we may identify $V_{\mathbb{C}}^k = \mathbb{C}^{mk}$. By Lemma 1.6.3, $\mathcal{O}$ is a connected open subset of $\mathbb{C}^{mk}$. Since $\mathcal{O}_{(r,s)}$ is a non-empty open subset of $\mathcal{O} \cap \mathbb{R}^{mk}$, we may use Lemma 1.6.3 to see that if the functions $f_{i,R}$ are constant on $\mathcal{O}_{(r,s)}$, then the functions $f_{i,R}$ are constant on $\mathcal{O}_{\mathbb{C}}$. This shows that the functions $f_{i,R}$ are constant on $\cup_{\tilde{r}+\tilde{s}=k}\mathcal{O}_{(\tilde{r},\tilde{s})}$. □

An algebraic curvature tensor R is said to be *Jordan Osserman of type* (r,s) if the Jordan normal form of $\mathcal{J}_R$ is constant on $\mathrm{Gr}_{r,s}(V)$. Clearly, if

R is Jordan Osserman of type (r, s), then R is Osserman of type (r, s) and hence is k Osserman, where $k = r + s$. Let $p \geq 2$ and let $q \geq 2$. In Lemma 3.2.2, we will construct an algebraic curvature tensor which is Osserman of type $(p, 0)$ but which is not Jordan Osserman of type $(p-1, 1)$; thus the value $r + s$ does not suffice to determine if R is Jordan Osserman. Furthermore, Osserman does not imply Jordan Osserman in the higher order context.

We use Lemma 1.10.2 to derive the following duality result of Gilkey, Stanilov, and Videv [85].

1.10.3 Lemma. *Let R be an algebraic curvature tensor on a vector space V of signature (p, q) and dimension $m = p + q \geq 2$. Suppose that R is k Osserman, where $1 \leq k \leq m - 1$.*

(1) *If v is a null vector, then* $\text{Tr}\{\mathcal{J}_R(v)^i\} = 0$ *for any i.*
(2) *The curvature tensor R is Einstein.*
(3) *The curvature tensor R is $m - k$ Osserman.*
(4) *If R is Jordan Osserman of type (r, s), then R is Jordan Osserman of type $(p - r, q - s)$.*

Proof. We complexify and let

$$\mathcal{N} := \{x \in V \otimes \mathbb{C} : (x, x) = 0\}$$

be the variety of complex null-vectors. Let $0 \neq x_1 \in \mathcal{N}$. Choose $x_2 \notin x_1^\perp$. Define linear functions T_i on $V \otimes \mathbb{C}$ by $T_i(x) := (x, x_i)$. As

$$T_1(x_1) = 0,\ T_1(x_2) \neq 0, \text{ and } T_2(x_1) \neq 0,$$

the two linear functions T_1 and T_2 are linearly independent. We define:

$$W := \ker(T_1) \cap \ker(T_2) = x_1^\perp \cap x_2^\perp.$$

If $a_1x_1 + a_2x_2 \in W$, then $0 = (a_1x_1 + a_2x_2, x_1) = a_2(x_2, x_1)$ so $a_2 = 0$ since $(x_2, x_1) \neq 0$. If $a_1x_1 \in W$, then $0 = (a_1x_1, x_2) = a_1(x_1, x_2)$ so $a_1 = 0$. Thus,

$$\text{span}\{x_1, x_2\} \cap W = \{0\}.$$

Consequently, we can choose a basis $\mathcal{B}$ for V of the form

$$\mathcal{B} = \{x_1, x_2, w_3, ..., w_m\}$$

so that the vectors $\{w_3, ..., w_m\}$ form a basis for W. Suppose that $w \in W$ and that $(w, w_i) = 0$ for $3 \leq i \leq m$. Since $(w, x_1) = (w, x_2) = 0$, this implies that $(w, v) = 0$ for all $v \in V$ and hence $w = 0$. This shows that the induced inner product on W is non-degenerate.

Since $k - 1 \leq m - 2 = \dim W$, we may choose a non-degenerate $k - 1$ plane $\sigma \subset W$. Let

$$x_t := x_1 + tx_2 \text{ and } g(t) := (x_t, x_t) = (x_2, x_2)t^2 + 2t(x_1, x_2).$$

Since $(x_1, x_2) \neq 0$, $g(t)$ is a non-trivial polynomial of degree at most 2. This shows that $g(t)$ has at most two roots; in particular, there exists $\varepsilon > 0$ so $g(t) \neq 0$ for $t \in (0, \varepsilon)$. We consider the non-degenerate k plane

$$\pi_t := \sigma \oplus \operatorname{span}\{x_t\} \text{ for } t \in (0, \varepsilon).$$

Since R is k Osserman, we apply Lemma 1.10.2 to see there are universal constants c_i so that

$$c_i = \operatorname{Tr}\{\mathcal{J}_R(\pi_t)^i\} \text{ for } 1 \leq i \leq m.$$

Since $\pi_t = \sigma \oplus \operatorname{span}\{x_t\}$ is an orthonormal direct sum,

$$\begin{aligned}
\mathcal{J}_R(\pi_t) &= \mathcal{J}_R(\sigma) + g(t)^{-1}\mathcal{J}_R(x_t), \\
c_i &= \operatorname{Tr}\{[\mathcal{J}_R(\sigma) + g(t)^{-1}\mathcal{J}_R(x_t)]^i\}, \text{ and} \\
g(t)^i c_i &= g(t)^i \operatorname{Tr}\{[\mathcal{J}_R(\sigma) + g(t)^{-1}\mathcal{J}_R(x_t)]^i\} \\
&= \operatorname{Tr}\{[g(t)\mathcal{J}_R(\sigma) + \mathcal{J}_R(x_t)]^i\}.
\end{aligned}$$

We now take the limit as $t \downarrow 0$ to prove Assertion (1) by computing:

$$\text{(1.10.3.a)} \qquad \operatorname{Tr}(\mathcal{J}_R(x_1)^i) = 0 \text{ for any } x_1 \in \mathcal{N}.$$

Let x_i be complex vectors with $g(x_i, x_i) = 1$ and $g(x_i, x_j) = 0$ for $i \neq j$. Let $z := x_1 + \sqrt{-1}x_2 \in \mathcal{N}$. We expand

$$\mathcal{J}_R(z) = \mathcal{J}_R(x_1) - \mathcal{J}_R(x_2) + 2\sqrt{-1}\mathcal{J}_R(x_1, x_2).$$

Let ρ_R be the complexification of the Ricci tensor. If $\{e_1, ..., e_m\}$ is an orthonormal basis for V, then

$$\rho_R(x, y) = \textstyle\sum_i (e_i, e_i) R(x, e_i, e_i, y).$$

We use Equation (1.10.3.a) to see that:

$$\begin{aligned}
0 &= \operatorname{Tr}(\mathcal{J}_R(z)) \\
\text{(1.10.3.b)} \qquad &= \operatorname{Tr}(\mathcal{J}_R(x_1)) - \operatorname{Tr}(\mathcal{J}_R(x_2)) + 2\operatorname{Tr}(\sqrt{-1}\mathcal{J}_R(x_1, x_2)) \\
&= \rho_R(x_1, x_1) - \rho_R(x_2, x_2) + 2\sqrt{-1}\rho_R(x_1, x_2).
\end{aligned}$$

We replace x_2 by $-x_2$ in Equation (1.10.3.b) to see

(1.10.3.c) $$0 =\rho_R(x_1,x_1) - \rho_R(x_2,x_2) - 2\sqrt{-1}\rho_R(x_1,x_2).$$

We combine Equations (1.10.3.b) and (1.10.3.c) to establish the identities:

$$0 = \rho_R(x_1,x_1) - \rho_R(x_2,x_2) \text{ and } \rho_R(x_1,x_2) = 0.$$

This shows that there exists a constant c_1 so $\rho_R(x,x) = c_1(x,x)$ for any complex vector x and, consequently, R is Einstein; this completes the proof of Assertion (2).

We complete the proof by establishing the duality results of Assertions (3) and (4). We have assumed that the curvature tensor R is k Osserman. Let τ be a non-degenerate $m-k$ plane and let $\tau^\perp$ be the non-degenerate complementary k plane. Let $\{e_1, ..., e_k\}$ be an orthonormal basis for $\tau^\perp$ and let $\{e_{k+1}, ..., e_m\}$ be an orthonormal basis for τ. We use Assertion (2) to see R is Einstein. We may then use Lemma 1.7.2 to see that there exists a constant c_1 so that $\sum_i(e_i,e_i)\mathcal{J}_R(e_i) = c_1 \operatorname{Id}$. Consequently, we have:

(1.10.3.d) $$c_1 \operatorname{Id} = \textstyle\sum_{i\le k}\varepsilon_i\mathcal{J}_R(e_i) + \sum_{i>k}\varepsilon_i\mathcal{J}_R(e_i) = \mathcal{J}_R(\sigma^\perp) + \mathcal{J}_R(\sigma).$$

Therefore, the eigenvalues and Jordan normal form of $\mathcal{J}_R(\sigma)$ are determined by the eigenvalues and Jordan normal form of $\mathcal{J}_R(\sigma^\perp)$. Assertions (3) and (4) now follow. □

In Lemma 1.10.3, we showed that any k Osserman algebraic curvature tensor was Einstein. We can now show that R is *zweistein* (2 stein) if we impose the restriction that $2k \ne \dim V$.

1.10.4 Lemma. *Let R be a k Osserman algebraic curvature tensor on a vector space of signature (p,q) and dimension $m := p+q$, for $1 \le k \le m-1$.*

(1) *There exists a constant C so that if σ is any non-degenerate k plane, then $\operatorname{Tr}\{\mathcal{J}_R(\sigma)\mathcal{J}_R(\sigma^\perp)\} = C$.*
(2) *If $2k \ne m$, then R is 2-stein.*

Proof. Let $\widetilde{\operatorname{Gr}}_k(V)$ be the set of non-degenerate k dimensional subspaces of V. Let R be a k Osserman algebraic curvature tensor on V. We apply Lemma 1.10.3 to see that R is Einstein. Thus, as noted above in Equation (1.10.3.d), there exists a constant c_1 so that

(1.10.4.a) $$\mathcal{J}_R(\sigma) + \mathcal{J}_R(\sigma^\perp) = c_1 \operatorname{Id} \text{ for all } \sigma \in \widetilde{\operatorname{Gr}}_k(V).$$

We use Lemma 1.10.3 to see that R is $m-k$ Osserman. Thus, by Lemma 1.10.2, the functions $\operatorname{Tr}\{\mathcal{J}_R(\sigma)^2\}$ and $\operatorname{Tr}\{\mathcal{J}_R(\sigma^\perp)^2\}$ are constant on $\widetilde{\operatorname{Gr}}_k(V)$. We use Equation (1.10.4.a) to see that the function $\operatorname{Tr}\left\{\{\mathcal{J}_R(\sigma)+\mathcal{J}_R(\sigma^\perp)\}^2\right\}$ is constant on $\widetilde{\operatorname{Gr}}_k(V)$. We establish Assertion (1) by computing

$$\begin{aligned}\operatorname{Tr}\{\mathcal{J}_R(\sigma)\mathcal{J}_R(\sigma^\perp)\} =&\tfrac{1}{2}\operatorname{Tr}\left\{\{\mathcal{J}_R(\sigma)+\mathcal{J}_R(\sigma^\perp\}^2\right\}\\ &-\tfrac{1}{2}\operatorname{Tr}\{\mathcal{J}_R(\sigma)^2\}-\tfrac{1}{2}\operatorname{Tr}\{\mathcal{J}_R(\sigma^\perp)^2\}.\end{aligned}$$

We follow the argument given in Gilkey, Stanilov, and Videv [85] to give a rather combinatorial proof of Assertion (2). We must show that if $2k\neq m=\dim(V)$ and if R is k Osserman, then R is 2-stein. The case $k=1$ follows from Lemma 1.10.1; the case $k=m-1$ then follows by duality from Lemma 1.10.3. We may therefore suppose $2\leq k\leq m-2$. Thus, $m\geq 4$; since $2k\neq m$, $m\geq 5$.

We suppose given unit vectors e_1 and e_2 so $e_1\perp e_2$. We can extend the set $\{e_1,e_2\}$ to an orthonormal basis $\mathcal{B}:=\{e_1,...,e_m\}$ for V; we then have:

$$(e_i,e_j)=0 \text{ for } i\neq j \text{ and } (e_i,e_i):=\varepsilon_i=\pm 1.$$

Let $I=\{i_2,...,i_k\}$ be an ordered collection of $k-1$ distinct indices which range from 2 to m and let $\mathcal{I}$ be the set of all such collections; the number of such collections is then given by

$$|\mathcal{I}|=(m-1)...(m+1-k).$$

For $I\in\mathcal{I}$, we define:

$$\begin{aligned}&\sigma_I:=\operatorname{span}\{e_1,e_{i_2},...,e_{i_k}\}\in\widetilde{\operatorname{Gr}}_k(V) \text{ and}\\ &\mathcal{E}:=\textstyle\sum_{I\in\mathcal{I}}\operatorname{Tr}\{\mathcal{J}_R(\sigma_I)\mathcal{J}_R(\sigma_I^\perp)\};\end{aligned}$$

by Assertion (1), $\mathcal{E}=|\mathcal{I}|C$ is independent of the basis $\mathcal{B}$. Let $J_i:=\varepsilon_i\mathcal{J}_R(e_i)$. We then have

$$\mathcal{J}_R(\sigma_I)=\mathcal{J}_1+\textstyle\sum_{i\in I}\mathcal{J}_i \text{ and } \mathcal{J}_R(\sigma_i^\perp)=\sum_{j\notin I}\mathcal{J}_j.$$

By taking the multiplicities involved into account, we have

$$\mathcal{E}=a_{k,m}\textstyle\sum_{i>1,j>1,i\neq j}\operatorname{Tr}\{J_iJ_j\}+b_{k,m}\sum_{i>1}\operatorname{Tr}\{J_1J_i\},$$

where the coefficients $a_{k,m}$ are given by:

$$\begin{aligned}a_{k,m}:=&|\{I\in\mathcal{I}:i\in I,j\notin I\}|\\ =&\begin{cases}1 & \text{if } k=2,\\ (k-1)(m-3)...(m-k) & \text{if } k\geq 3,\end{cases}\end{aligned}$$

and the coefficients $b_{k,m}$ are given by:

$$b_{k,m} :=|\{I \in \mathcal{I} : i \notin I\}| = \begin{cases} m-2 & \text{if } k=2, \\ (m-2)(m-3)...(m-k) & \text{if } k \geq 3. \end{cases}$$

We use the Einstein condition to see:

$$J_1 + J_2 + ... + J_m = c_1 \operatorname{Id}, \text{ and } \operatorname{Tr}\{J_1\} = c_1.$$

Consequently, we may express

$$\begin{aligned} \mathcal{E} =& a_{k,m} \textstyle\sum_{i,j} \operatorname{Tr}\{J_i J_j\} + (b_{k,m} - 2a_{k,m}) \sum_i \operatorname{Tr}\{J_1 J_i\} \\ & - a_{k,m} \textstyle\sum_i \operatorname{Tr}\{J_i^2\} + (2a_{k,m} - b_{k,m}) \operatorname{Tr}\{J_1^2\} \\ =& m a_{k,m} c_1^2 + (b_{k,m} - 2a_{k,m}) c_1^2 - a_{k,m} \textstyle\sum_i \operatorname{Tr}\{J_i^2\} \\ & + (2a_{k,m} - b_{k,m}) \operatorname{Tr}\{J_1^2\}. \end{aligned}$$

A priori, the term $\sum_i \operatorname{Tr}\{J_i^2\}$ depends on the basis $\mathcal{B}$ but not on the order of the elements in the basis. We interchange the roles of e_1 and e_2 to see:

$$(1.10.4.b) \qquad (2a_{k,m} - b_{k,m})\{\operatorname{Tr}\{J_1^2\} - \operatorname{Tr}\{J_2^2\}\} = 0.$$

We study the coefficient $2a_{k,m} - b_{k,m}$ appearing in Equation (1.10.4.b):

$$2a_{k,m} - b_{k,m} = \begin{cases} 2-(m-2) = 4-m = 2k-m \neq 0 & \text{if } k=2, \\ (2k-2-m+2)(m-3)...(m-k) \neq 0 & \text{if } k \geq 3. \end{cases}$$

Thus, by Equation (1.10.4.b), $\operatorname{Tr}\{\mathcal{J}_R(v_1)^2\} = \operatorname{Tr}\{\mathcal{J}_R(v_2)^2\}$ if $\{v_1, v_2\}$ is an orthonormal subset of V. Let x and y be arbitrary unit vectors. We can choose a unit vector z so $z \perp x$ and $z \perp y$. We may then complete the proof by computing:

$$\operatorname{Tr}\{\mathcal{J}_R(x)^2\} = \operatorname{Tr}\{\mathcal{J}_R(z)^2\} = \operatorname{Tr}\{\mathcal{J}_R(y)^2\}. \quad \square$$

Let R be an algebraic curvature tensor which is defined on a vector space V of signature (p, q). Let $\{x, y\}$ be an oriented basis for a non-degenerate 2 plane π. In Section 1.5, we defined the skew-symmetric curvature operator

$$R(\pi) := |(x,x)(y,y) - (x,y)^2|^{-1/2} R(x,y);$$

we showed in Lemma 1.9.1 that this was independent of the particular oriented basis which was chosen. Recall that R is *spacelike IP*, *timelike IP*, or *mixed IP* if the eigenvalues of $R(\pi)$ are constant on the appropriate Grassmannian.

1.10.5 Lemma. *Let R be an algebraic curvature tensor on a vector space V of signature (p,q). The following assertions are equivalent and all define the notion of an IP algebraic curvature tensor:*

(1) *If $q \geq 2$, then R is timelike IP.*

(2) *If $p \geq 2$, then R is spacelike IP.*

(3) *If $p \geq 1$ and $q \geq 1$, then R is mixed IP.*

Proof. We adapt the proof given for Lemma 1.10.2 to this setting. We let

$$\mathcal{O} := \{(v_1, v_2) \in V \times V : (v_1, v_1)(v_2, v_2) - (v_1, v_2)^2 \neq 0\}$$

be the set of all non-degenerate oriented 2 frames. We define

$$f_{i,R}(v_1, v_2) = \operatorname{Tr}\{R(\operatorname{span}\{v_1, v_2\})^i\}^2 = \frac{\operatorname{Tr}\{R(v_1, v_2)^i\}^2}{((v_1, v_1)(v_2, v_2) - (v_1, v_2)^2)^i}.$$

Clearly, the functions $\operatorname{Tr}\{R(\operatorname{span}\{v_1, v_2\})^i\}$ are constant if and only if the functions $\operatorname{Tr}\{R(\pi)^i\}^2$ are constant so we have not lost any information. But with this definition, we have eliminated the absolute value sign so the functions $f_{i,R}$ extend holomorphically to the space of complex non-degenerate 2 frames. The desired result now follows using the identity theorem in exactly the same way as was used to prove Lemma 1.10.2. □

We can generalize Assertion (1) of Lemma 1.10.3 to this setting.

1.10.6 Lemma. *Let R be an IP algebraic curvature tensor on a vector space V of signature (p,q), where $q \geq 3$. Let $\{x, y\}$ span a degenerate 2 plane π. Then $R(x, y)$ is nilpotent.*

Proof. We may choose a basis $\{v_1, v_2\}$ for π so that $(v_1, v_1) = 0$ and so that $v_1 \perp v_2$. Then $R(x, y)$ is a non-zero multiple of $R(v_1, v_2)$ so it suffices to show $R(v_1, v_2)$ is nilpotent or equivalently, using Lemmas 1.2.6 and 1.2.8, that $\operatorname{Tr}\{R(v_1, v_2)^i\} = 0$ for any i. By perturbing v_2 slightly if necessary, we may suppose without loss of generality that v_2 is not a null vector and thus we may assume $v_2 \in S^{\pm}(V)$.

Since $q \geq 3$, we can choose v_3 a spacelike unit vector so $v_3 \perp \pi$. Let $\{t^{-1}v_1 + v_3, v_2\}$ be an orthonormal basis for a non-degenerate 2 plane π_t. Since R is IP, there exist universal constants c_i so

$$c_i := \operatorname{Tr}\{R(t^{-1}v_1 + v_3, v_2)^i\}$$

for any $t > 0$. Thus, taking into account the homogeneity involved, we have:

$$t^i c_i = \operatorname{Tr}\{R(v_1 + tv_3, v_2)^i\} \text{ for } t > 0.$$

We take the limit as $t \downarrow 0$ to complete the proof. $\square$

Let R be an algebraic curvature tensor which is defined on a vector space V of signature (p, q). We suppose given a pseudo-Hermitian almost complex structure J so that $J^*R = R$. We use Lemma 1.6.2 to see that $R(\pi)$ is a complex linear map for any non-degenerate complex line π. Recall that R is said to be *spacelike almost complex IP* or *timelike almost complex IP* if the eigenvalues of $R(\pi)$, regarded as a complex linear map, are constant on the Grassmannians of spacelike or timelike complex lines; there are no mixed complex lines.

1.10.7 Lemma. *Let R be an almost complex algebraic curvature tensor on a vector space V of signature (p, q) and dimension $m = p+q$. The following assertions are equivalent and all define the notion of an almost complex IP algebraic curvature tensor:*

(1) *There exist constants c_i so that $\operatorname{Tr}\{(JR(x, Jx))^i\} = c_i(x, x)^i$ for any $x \in V$ and for $1 \leq i \leq m$.*

(2) *If $q \geq 2$, then R is almost complex timelike IP.*

(3) *If $p \geq 2$, then R is almost complex spacelike IP.*

Proof. We use Lemma 1.2.6 to see that Assertion (1) implies Assertion (2). Suppose that Assertion (2) holds. Since the eigenvalues of $JR(x, Jx)$ are constant on $S^-(V)$, we may use Lemma 1.2.6 to see that there exist constants so the functions $f_i(x) := \operatorname{Tr}(JR(x, Jx)^i) - c_i(x, x)^i$ vanish on $S^-(V)$. Taking into account the homogeneity involved, we see that $f_i(x)$ vanishes on all timelike vectors. The proof of Lemma 1.10.1 then extends to this setting to yield that f_i vanishes identically and Assertion (1) holds. The proof that Assertion (3) implies Assertion (1) is similar and therefore omitted. $\square$

Let R be an algebraic curvature tensor which is defined on a vector space V of signature (p, q). Let $\mathcal{B} := \{e_1, ..., e_k\}$ be a basis for a non-degenerate k plane $\sigma \subset V$. In Section 1.5, we defined the *higher order skew-symmetric curvature operator*

$$\mathfrak{S}_R(\sigma) := \textstyle\sum_{i,j,k,l} h^{ik} h^{jl} R(e_i, e_j) R(e_k, e_l);$$

we showed in Lemma 1.9.4 that this operator was independent of the basis $\mathcal{B}$ chosen for σ. In analogy to the higher order Osserman operator, we say that an algebraic curvature tensor R is *IP of type* (r, s) if the eigenvalues of $\mathfrak{S}_R(\sigma)$ are constant on the Grassmannian $\operatorname{Gr}_{r,s}(V)$ of all unoriented planes of type (r, s). The proof of Lemma 1.10.2 extends immediately to this setting to yield:

1.10.8 Lemma. *Let R be an algebraic curvature tensor on a vector space V of signature (p,q) and dimension $m = p+q$. Let $1 \le k \le m-1$. The following assertions are equivalent and all define the notion of a k-IP algebraic curvature tensor:*

(1) *There exists an admissible pair (r,s) with $r+s=k$, $0 \le r \le p$, and $0 \le s \le q$ so that the eigenvalues of $\mathfrak{S}_R$ are constant on $\operatorname{Gr}_{r,s}(V)$.*

(2) *The eigenvalues of $\mathfrak{S}_R$ are constant on $\widetilde{\operatorname{Gr}}_k(V)$.*

(3) *If $1 \le i \le m$, then $\operatorname{Tr}\{\mathfrak{S}_R(\cdot)^i\}$ is constant on $\widetilde{\operatorname{Gr}}_k(V)$.*

Let ∇R be an *algebraic covariant derivative*; this means that $\nabla R \in \otimes^5 V^*$ is a 5 tensor which satisfies the symmetries of Equations (1.5.4.b)-(1.5.4.d):

$$\begin{aligned}
&\nabla R(a,b,c,d;e) = -\nabla R(b,a,c,d;e) = \nabla R(c,d,a,b;e)\\
&\nabla R(a,b,c,d;e) + \nabla R(a,c,d,b;e) + \nabla R(a,d,b,c;e) = 0\\
&\nabla R(a,b,c,d;e) + \nabla R(a,b,d,e;c) + \nabla R(a,b,e,c;d) = 0.
\end{aligned}$$

Here ∇R is to be understood in a purely formal sense. In Section 1.5, we defined the Szabó operator $\mathcal{S}_{\nabla R}(x)$ by the identity:

$$(\mathcal{S}_{\nabla R}(x)y, z) = \nabla R(y,x,x,z;x).$$

Recall that ∇R is said to be *spacelike Szabó* or *timelike Szabó* if the eigenvalues of $\mathcal{S}_{\nabla R}$ are constant on the pseudo-spheres $S^{\pm}(V)$.

1.10.9 Lemma. *Let ∇R be a covariant derivative algebraic curvature tensor on a vector space V of signature (p,q) and dimension m. The following assertions are equivalent and all define the notion of a Szabó covariant derivative algebraic curvature tensor:*

(1) *There exist constants c_i so that for any $x \in S^{\pm}(V)$ and for any i with $2i \le m$, $\operatorname{Tr}\{\mathcal{S}_{\nabla R}(x)^{2i}\} = c_i(x,x)^{3i}$ and $\operatorname{Tr}\{\mathcal{S}_{\nabla R}(x)^{2i-1}\} = 0$.*

(2) *If $p \ge 1$, then R is timelike Szabó.*

(3) *If $q \ge 1$, then R is spacelike Szabó.*

Proof. We use Lemma 1.2.6 to see that Assertion (1) implies Assertions (2) and (3). Conversely, suppose Assertion (2) holds. Then there exist constants b_i so that

$$\operatorname{Tr}\{\mathcal{S}_{\nabla R}(x)^i\} = b_i \text{ if } x \in S^-(V).$$

Since $\mathcal{S}_{\nabla R}(-x) = -\mathcal{S}_{\nabla R}(x)$, the constants b_i vanish for i odd. Taking the homogeneity involved into account, we see therefore that

$$\text{(1.10.9.a)} \qquad \operatorname{Tr}\{\mathcal{S}_{\nabla R}(x)^{2i-1}\} = 0, \text{ and } \operatorname{Tr}\{\mathcal{S}_{\nabla R}(x)^{2i}\} = c_i(x,x)^{3i}$$

for any timelike vector x and $1 \leq i \leq m$. The same argument used to prove Lemma 1.10.1 now can be used to extend Equation (1.10.9.a) from the open subset of timelike vectors to all of V and thereby to show that Assertion (2) implies Assertion (1); similarly Assertion (3) implies Assertion (1). □

1.11 The exponential map and Jacobi vector fields

We begin with a technical observation. Let ∇ be the Levi-Civita connection of a pseudo-Riemannian manifold (M, g). Let $x = (x_1, ..., x_m)$ be a system of local coordinates on M. Let

$$g_{ij} := (\partial_i, \partial_j) \text{ and } \Gamma_{ijk} := (\nabla_{\partial_i}\partial_j, \partial_k) = \tfrac{1}{2}\{\partial_i g_{jk} + \partial_j g_{ik} - \partial_k g_{ij}\}$$

give the components of the metric and of the Christoffel symbols relative to the coordinate frame $\partial_i := \frac{\partial}{\partial x_i}$. If we decompose $\nabla_{\partial_i}\partial_j = \sum_k \Gamma_{ij}{}^k \partial_k$, then:

$$\Gamma_{ijk} = \textstyle\sum_l g_{kl}\Gamma_{ij}{}^l, \text{ and } \Gamma_{ij}{}^k = \sum_l g^{kl}\Gamma_{ijl}.$$

Let ${}^gR_{ijkl} := {}^gR(\partial_i, \partial_j, \partial_k, \partial_l)$ and ${}^gR_{ijkl;n} := \nabla_{\partial_n}{}^gR(\partial_i, \partial_j, \partial_k, \partial_l)$ be the components of ${}^gR$ and $\nabla^g R$ relative to the coordinate frame.

1.11.1 Lemma. *Let (M, g) be a pseudo-Riemannian manifold. Assume given a coordinate system on M centered at P so that $\partial_i g_{jk}(P) = 0$ for all i, j, k. Then:*

(1) ${}^gR_{ijkl}(P) = \frac{1}{2}\{\partial_i\partial_k g_{jl} + \partial_j\partial_l g_{ik} - \partial_i\partial_l g_{jk} - \partial_j\partial_k g_{il}\}(P)$.

(2) ${}^gR_{ijkl;n}(P) = \frac{1}{2}\{\partial_n\partial_i\partial_k g_{jl} + \partial_n\partial_j\partial_l g_{ik} - \partial_n\partial_i\partial_l g_{jk} - \partial_n\partial_j\partial_k g_{il}\}(P)$.

Proof. Since the first derivatives of the metric vanish at P, $\Gamma = O(|x|)$ so we may compute:

$$\begin{aligned}
{}^gR_{ijkl} =& \textstyle\sum_n (\partial_i\Gamma_{jk}{}^n + \sum_p \Gamma_{ip}{}^n\Gamma_{jk}{}^p \\
& -\partial_j\Gamma_{ik}{}^n - \textstyle\sum_p \Gamma_{jp}{}^n\Gamma_{ik}{}^p)\partial_n, \partial_\ell) + O(|x|^2) \\
=& (\partial_i\Gamma_{jk\ell} - \partial_j\Gamma_{ik\ell}) + O(|x|^2) \\
=& \tfrac{1}{2}\{\partial_i\partial_k g_{j\ell} - \partial_i\partial_\ell g_{jk} - \partial_j\partial_k g_{i\ell} + \partial_j\partial_\ell g_{ik}\} + O(|x|^2).
\end{aligned}$$

This proves Assertion (1). Since the Christoffel symbols vanish at P, we have ${}^gR_{ijkl;n} = \partial_n{}^gR_{ijkl} + O(|x|)$. Assertion (2) now follows from the calculation performed above. □

Let (M, g) be a pseudo-Riemannian manifold. A curve σ in M is said to be a *geodesic* if $\nabla_{\sigma'}\sigma' = 0$.

1.11.2 Lemma. *Let P be a point of a pseudo-Riemannian manifold (M, g). There exists a convex open neighborhood $\mathcal{O}$ of $0 \in T_PM$ and a smooth map* $\exp_P$ *from $\mathcal{O}$ to M so that the curves $t \rightarrow \exp_P(tv)$ for $v \in \mathcal{O}$ and $t \in [0, 1]$ are geodesics in M starting at P with initial direction v.*

Proof. We can use the fundamental theorem of ordinary differential equations to see that there is a convex open neighborhood $\mathcal{O}$ of 0 in T_PM, that there is $\varepsilon > 0$, and that there is a smooth map

$$\mathcal{E} : \mathcal{O} \times [0, \varepsilon) \rightarrow M$$

so that the curves $t \rightarrow \mathcal{E}(v, t)$ are geodesics starting at P with initial direction v. We may reparametrize any geodesic σ to define a new geodesic $\sigma_\varepsilon(t) := \sigma(\varepsilon t)$ starting at P with initial direction εv. Thus, by uniqueness, we have:

$$\mathcal{E}(v, \varepsilon t) = \mathcal{E}(\varepsilon v, t).$$

Consequently, by replacing $\mathcal{O}$ by $\varepsilon\mathcal{O}$, we may assume that $\mathcal{E} : \mathcal{O} \times [0, 1] \rightarrow M$; the desired map is then obtained by setting $\exp_P(v) = \mathcal{E}(v, 1)$. □

1.11.3 Geodesic coordinates. Let $\{e_1, ..., e_m\}$ be a basis for T_PM and let $\{x_1, ..., x_m\}$ be the associated dual coordinates on V. Let

$$\psi(x_1, ..., x_m) := \exp_P(x_1 e_1 + ... + x_m e_m)$$

be the associated exponential *geodesic coordinates*; the coordinate system is defined for x such that $x_1 e_1 + ... + x_m e_m \in \mathcal{O}$, where $\mathcal{O}$ is a suitably chosen convex open neighborhood of $0 \in T_PM$. This system of coordinates is characterized by the fact that the curves

$$\sigma : t \rightarrow \psi(tx) \text{ for } t \in [0, \varepsilon]$$

are geodesics starting at P with initial direction $\sigma'(0) = x_1 e_1 + ... + x_m e_m$.

In a system of geodesic coordinates, we can express the components of the curvature tensor in terms of the 2 jets of the metric, and we can express the 2 jets of the metric in terms of the components of the curvature tensor at the center of the coordinate system. More precisely:

1.11.4 Lemma. *Let P be a point of a pseudo-Riemannian manifold (M, g) of dimension m. Let $\{e_1, ..., e_m\}$ be an orthonormal basis for T_PM and let $x \rightarrow \exp_P(x_1 e_1 + ... + x_m e_m)$ define geodesic coordinates centered at P. Let i, j, k, l be indices with $1 \le i, j, k, l \le m$. Let $\partial_i := \frac{\partial}{\partial x_i}$ and let:*

$$g_{ij} := g(\partial_i, \partial_j), \quad g_{ij/k} := \partial_k g_{ij},$$
$$g_{ij/kl} := \partial_l \partial_k g_{ij}, \quad {}^gR_{ijkl} := {}^gR(\partial_i, \partial_j, \partial_k, \partial_l).$$

(1) *We have* $g_{ij}(P) = 0$ *for* $i \neq j$ *and* $g_{ii}(P) = (e_i, e_i)$.
(2) *We have* $g_{ij/k}(P) = 0$.
(3) *We have* $g_{ik/jl}(P) = -(g_{kj/il} + g_{kl/ij})(P) = g_{jl/ik}(P)$.
(4) *We have* ${}^gR_{ijkl}(P) = g_{ik/jl}(P) - g_{il/jk}(P)$.
(5) *We have* $3g_{ik/jl}(P) = {}^gR_{ijkl}(P) - {}^gR_{iljk}(P)$.

Proof. Let $\psi(x) = \exp_P(x_1e_1 + ... + x_me_m)$ identify a small neighborhood of 0 in $T_PM = \mathbb{R}^m$ with a small neighborhood of P in M. Since $\psi_*\partial_i = e_i$ at P, Assertion (1) follows. Let u, v, w, and x be vectors in $\mathbb{R}^m$ and let

$$U(\cdot) = u,\ V(\cdot) = v,\ W(\cdot) = w,\ \text{and } X(\cdot) = x$$

be the corresponding constant vector fields. The bracket of any pair of these vector fields vanishes. We have

$$g(\nabla_V V, W) = Vg(V, W) - \tfrac{1}{2}Wg(V, V).$$

The curve $t \to tv$ is a geodesic so $g(\nabla_V V, W)(tv) = 0$ for $t \in [0, \epsilon(v))$. Thus,

(1.11.4.a) $0 = 2Vg(V, W)(P) - Wg(V, V)(P)$ and
(1.11.4.b) $0 = 2VVg(V, W)(P) - VWg(V, V)(P)$ for any V, W.

We set $V = W$ in Equation (1.11.4.a) to see $Vg(V, V)(P) = 0$. We polarize this identity. Let $V(\varepsilon) := V + \varepsilon W$. We differentiate the identity $Vg(V, V)(P) = 0$ with respect to ε and set $\varepsilon = 0$ to see that

(1.11.4.c) $\qquad 0 = Wg(V, V)(P) + 2Vg(V, W)(P)$ for any V, W.

We use Equations (1.11.4.a) and (1.11.4.c) to see that

(1.11.4.d) $\qquad 0 = Wg(V, V)(P)$ for any V, W.

We polarize this identity. Let $V(\varepsilon) := V + \varepsilon U$. We differentiate the identity of Equation (1.11.4.d), and set $\varepsilon = 0$ to prove Assertion (2) by showing that

$$0 = Wg(U, V)(P) \text{ for any } U, V, W.$$

We set $V = W$ in Equation (1.11.4.b) to see that

$$0 = VVg(V, V)(P) \text{ for any } V.$$

We polarize this identity. Let $V(\varepsilon) := V + \varepsilon W$. We differentiate the identity $0 = VVg(V,V)(P)$ with respect to ε and set $\varepsilon = 0$ to see

$$0 = VVg(V,W)(P) + VWg(V,V)(P) \text{ for any } V, W. \tag{1.11.4.e}$$

We use Equations (1.11.4.b) and (1.11.4.e) to see that

$$\begin{aligned} 0 =& VVg(V,W)(P) \text{ and} \\ 0 =& VWg(V,V)(P) \text{ for all } V, W. \end{aligned} \tag{1.11.4.f}$$

We polarize these identities. Let $V(\varepsilon) := V + \varepsilon U$. We differentiate the identities of Display (1.11.4.f) with respect to ε and set $\varepsilon = 0$ to see that

$$\begin{aligned} 0 =& 2UVg(V,W)(P) + VVg(U,W)(P) \text{ and} \\ 0 =& UWg(V,V)(P) + 2VWg(U,V)(P) \text{ for all } U, V, W. \end{aligned} \tag{1.11.4.g}$$

Note that $VVg(U,W)$ is symmetric in U and W. We use the relations of Display (1.11.4.g) twice to see that

$$\begin{aligned} VVg(U,W)(P) =& -2UVg(V,W)(P) = -2WVg(U,V)(P) \\ =& UWg(V,V)(P) \text{ for all } U, V, W. \end{aligned} \tag{1.11.4.h}$$

We polarize these identities. Let $V = V + \varepsilon X$. We differentiate with respect to ε and set $\varepsilon = 0$ to prove Assertion (3) by computing

$$\begin{aligned} VXg(U,W)(P) =& -UXg(W,V)(P) - UVg(W,X)(P) \\ =& UWg(V,X)(P) \text{ for all } U, V, W, X. \end{aligned} \tag{1.11.4.i}$$

Assertion (4) now follows from Assertion (3) and from Lemma 1.11.1. We prove Assertion (5) by using Assertions (3) and (4) to compute:

$${}^gR_{ijkl} - {}^gR_{iljk} = g_{ik/jl} - g_{il/jk} - g_{ij/kl} + g_{ik/jl} = 3g_{ik/jl}. \qquad \square$$

We remark that all the derivatives of the metric can be expressed in terms of the covariant derivatives of the curvature tensor at P; we omit the proof of this fact as we shall not need it and instead refer to Atiyah, Bott, and Patodi [8, 9].

Let $\mathcal{J}_R(x) : y \to R(y,x)x$ be the *Jacobi operator*. Let σ be a geodesic. If Y is a vector field along σ, then we define

$$Y' := \nabla_{\sigma'} Y \text{ and } Y'' := \nabla_{\sigma'} \nabla_{\sigma'} Y.$$

We say that Y is a *Jacobi vector field along* σ if Y satisfies the *Jacobi equation*

$$Y'' + \mathcal{J}_R(\sigma')Y = 0.$$

1.11.5 Lemma. *Let Y be a Jacobi vector field along a geodesic σ such that $Y(0) \perp \sigma'(0)$ and such that $Y'(0) \perp \sigma'(0)$. Then $Y(t) \perp \sigma'(t)$ for all t.*

Proof. Let Y be a Jacobi vector field along a geodesic σ with $Y(0) \perp \sigma'$ and $Y'(0) \perp \sigma'$. Set

$$f(t) := (Y(t), \sigma'(t)).$$

We use the geodesic equation $\sigma'' = 0$ and the fact that Y is a Jacobi vector field to compute:

$$\begin{aligned} f'(t) =& \partial_t(Y(t), \sigma'(t)) = (Y'(t), \sigma'(t)) + (Y(t), \sigma''(t)) \\ =& (Y'(t), \sigma'(t)) \\ f''(t) =& (Y''(t), \sigma'(t)) + (Y'(t), \sigma''(t)) = (Y''(t), \sigma'(t)) \\ =& -(\mathcal{J}_R(\sigma'(t))Y(t), \sigma'(t)) = (Y(t), \mathcal{J}_R(\sigma'(t))\sigma'(t)) \\ =& 0. \end{aligned}$$

This shows that $f(t) = A + Bt$ is a linear function of t. Since $Y(0) \perp \sigma'(0)$ and $Y'(0) \perp \sigma'(0)$, we have $A = B = 0$. □

Jacobi vector fields arise in the study of geodesic sprays.

1.11.6 Lemma. *Let (M, g) be a pseudo-Riemannian manifold. Fix a point $P \in M$. Let $v, w \in T_PM$ be linearly independent. Define the geodesic spray*

$$F(t, s) := \exp_P(t(v + sw)).$$

Let $\sigma(t) = F(t, 0)$ and let $Y = F_*(\partial_s)|_{s=0}$ define a smooth vector field along σ. Then Y is a Jacobi vector field along σ with $Y(0) = 0$ and $Y'(0) = w$.

Proof. Let $\{e_1, ..., e_m\}$ be a basis for T_PM, where $e_1 = v$ and $e_2 = w$. Let

$$\psi : x \to \exp_P(x_1e_1 + ... + x_me_m)$$

be the associated geodesic coordinates. Since $F(t, s) = (t, st, 0, ..., 0)$,

(1.11.6.a) $$X := F_*(\partial_t) = \partial_1 + s\partial_2 \text{ and } Y := F_*(\partial_s) = t\partial_2.$$

Extend X and Y to smooth vector fields on $\mathcal{O}_1 = \{x \in \mathbb{R}^m : x_1 > 0\}$ by:

$$X(x) := \partial_1 + \frac{x_2}{x_1}\partial_2 \text{ and } Y(x) = x_1\partial_2.$$

We then have $[X, Y] = \partial_2 - \partial_2 = 0$. We can now use the properties of the Levi-Civita connection to compute:

(1.11.6.b) $$\begin{aligned} Y'' :=& \nabla_X\nabla_X Y = \nabla_X\nabla_Y X = \nabla_Y\nabla_X X + R(X, Y)X \\ =& \nabla_Y\nabla_X X - \mathcal{J}_R(X)Y. \end{aligned}$$

Fix s. Since the t curves are straight lines from the origin, they are geodesics. Consequently, we have that:

$$(\nabla_X X)(F(t,s)) \equiv 0 \text{ for any } s \text{ and } t > 0.$$

Since the s curves are integral curves for Y,

$$(\nabla_Y \nabla_X X)(F(t,s)) \equiv 0 \text{ for any } s \text{ and } t > 0. \quad (1.11.6.c)$$

We use Equations (1.11.6.b) and (1.11.6.c) to see therefore that

$$Y''(\sigma(t)) + \mathcal{J}_R(\sigma'(t))\{Y(\sigma(t))\} = 0 \text{ for } t > 0. \quad (1.11.6.d)$$

We use continuity to see that Equation (1.11.6.d) also holds when $t = 0$. This shows that Y is a Jacobi vector field along σ.

We use Equation (1.11.6.a) to see that $Y(0) = 0$. We use Lemma 1.11.4 to see that the 1 jets of the metric vanish at the origin for a system of exponential geodesic coordinates; this implies the Christoffel symbols vanish at the origin. Consequently, we have that:

$$\nabla_X Y(0) = \{X(Y)\}(0) = \partial_2 = w. \quad \square$$

1.12 Geometric realizations of algebraic curvature tensors

We can draw the following consequence from Lemmas 1.11.1 and 1.11.4:

1.12.1 Lemma. *Let P be a point of a pseudo-Riemannian manifold (M, g). Let X, Y, Z, W be tangent vectors at P.*

(1) *$R := {}^gR(P)$ is an algebraic curvature tensor on T_PM, i.e. we have:*

1a) $R(X,Y,Z,W) = -R(Y,X,Z,W) = -R(X,Y,W,Z)$.

1b) $R(X,Y,Z,W) = R(Z,W,X,Y)$.

1c) $R(X,Y,Z,W) + R(Y,Z,X,W) + R(Z,X,Y,W) = 0$.

(2) *$\nabla R := \nabla^g R_P$ is a algebraic covariant derivative curvature tensor on T_PM, i.e. we have:*

2a) $\nabla R(X,Y,Z,W;V) = -\nabla R(Y,X,Z,W;V)$
$= \nabla R(Z,W,X,Y;V)$.

2b) $0 = \nabla R(X,Y,Z,W;V) + \nabla R(X,Z,W,Y;V)$
$+\nabla R(X,W,Y,Z;V)$.

2c) $0 = \nabla R(X,Y,Z,W;V) + \nabla R(X,Y,W,V;Z)$
$+\nabla R(X,Y,V,Z;W)$.

Remark. Curvature symmetry (1c) is the *first Bianchi identity*, curvature symmetry (2b) is the *covariant derivative of the first Bianchi identity*, and curvature symmetry (2c) is the *second Bianchi identity*.

Proof. Fix a point P of M. By Lemma 1.11.4, we can find local coordinates so $g_{ij}(P) = \varepsilon_i\delta_{ij}$ and so $\partial_k g_{ij}(P) = 0$. Let $R := {}^gR_P$ and $\nabla R = {}^g\nabla R_P$. We use Lemma 1.11.1 to see that at the distinguished point P, we have:

$$\begin{aligned}R_{ijkl} &= \tfrac{1}{2}\{\partial_i\partial_k g_{jl} + \partial_j\partial_l g_{ik} - \partial_i\partial_l g_{jk} - \partial_j\partial_k g_{il}\}\\ R_{ijkl;n} &= \tfrac{1}{2}\{\partial_n\partial_i\partial_k g_{jl} + \partial_n\partial_j\partial_l g_{ik} - \partial_n\partial_i\partial_l g_{jk} - \partial_n\partial_j\partial_k g_{il}\}.\end{aligned}$$

Symmetries (1a), (1b), and (2a) are now immediate. Symmetry (1c) will follow from symmetries (1a) and (1b) and from the following computation:

$$\begin{aligned}&R_{ijkl} + R_{iklj} + R_{iljk}\\ &= \tfrac{1}{2}\{\partial_i\partial_k g_{jl} + \partial_j\partial_l g_{ik} - \partial_i\partial_l g_{jk} - \partial_j\partial_k g_{il}\\ &\quad +\partial_i\partial_l g_{kj} + \partial_k\partial_j g_{il} - \partial_i\partial_j g_{kl} - \partial_k\partial_l g_{ij}\\ &\quad +\partial_i\partial_j g_{lk} + \partial_l\partial_k g_{ij} - \partial_i\partial_k g_{lj} - \partial_l\partial_j g_{ik}\}\\ &= 0.\end{aligned}$$

Symmetry (2b) is proved similarly. We prove symmetry (2c) by computing:

$$\begin{aligned}&R_{ijkl;n} + R_{ijln;k} + R_{ijnk;l}\\ &=\tfrac{1}{2}\{\partial_n\partial_i\partial_k g_{jl} + \partial_n\partial_j\partial_l g_{ik} - \partial_n\partial_i\partial_l g_{jk} - \partial_n\partial_j\partial_k g_{il}\\ &\quad + \partial_k\partial_i\partial_l g_{jn} + \partial_k\partial_j\partial_n g_{il} - \partial_k\partial_i\partial_n g_{jl} - \partial_k\partial_j\partial_l g_{in}\\ &\quad + \partial_l\partial_i\partial_n g_{jk} + \partial_l\partial_j\partial_k g_{in} - \partial_l\partial_i\partial_k g_{jn} - \partial_l\partial_j\partial_n g_{ik}\}\\ &=0. \quad \square\end{aligned}$$

Let V be a vector space of signature (p, q) and let R be an algebraic curvature tensor on V. Recall that a pseudo-Riemannian manifold (M, g) is said to be a *geometric realization* of R at a point $P \in M$ if there exists an isometry $\Psi : T_PM \to V$ so that $\Psi^*R = {}^gR_P$ - i.e. for all tangent vector fields X, Y, Z, and W in T_PM, we have:

$$\begin{aligned}&(\Psi X, \Psi Y) = (X, Y) \text{ and}\\ &R(\Psi X, \Psi Y, \Psi Z, \Psi W) = {}^gR(X, Y, Z, W).\end{aligned}$$

1.12.2 Theorem. *Let R be an algebraic curvature tensor on a vector space V of signature (p,q). Then R is geometrically realizable.*

Proof. Let R be an algebraic curvature tensor on a vector space of signature (p,q). Let $m := \dim V = p+q$ and let $\{e_1,...,e_m\}$ be an orthonormal basis for V. Let $\{x_1,...,x_m\}$ be the associated dual coordinates on V. Motivated by Lemma 1.11.4, we define the germ of a pseudo-Riemannian metric on V by setting:

$$g_{ik} = (e_i, e_k) - \tfrac{1}{6}\textstyle\sum_{jl}(R_{ijlk} + R_{iljk})x_j x_l.$$

The coefficients are symmetric in the pair (j,l). Thus,

$$\partial_j\partial_l g_{ik} = -\tfrac{1}{3}(R_{ijlk} + R_{iljk}). \tag{1.12.2.a}$$

Curvature symmetries (1.5.1.e) and (1.5.1.f) then show that $g_{ik} = g_{ki}$. Consequently, the inner product is symmetric; since the given metric on V is non-degenerate and since the set of non-degenerate metrics is open, g will be non-degenerate for $|x|$ small. Since $\partial_i g_{jk}(P) = 0$, we may use Lemma 1.11.1 to express ${}^gR$ in terms of the second derivatives of the metric at the center of the coordinate system. We then use Equation (1.12.2.a) to complete the proof by computing:

$$\begin{aligned}{}^gR_{ijkl}(P) =&\tfrac{1}{2}\{\partial_i\partial_k g_{jl} + \partial_j\partial_l g_{ik} - \partial_i\partial_l g_{jk} - \partial_j\partial_k g_{il}\}(P)\\ =&\tfrac{1}{6}\{-(R_{jikl} + R_{jkil}) - (R_{ijlk} + R_{iljk})\\ &\quad + (R_{jilk} + R_{jlik}) + (R_{ijkl} + R_{ikjl})\}\\ =&\tfrac{1}{6}\{4R_{ijkl} - 2R_{iljk} - 2R_{iklj}\}\\ =&R_{ijkl}. \quad \square\end{aligned}$$

Let F be an immersion of a manifold M of dimension $m = r+s-1$ into a vector space W of signature (r,s). The *first fundamental form* $g := F^*(\cdot,\cdot)_W$ is the pull-back metric defined by setting:

$$g(v_1, v_2) := (F_*v_1, F_*v_2)_W = (v_1F, v_2F).$$

We say F is a *non-degenerate immersion* if the first fundamental form is non-degenerate. We shall assume this condition henceforth. Locally, we may choose a normal ν along M so that

$$(\nu,\nu) = \pm 1 \text{ and } (dF,\nu) = 0.$$

The metric g is then a non-degenerate metric of signature $(r, s-1)$ if ν is spacelike or signature $(r-1, s)$ if ν is timelike. Let L be the second fundamental form of the immersion and let S be the associated *shape operator;* S

is also sometimes called the *Weingarten operator.* L is the symmetric bilinear form and S is the self-adjoint map which are defined by the relations:

$$L(v_1, v_2) := (v_1 v_2(F), \nu) = g(Sv_1, v_2) \text{ for } v_1, v_2 \in TM.$$

We compute the curvature tensor of such a hypersurface and establish the *Codazzi equation* by showing that ∇S is a totally symmetric 3 tensor.

1.12.3 Lemma. *Let F be an immersion of a manifold M into a vector space W of signature (r, s) as a non-degenerate hypersurface. Let ν be a unit normal and let $\varepsilon := (\nu, \nu) = \pm 1$. Let S be the associated shape operator and let g be the induced metric on M.*

(1) ${}^gR = \varepsilon R_S$.

(2) *Let $S_{ij;k}$ be the components of ∇S. Then $S_{ij;k} = S_{ji;k} = S_{ik;j}$.*

Proof. Fix a point $P \in M$. By translating F, we may suppose without loss of generality that $F(P) = 0$. Choose a basis $\{e_1, ..., e_{m+1}\}$ for W so that $\{e_1, ..., e_m\}$ is a basis for T_PM and so that $e_{m+1} = \nu(P)$. Let $\{x_1, ..., x_{m+1}\}$ be the dual basis for W^*; these give coordinates on W. Then

$$\vec{y} := (F^* x_1, ..., F^* x_m)$$

is a system of local coordinates on M and the immersion takes the form

$$F(\vec{y}) = y_1 e_1 + ... + y_m e_m + f(\vec{y}) e_{m+1},$$

where f is a smooth scalar valued function with $df(P) = 0$; this represents the immersion as a graph over the tangent plane at P. Let

$$\varepsilon = (e_{m+1}, e_{m+1}) = \pm 1.$$

We have $F_*(\partial_i^y) = e_i + \partial_i f e_{m+1}$. Let indices i, j, k, l range from 1 to m. The shape operator and the second fundamental form at P are given by:

$$(S\partial_i, \partial_j)(P) = L(\partial_i, \partial_j)(P) = (\partial_i \partial_j F, e_{m+1})(P) = \varepsilon \partial_i \partial_j f(P).$$

This shows that

$$\begin{aligned} (R_S)_{ijkl}(P) =& \{(S\partial_i, \partial_l)(S\partial_j, \partial_k) - (S\partial_i, \partial_k)(S\partial_j, \partial_l)\}(P) \\ =& \{L(\partial_i, \partial_l)L(\partial_j, \partial_k) - L(\partial_i, \partial_k)L(\partial_j, \partial_l)\}(P) \qquad (1.12.3.a) \\ =& \{\partial_i\partial_l(f) \cdot \partial_j\partial_k(f) - \partial_i\partial_k(f) \cdot \partial_j\partial_l(f)\}(P). \end{aligned}$$

The metric takes the form:

$$g_{ij} = (e_i, e_j) + \varepsilon\partial_i(f)\partial_j(f).$$

Since $\partial_k(g_{ij})(P) = 0$, we may use Lemma 1.11.1 to see that:

$$\begin{aligned} R_{ijkl}(P) =& \tfrac{1}{2}\{\partial_i\partial_k g_{jl} + \partial_j\partial_l g_{ik} - \partial_i\partial_l g_{jk} - \partial_j\partial_k g_{il}\}(P) \\ =& \varepsilon\{\partial_i\partial_l(f)\cdot\partial_j\partial_k(f) - \partial_i\partial_k(f)\cdot\partial_j\partial_l(f)\}(P). \end{aligned} \tag{1.12.3.b}$$

Assertion (1) now follows from Equations (1.12.3.a) and (1.12.3.b).

Let $\varepsilon_i := (e_i, e_i)$. We compute:

$$\begin{aligned} g_{ij} &= (e_i, e_j) + \varepsilon\partial_i f\cdot\partial_j f \\ \nu &= \frac{e_{m+1} - \varepsilon\sum_k \varepsilon_k\partial_k f\cdot e_k}{\sqrt{1+\varepsilon\sum_k\varepsilon_k(\partial_k f)^2}}, \text{ and} \\ L(\partial_i, \partial_j) &= \varepsilon\frac{\partial_i\partial_j f}{\sqrt{1+\varepsilon\sum_k\varepsilon_k(\partial_k f)^2}}. \end{aligned}$$

Since $g_{ij} = (e_i, e_j) + O(|y|^2)$, $\Gamma_{ijk} = O(|y|)$ and thus

$$\begin{aligned} S_{ij;k} =& \partial_k(Se_i, e_j) - (S\nabla_{\partial_k}\partial_i, \partial_j) - (S\partial_i, \nabla_{\partial_k}\partial_j) \\ =& \varepsilon\partial_i\partial_j\partial_k f + O(|y|). \quad \square \end{aligned}$$

If ϕ is self-adjoint, then we defined an algebraic curvature tensor R_ϕ in Equation (1.1.1.b) by setting:

$$\begin{aligned} &R_\phi(x, y)z := (\phi(y), z)\phi(x) - (\phi(x), z)\phi(y) \text{ and} \\ &R_\phi(x, y, z, w) := (\phi(y), z)(\phi(x), w) - (\phi(x), z)(\phi(y), w). \end{aligned}$$

We use Lemma 1.12.3 to show that these tensors R_ϕ, where ϕ is self-adjoint, are geometrically realizable by hypersurfaces.

1.12.4 Theorem. *Let ϕ be a self-adjoint linear map of a vector space V of signature (p, q). Let $\varepsilon = \pm 1$ be given. There exists a non-degenerate hypersurface M and a point $P \in M$ so $S(P) = \phi$ and so $\varepsilon R_\phi = {}^g R_P$.*

Proof. We extend the given innerproduct on V to an innerproduct on the vector space $W := V \oplus e_{m+1}\cdot\mathbb{R}$ by setting:

$$(v_1 \oplus r_1 e_{m+1}, v_2 \oplus r_2 e_{m+1}) = (v_1, v_2) + \varepsilon r_1 r_2.$$

Let $\{e_1, ..., e_{\dim V}\}$ be a basis for V and let $(y_1, ..., y_m)$ be the corresponding dual coordinates on V. We define an embedding F of V in W by setting:

(1.12.4.a) $F(y_1, ..., y_m) := y_1e_1 + ... + y_me_m + \frac{\varepsilon}{2}\sum_{i,j}(\phi e_i, e_j)y_iy_je_{m+1}$.

Since $g_{ij}(0) = (e_i, e_j)$ is non-degenerate and since the set of non-degenerate metrics forms an open subset of the space of all metrics, g is the germ of a pseudo-Riemannian metric. Furthermore, $L_{ij}(0) = (\phi e_i, e_j)$ so $S = \phi$ and thus (M, g) is a geometric realization of εR_ϕ by Lemma 1.12.3. □

1.13 Schur problems

Let (M, g) be a pseudo-Riemannian manifold. We shall denote the inner product on the tangent bundle by $(\cdot,\cdot)$ in the interests of notational simplicity. We say that (M, g) is said to be *k-stein* if the associated algebraic curvature tensor ${}^gR$ is k-stein at each point of M; this means that there exist smooth functions $c_i(\cdot)$ so that

$$\mathrm{Tr}\{\mathcal{J}_R(x)^i\} = c_i(P)(x,x)^i \text{ for any } x \in T_PM \text{ and for } 1 \leq i \leq k.$$

We wish to study if the coefficients c_i are necessarily constant. Such problems are called Schur-like problems. We focus on the cases $k = 1$ (Einstein) and $k = 2$ (zweistein) in this section and establish the following result:

1.13.1 Theorem. *Let (M, g) be a connected pseudo-Riemannian manifold of signature (p, q). Let $m := p + q = \dim(M)$.*

(1) *If $m \neq 2$ and if (M, g) is Einstein, then c_1 is constant.*

(2) *If $m \neq 2, 4$ and if (M, g) is 2-stein, then c_2 is constant.*

Proof. Let (M, g) be a connected pseudo-Riemannian manifold of signature (p, q) which is Einstein. Let $R(x, y, z, w; v) = \nabla_vR(x, y, z, w)$ be the covariant derivative of the curvature tensor. We proved in Lemma 1.12.1 that we have the symmetries:

$$\begin{aligned}
&R(a, b, c, d; e) = -R(b, a, c, d; e) = R(c, d, a, b; e),\\
&R(a, b, c, d; e) + R(a, c, d, b; e) + R(a, d, b, c; e) = 0,\\
&R(a, b, c, d; e) + R(a, b, d, e; c) + R(a, b, e, c; d) = 0.
\end{aligned}$$

Suppose that (M, g) is Einstein and that $m \neq 2$. We follow the argument given in Besse [16, Theorem 1.97] to show that c_1 is constant and thereby prove Assertion (1).

Let $\{e_1, ..., e_m\}$ be a local orthonormal frame for the tangent bundle of (M, g). Define $\varepsilon_i := (e_i, e_i) = \pm 1$. Since R is Einstein,

$$\rho(x, y) = c_1(P) g_P(x, y) \text{ for all } x, y \in T_P M$$

by Lemma 1.7.2. The *scalar curvature* τ of the manifold is defined by:

$$\tau := \textstyle\sum_{i,j} \varepsilon_i \varepsilon_j R_{ijji}.$$

Let $\rho_{ij;k}$ be the components of the covariant derivative of the Ricci tensor. We covariantly differentiate and use the second Bianchi identity to compute:

$$\begin{aligned}
\tau_{;k} =& \{\textstyle\sum_{i,j} \varepsilon_i \varepsilon_j R_{ijji}\}_{;k} = \{\sum_{i,j} \varepsilon_i \rho_{ii}\}_{;k} = \{\sum_{i,j} \varepsilon_i c_1 g_{ii}\}_{;k} \\
=& m c_{1;k} \\
=& -\textstyle\sum_{ij} \varepsilon_i \varepsilon_j R_{ijji;k} = -\sum_{i,j} \varepsilon_i \varepsilon_j (R_{ijik;j} + R_{ijkj;i}) \\
=& 2 \textstyle\sum_{i,j} \varepsilon_i \varepsilon_j R_{ijki;j} = 2 \sum_j \varepsilon_j \rho_{jk;j} = 2 \sum_j \varepsilon_j \{c_1 g_{jk}\}_{;j} \\
=& 2 c_{1;k}.
\end{aligned}$$

Since $m \neq 2$ by assumption, c_1 is constant. This proves Assertion (1).

To prove Assertion (2), we suppose that $m \neq 2, 4$ and that (M, g) is 2-stein. We must show that c_2 is constant. We follow Gilkey, Swann, and Vanhecke [87] making appropriate changes in the argument given there as we are dealing with an indefinite metric. We also refer to Besse [16, p. 165] for another proof.

Let x be an arbitrary tangent vector. As R is 2-stein, we have that:

$$c_2(x, x)^2 = \operatorname{Tr}\{\mathcal{J}_R(x)^2\} = \textstyle\sum_{i,j} \varepsilon_i \varepsilon_j R(e_i, x, x, e_j) R(e_j, x, x, e_i). \tag{1.13.1.a}$$

We polarize Equation (1.13.1.a). We set $x(t) = x + ty$ and study the terms which are quadratic in t to see

$$\begin{aligned}
& c_2\{g(x, x) g(y, y) + 2 g(x, y) g(x, y)\} \\
&= \textstyle\sum_{i,j} \varepsilon_i \varepsilon_j \{R(e_i, x, x, e_j) R(e_j, y, y, e_i) \\
&\quad + R(e_i, x, y, e_j) R(e_j, x, y, e_i) + R(e_i, x, y, e_j) R(e_j, y, x, e_i)\}.
\end{aligned}$$

We set $y = e_k$. We multiply by ε_k and sum over the index k to see:

$$\begin{aligned}
& c_2(m+2)(x, x) \\
&= c_2 \textstyle\sum_k \varepsilon_k \{g(x, x) g(e_k, e_k) + 2 g(x, e_k) g(x, e_k)\} \\
&= \textstyle\sum_{i,j,k} \varepsilon_i \varepsilon_j \varepsilon_k \{R(e_i, x, x, e_j) R(e_j, e_k, e_k, e_i) \\
&\quad + R(e_i, x, e_k, e_j) R(e_j, x, e_k, e_i) + R(e_i, x, e_k, e_j) R(e_j, e_k, x, e_i)\}.
\end{aligned} \tag{1.13.1.b}$$

Since (M,g) is Einstein, we have $\rho = c_1 g$. Consequently:

$$\begin{aligned}
&\textstyle\sum_{i,j,k}\varepsilon_i\varepsilon_j\varepsilon_k R(e_i,x,x,e_j)R(e_j,e_k,e_k,e_i)\\
&=\textstyle\sum_{i,j}\varepsilon_i\varepsilon_j R(e_i,x,x,e_j)\rho(e_i,e_j)\\
&=c_1\textstyle\sum_{i,j}\varepsilon_i\varepsilon_j R(e_i,x,x,e_j)g(e_i,e_j)\\
&=c_1\textstyle\sum_{i}\varepsilon_i R(e_i,x,x,e_i) = c_1^2(x,x).
\end{aligned} \tag{1.13.1.c}$$

The Ricci tensor is a bilinear form which arises from contracting the curvature tensor. There are two different contractions of $R\otimes R\in\otimes^8 T^*M$ which are not expressible in terms of the Ricci tensor, that we could use to define symmetric bilinear forms. We define:

$$\begin{aligned}
\xi(x,y) &:= \textstyle\sum_{i,j,k}\varepsilon_i\varepsilon_j\varepsilon_k R(x,e_i,e_j,e_k)R(y,e_i,e_j,e_k)\\
\tilde\xi(x,y) &:= \textstyle\sum_{i,j,k}\varepsilon_i\varepsilon_j\varepsilon_k R(x,e_i,e_j,e_k)R(y,e_j,e_i,e_k)
\end{aligned}.$$

We use the first Bianchi identity to see that these two quadratic forms are equivalent by studying the associated quadratic functions:

$$\begin{aligned}
\tilde\xi(x,x) &= \textstyle\sum_{i,j,k}\varepsilon_i\varepsilon_j\varepsilon_k R(x,e_i,e_j,e_k)R(x,e_j,e_i,e_k)\\
&= -\textstyle\sum_{i,j,k}\varepsilon_i\varepsilon_j\varepsilon_k R(x,e_i,e_j,e_k)R(x,e_i,e_k,e_j)\\
&\quad -\textstyle\sum_{i,j,k}\varepsilon_i\varepsilon_j\varepsilon_k R(x,e_i,e_j,e_k)R(x,e_k,e_j,e_i)\\
&= \xi(x,x) - \tilde\xi(x,x).
\end{aligned} \tag{1.13.1.d}$$

Consequently, $\tilde\xi = \frac{1}{2}\xi$. We use Equations (1.13.1.b) and (1.13.1.c) to see:

$$\{c_2(m+2) - c_1^2\}(x,x) = \xi(x,x) + \tilde\xi(x,x) \text{ so}$$

$$\xi(x,x) = \tfrac{2}{3}\{c_2(m+2) - c_1^2\}(x,x). \tag{1.13.1.e}$$

We set $x = e_l$, multiply by ε_l and sum Equation (1.13.1.e) to see:

$$\begin{aligned}
\|R\|^2 &:= \textstyle\sum_{i,j,k,l}\varepsilon_i\varepsilon_j\varepsilon_k\varepsilon_l R_{ijkl}R_{ijkl} = \sum_l \varepsilon_l\xi_{ll}\\
&= \textstyle\sum_l \varepsilon_l \frac{2}{3}\{c_2(m+2) - c_1^2\}(e_l,e_l)\\
&= m\tfrac{2}{3}\{c_2(m+2) - c_1^2\}.
\end{aligned}$$

Consequently, we may rewrite Equation (1.13.1.e) in the form

$$\xi(x,x) = \tfrac{2}{3}\{c_2(m+2) - c_1^2\}(x,x) = \tfrac{1}{m}\|R\|^2(x,x). \tag{1.13.1.f}$$

This means that (M,g) is *super-Einstein* in the terminology of Gray and Willmore [92]; see also Besse [16, p. 165]. We have previously shown that c_1 is constant. In light of Equation (1.13.1.f), to show that c_2 is constant and prove Assertion (2), it suffices to show that $\|R\|^2$ is constant.

We polarize Equation (1.13.1.f) to see that:

$$\text{(1.13.1.g)} \qquad \begin{aligned} \xi(x,y) &= \textstyle\sum_{i,j,k} \varepsilon_i\varepsilon_j\varepsilon_k R(x,e_i,e_j,e_k)R(y,e_i,e_j,e_k) \\ &= \tfrac{1}{m}||R||^2(x,y). \end{aligned}$$

Let $\xi_{ij;k}$ and $\rho_{ij;k}$ denote the components of the covariant derivative of the tensors ξ and ρ. We use Equation (1.13.1.g) to compute:

$$\textstyle\sum_b \varepsilon_b \xi_{ab;b} = \sum_{b,i,j,k} \varepsilon_i\varepsilon_j\varepsilon_k\varepsilon_b\{R_{aijk;b}R_{bijk} + R_{aijk}R_{bijk;b}\}.$$

Since M is Einstein, we may use the second Bianchi identity to compute:

$$\textstyle\sum_b \varepsilon_b R_{bijk;b} = -\sum_b \varepsilon_b\{R_{bikb;j} + R_{bibj;k}\} = -\{\rho_{ik;j} - \rho_{ij;k}\} = 0.$$

We also use the second Bianchi identity to compute:

$$\begin{aligned} &\textstyle\sum_{i,j,k,b} \varepsilon_i\varepsilon_j\varepsilon_k\varepsilon_b R_{aijk;b}R_{bijk} \\ &= -\textstyle\sum_{i,j,k,b} \varepsilon_i\varepsilon_j\varepsilon_k\varepsilon_b(R_{ibjk;a}R_{bijk} + R_{bajk;i}R_{bijk}) \\ &= \tfrac{1}{2}||R||^2_{;a} - \textstyle\sum_{i,j,k,b} \varepsilon_i\varepsilon_j\varepsilon_k\varepsilon_b R_{iajk;b}R_{ibjk} \end{aligned}$$

so $\sum_{i,j,k,b} \varepsilon_i\varepsilon_j\varepsilon_k\varepsilon_b R_{aijk;b}R_{bijk} = \frac{1}{4}||R||^2_{;a}$ and thus

$$\text{(1.13.1.h)} \qquad \textstyle\sum_b \varepsilon_b \xi_{ab;b} = \frac{1}{4}||R||^2_{;a}.$$

On the other hand, we may use the second identity in Display (1.13.1.g) to see that $\xi_{ab} = \frac{1}{m}||R||^2 g_{ab}$ and, consequently, that:

$$\textstyle\sum_b \varepsilon_b \xi_{ab;b} = \frac{1}{m}\sum_b \varepsilon_b ||R||^2_{;b} g_{ab} = \frac{1}{m}||R||^2_{;a}.$$

Since $\frac{1}{4} \neq \frac{1}{m}$, we may use Equations (1.13.1.h) and (1.13.1.i) to see that $||R||^2_{;a} = 0$ and thereby complete the proof of Assertion (2). □

1.14 Space forms

Let V be a vector space of signature (p,q). If R is an algebraic curvature tensor on R and if the sectional curvature $\kappa(\pi) = \kappa$ for every non-degenerate 2 plane, then we use Lemma 1.6.4 to see that $R = \kappa R_{\mathrm{Id}}$, where

$$R_{\mathrm{Id}}(x,y)z := (y,z)x - (x,z)y.$$

Thus, R_{Id} can properly be called the algebraic curvature tensor of *constant sectional curvature* +1. Let (M, g) be a connected pseudo-Riemannian manifold of dimension m and signature (p, q). We say that (M, g) is a *space form* if for every $P \in M$, there exists a constant $\kappa(P)$ so that

$$^gR_P = \kappa(P) R_{\mathrm{Id}}.$$

In Lemma 1.14.1, we shall show that space forms are m-stein and that the sectional curvature κ is constant if $m \neq 2$. In Lemma 1.14.2, we show that the local isometry type of a space form is determined by (p, q, κ). In Lemma 2.6.1, we will show the pseudo-spheres

$$S^{r,s}(\rho) := \{v \in \mathbb{R}^{r,s} : (v, v) = \rho\}$$

have constant sectional curvature. Thus, by combining Lemmas 1.14.2 and 2.6.1, we see that any pseudo-Riemannian manifold of constant sectional curvature is locally isometric to some pseudo-sphere; see Theorem 3.6.1 for details.

1.14.1 Lemma. *Let (M, g) be a connected space form of dimension m and signature (p, q). Then (M, g) is m-stein. If $m \neq 2$, then κ is constant.*

Proof. To show that a pseudo-Riemannian manifold (M, g) is m-stein, we must find constants $c_i(P)$ so that

$$\text{(1.14.1.a)} \qquad \mathrm{Tr}\{\mathcal{J}_R(x)^i\} = c_i(P)(x, x)^i \text{ for } 1 \leq i \leq m \text{ and } x \in T_PM.$$

By Lemma 1.7.3, it suffices to check that this identity holds for all non-degenerate tangent vectors x.

Suppose that (M, g) is a space form. Fix $P \in M$ and let $R = {}^gR_P$. Since $R = \kappa(P) R_{\mathrm{Id}}$,

$$\text{(1.14.1.b)} \qquad R(x, y)z = \kappa(P)\{(y, z)x - (x, z)y\} \text{ on } T_PM.$$

We may decompose $T_PM = \mathrm{span}\{x\} \oplus x^\perp$ if x is non-degenerate. We show that (M, g) is k-stein for all k by using Equation (1.14.1.b) to compute:

$$\text{(1.14.1.c)} \qquad \begin{aligned} &\mathcal{J}_R(x)x = 0, \\ &\mathcal{J}_R(x)y = \kappa(P)(x, x)y \text{ if } y \perp x, \text{ and} \\ &\mathrm{Tr}\{\mathcal{J}_R(x)^i\} = (m-1)\kappa(P)^i(x, x)^i. \end{aligned}$$

This shows that $c_1(P) = (m-1)\kappa(P)$. Since $m \neq 2$, we apply Theorem 1.13.1 to see that c_1 is constant; this implies that the sectional curvature κ is constant. □

Let $t \to \sigma(t)$ be a curve in a pseudo-Riemannian manifold (M, g). If Y is a vector field along σ, then we define $Y' := \nabla_{\sigma'} Y$ and $Y'' := \nabla_{\sigma'}\nabla_{\sigma'} Y$. A vector field Y along σ is said to be a *parallel vector field* if $Y' = 0$; σ' is a parallel vector field if and only if σ is a geodesic. If Y_1 and Y_2 are parallel vector fields along σ, then

$$\text{(1.14.1.d)} \qquad \partial_t(Y_1, Y_2) = (Y_1', Y_2) + (Y_1, Y_2') = 0$$

so (Y_1, Y_2) is constant. In particular, if Y is a parallel vector field, then (Y, Y) is constant.

We now show that the condition that (M, g) has constant sectional curvature κ and signature (p, q) completely characterizes (M, g) up to local isometry.

1.14.2 Lemma. *Assume that (M, g) and $(\tilde{M}, \tilde{g})$ are space forms with the same signature (p, q) and constant sectional curvature κ. Choose points $P \in M$ and $\tilde{P} \in \tilde{M}$ and choose a linear isometry $\Psi : T_PM \to T_{\tilde{P}}\tilde{M}$. Then there exists a local isometry ψ from (M, g) to $(\tilde{M}, \tilde{g})$ so that $\psi(P) = \tilde{P}$ and so that $\psi_*(P) = \Psi$.*

Proof. We use the exponential map to define a diffeomorphism

$$\psi := \exp_{\tilde{P}} \circ \Psi \circ \exp_P^{-1}$$

from a neighborhood of P in M to a neighborhood of $\tilde{P}$ in $\tilde{M}$, such that:

$$\psi(P) = \tilde{P} \text{ and } \psi_*(P) = \Psi.$$

Let v be a spacelike vector in T_PM and let $\tilde{v} := \Psi v$ be the corresponding spacelike vector in $T_{\tilde{P}}\tilde{M}$. Let $c^2 := (v, v) = (\tilde{v}, \tilde{v})$. Let

$$\sigma(t) := \exp_P(tv) \text{ and } \tilde{\sigma}(t) := \psi\sigma(t) = \exp_{\tilde{P}}(t\tilde{v})$$

be corresponding geodesics in M and $\tilde{M}$. We restrict to values of t so that we are in the domain of the exponential maps in question. Let t_0 be in the admissible parameter range. Let

$$Q := \sigma(t_0) \text{ and } \tilde{Q} := \psi Q = \tilde{\sigma}(t_0)$$

be corresponding points on the geodesics σ and $\tilde{\sigma}$. Let

$$\Psi_Q := \psi_*(Q) : T_QM \to T_{\tilde{Q}}\tilde{M}.$$

Since σ' and $\tilde{\sigma}'$ are parallel vector fields, we use Equation (1.14.1.d) to see

$$\begin{aligned}&(\Psi_Q\sigma'(t_0), \Psi_Q\sigma'(t_0)) = (\tilde{\sigma}'(t_0), \tilde{\sigma}'(t_0))\\ =&(\tilde{v}, \tilde{v}) = (v, v) = (\sigma'(t_0), \sigma'(t_0)).\end{aligned}$$

We use Jacobi vector fields to study the action of Ψ_Q on $\sigma'(t_0)^\perp$. Let $v \perp w \in T_PM$ and let $\tilde{v} \perp \tilde{w} := \Psi w \in T_{\tilde{P}}\tilde{M}$. Let

$$\begin{aligned}T(t,s) =& \exp_P(t(v+sw)),\ Y(t) := T_*(\partial_s)|_{s=0},\ Y_Q := Y(t_0),\\ \tilde{T}(t,s) =& \exp_{\tilde{P}}(t(\tilde{v}+s\tilde{w})),\ \tilde{Y}(t) := \tilde{T}_*(\partial_s)|_{s=0},\ \tilde{Y}_Q := \tilde{Y}(t_0).\end{aligned}$$

We then have $\tilde{T} = \psi T$ and $\tilde{Y}(\tilde{\sigma}(t)) = \psi_*(\sigma(t))Y(\sigma(t))$. We use Lemma 1.11.6 to see that Y and $\tilde{Y}$ are Jacobi vector fields along σ and $\tilde{\sigma}$. Consequently, we have that:

$$\begin{aligned}&Y(0) = 0 \perp \sigma'(0),\ Y'(0) = w \perp \sigma'(0),\\ &\tilde{Y}(0) = 0 \perp \tilde{\sigma}'(0),\ \tilde{Y}'(0) = \tilde{w} \perp \tilde{\sigma}'(0).\end{aligned}$$

Thus, we may use Lemma 1.11.5 to see that $Y \perp \sigma'$ and $\tilde{Y} \perp \tilde{\sigma}'$ for all t. Because (M, g) and $(\tilde{M}, \tilde{g})$ have constant sectional curvature κ and because $(\sigma', \sigma') = (\tilde{\sigma}', \tilde{\sigma}') = c^2 > 0$, we have that:

$$\mathcal{J}_R(\sigma')Y = c^2\kappa Y \text{ and } \mathcal{J}_R(\tilde{\sigma}')\tilde{Y} = c^2\kappa\tilde{Y}.$$

This shows that Y and $\tilde{Y}$ satisfy the ordinary differential equations with initial conditions:

$$\text{(1.14.2.a)}\qquad \begin{aligned}&Y''(t) = -c^2\kappa Y(t),\ Y(0) = 0,\ Y'(0) = w,\\ &\tilde{Y}''(t) = -c^2\kappa\tilde{Y}(t),\ \tilde{Y}(0) = 0,\ \tilde{Y}'(0) = \tilde{w}.\end{aligned}$$

Let e and $\tilde{e}$ be parallel vector fields along σ and $\tilde{\sigma}$ with $e(0) = w$ and $\tilde{e}(0) = \tilde{w}$. As σ and $\tilde{\sigma}$ are geodesics, σ' and $\tilde{\sigma}'$ are parallel vector fields. Since $e(0) \perp v$ and $\tilde{e}(0) \perp \tilde{v}$,

$$\begin{aligned}&e(t) \perp \sigma'(t),\ (e(t), e(t)) = (w, w),\\ &\tilde{e}(t) \perp \tilde{\sigma}'(t),\ (\tilde{e}(t), \tilde{e}(t)) = (\tilde{w}, \tilde{w}).\end{aligned}$$

Let $\delta := \sqrt{|\kappa|}$. We define:

$$\text{(1.14.2.b)}\qquad Z(t) := \begin{cases}\frac{1}{c\delta}\sin(c\delta t)e(t) & \text{if } \kappa = \delta^2,\\ te(t) & \text{if } \kappa = 0,\\ \frac{1}{c\delta}\sinh(c\delta t)e(t), & \text{if } \kappa = -\delta^2,\end{cases}$$

$$\tilde{Z}(t) := \begin{cases} \frac{1}{c\delta}\sin(c\delta t)\tilde{e}(t) & \text{if } \kappa = \delta^2, \\ t\tilde{e}(t) & \text{if } \kappa = 0, \\ \frac{1}{c\delta}\sinh(c\delta t)\tilde{e}(t) & \text{if } \kappa = -\delta^2. \end{cases} \tag{1.14.2.c}$$

The vector fields $Z(t)$ and $\tilde{Z}(t)$ satisfy the ordinary differential equation with initial condition given in Display (1.14.2.a). Thus, $Z = Y$ and $\tilde{Z} = \tilde{Y}$. Since $|Z| = |\tilde{Z}|$, we conclude that ψ_Q is an isometry *except* possibly if $\kappa > 0$ at the exceptional values of t, where $\sin(c\delta t) = 0$. Since these values are a discrete subset of σ, continuity then implies that ψ_* is an isometry at these values as well.

We have shown that Ψ_Q is an isometry from $T_Q M$ to $T_{\tilde{Q}}\tilde{M}$ if $Q = \exp_P tv$ for v spacelike. We replace the roles of sin and sinh to see Ψ_Q is an isometry if v is timelike. Let

$$\mathcal{N} := \{v \in T_P M : (v, v) = 0\}$$

be the set of null tangent vectors. We have shown that Ψ_Q is an isometry if $Q = \exp_P(v)$ for v not null. Since $\mathcal{N}$ is a nowhere dense subset of $T_P M$, continuity then implies that Ψ_Q is an isometry for all Q. □

1.15 Complex and para-complex space forms

Let V be a vector space of signature (p, q). Let $\phi \in \mathfrak{so}(V)$ be a skew-symmetric linear map of V with $\phi^2 = \varepsilon\,\mathrm{Id}$, where $\varepsilon = \pm 1$. Let

$$\begin{aligned} R_{\mathrm{Id}}(x, y)z &:= (y, z)x - (x, z)y \text{ and} \\ R_\phi(x, y)z &:= (\phi y, z)\phi x - (\phi x, z)\phi y - 2(\phi x, y)\phi z. \end{aligned}$$

We showed in Lemma 1.8.1 that R_{Id} and R_ϕ are algebraic curvature tensors; R_{Id} has constant sectional curvature $+1$.

Let (M, g) be a connected pseudo-Riemannian manifold of dimension m and signature (p, q), where $m \neq 2, 4$. We suppose that (M, g) is not a space form. We also suppose that for every $P \in M$, there exist constants $\lambda_0(P)$ and $\lambda_1(P)$ and a skew-adjoint map $\phi(P)$ of $T_P M$ with $\phi(P)^2 = \varepsilon\,\mathrm{Id}$ so that

$${}^g R_P = \lambda_0(P) R_{\mathrm{Id}} + \lambda_1(P) R_{\phi(P)}.$$

If $\varepsilon = -1$, then (M, g) is said to be a *complex space form*; if $\varepsilon = +1$, then (M, g) is said to be a *para-complex space form.* Since (M, g) is assumed not to be a space form, λ_1 does not vanish identically. If (M, g) is a complex space form, then p and q are both even; if (M, g) is a para-complex space form, then $p = q$. The constant $\varepsilon = \pm 1$ is assumed to be independent of the point $P \in M$.

Let (M, g) be a manifold of dimension $m \neq 2, 4$ which is either a para-complex space form or a complex space form. In Lemma 1.15.1, we shall show that (M, g) is m-stein, that λ_0 and λ_1 are constant, that locally the map $P \to \phi(P)$ can be chosen to be a smooth function of P, that $\nabla\phi = 0$, that $\nabla^g R = 0$, and that $\lambda_0 + \varepsilon\lambda_1 = 0$. In Lemma 1.15.3, we prove that (M, g) is characterized up to local isometry by (p, q, λ_0). We postpone until Section 3.6 the complete classification of such manifolds; see Theorem 3.6.5 and Theorem 3.6.7 for details.

The dimensions $m = 2$ and $m = 4$ are exceptional, and as noted above, and we shall avoid these dimensions in our discussion of (para-)complex space forms. However, many of the results we shall establish continue to hold in these dimensions modulo some additional hypotheses.

Let $\mathfrak{so}(M)$ be the bundle of skew-adjoint linear maps of the tangent bundle. In the following Lemma, we are primarily interested in the local properties of (para-)complex space forms so we shall assume that M is a contractible topological space to avoid topological complications.

1.15.1 Lemma. *Let (M, g) be a contractible (para-)complex space form of dimension $m \neq 2, 4$ and signature (p, q). Then:*

1. *(M, g) is m-stein and the functions λ_0 and λ_1 are constant.*
2. *There exists a smooth map $\phi : M \to \mathfrak{so}(M)$ so ${}^gR = \lambda_0 R_{\mathrm{Id}} + \lambda_1 R_\phi$.*
3. *$\nabla\phi = 0$, $\nabla^g R = 0$, and $\lambda_0 + \varepsilon\lambda_1 = 0$.*

Proof. To show that a pseudo-Riemannian manifold (M, g) is m-stein, we must find constants $c_i(P)$ so that

$$\mathrm{Tr}\{\mathcal{J}_R(x)^i\} = c_i(P)(x, x)^i \text{ for } 1 \leq i \leq m \text{ and } x \in T_P M.$$

By Lemma 1.7.3, it suffices to check that this identity holds for all non-degenerate tangent vectors x.

Fix $P \in M$ and let $R = {}^gR_P$. We have $R = \lambda_0(P)R_{\mathrm{Id}} + \lambda_1(P)R_\phi$, where ϕ is a skew-adjoint map of $T_P M$ with $\phi^2 = \varepsilon\,\mathrm{Id}$ for $\varepsilon = \pm 1$. We may express:

$$\begin{aligned} R(x, y)z =& \lambda_0(P)\{(y, z)x - (x, z)y\} \\ &+ \lambda_1(P)\{(\phi y, z)\phi x - (\phi x, z)\phi y - 2(\phi x, y)\phi z\}. \end{aligned}$$

Let x be a non-degenerate vector in $T_P M$. Then $\{x, \phi x\}$ spans a non-degenerate 2 plane. Decompose

$$T_P M = \mathrm{span}\{x, \phi x\} \oplus \{y : y \perp \{x, \phi x\}\}.$$

We show that R is k-stein for any k by computing:

$$\begin{aligned}
&\mathcal{J}_R(x)x = 0,\\
&\mathcal{J}_R(x)\phi x = (x,x)\{\lambda_0(P) - 3\varepsilon\lambda_1(P)\}\phi x,\\
&\mathcal{J}_R(x)y = (x,x)\lambda_0(P)y \text{ if } y \perp \{x, \phi x\},\\
&\operatorname{Tr}\{\mathcal{J}_R(x)^i\} = (x,x)^i\{(m-2)\lambda_0(P)^i + (\lambda_0(P) - 3\varepsilon\lambda_1(P))^i\}.
\end{aligned} \tag{1.15.1.a}$$

Furthermore, we may express:

$$\begin{aligned}
c_1(P) &= (m-2)\lambda_0(P) + (\lambda_0(P) - 3\varepsilon\lambda_1(P)) \text{ and}\\
c_2(P) &= (m-2)\lambda_0(P)^2 + (\lambda_0(P) - 3\varepsilon\lambda_1(P))^2.
\end{aligned}$$

Since $m \neq 2, 4$, we may use Theorem 1.13.1 to see that c_1 and c_2 are constant functions on M. Let

$$\begin{aligned}
\mathcal{E} &:= \{(u,v) \in \mathbb{R}^2 : (m-2)u^2 - 3\varepsilon v^2 = c_1\} \text{ and}\\
\mathcal{L} &:= \{(u,v) \in \mathbb{R}^2 : (m-2)u - 3\varepsilon v = c_2\}.
\end{aligned}$$

Since $m - 2 \neq 0$, $\mathcal{E}$ is either an ellipse or a hyperbola in $\mathbb{R}^2$ and $\mathcal{L}$ is a line in $\mathbb{R}^2$. Consequently, $\mathcal{E} \cap \mathcal{L}$ is a finite set. Let

$$\alpha(P) := (\lambda_0(P), \lambda_0(P) - 3\varepsilon\lambda_1(P))$$

be the two distinct eigenvalues of $(x,x)^{-1}\mathcal{J}_R$ for x non-degenerate. By Lemma 1.3.5, α is a continuous function on the space of non-degenerate vectors. Since $\alpha(P) \in \mathcal{E} \cap \mathcal{L}$, α takes discrete values and hence is locally constant. Since M is connected, we may conclude that λ_0 and λ_1 are constant. This completes the proof of Assertion (1).

For each P, we may express ${}^gR_P = \lambda_0 R_{\mathrm{Id}} + \lambda_1 R_{\phi(P)}$; the only question at issue in the proof of Assertion (2) is whether ϕ can be chosen to vary smoothly with P. We now use the assumption that M is contractible. If $x \in S^{\pm}(M,g)$, then we shall set

$$\mathcal{W}(x) := \{y : y \perp x \text{ and } \mathcal{J}_R(x)y = (x,x)(\lambda_0 - 3\varepsilon\lambda_1)y\}.$$

Since $\lambda_1 \neq 0$, we use Display (1.15.1.a) to see that

$$\dim \mathcal{W}(x) = 1.$$

Consequently, we may use Lemma 4.2.5 to see that $\mathcal{W}$ is a smooth line bundle over the pseudosphere bundles $S^{\pm}(M,g)$.

Fix $P \in M$. Since $s(x, P) := \phi(P)(x)$ is a unit section to the restriction of $\mathcal{W}$ to $S^{\pm}(T_PM)$, the line bundles $\mathcal{W}$ are trivial over $S^{\pm}(T_PM)$ and $\phi(P)$ is uniquely defined modulo a sign ambiguity. As M is contractible, we can use Lemma 4.2.4 to find an orthonormal frame field for TM and express

$$S^{\pm}(M, g) = M \times S^{\pm}(\mathbb{R}^{(p,q)}).$$

We may now use Lemma 4.2.4 to see that $\mathcal{W}$ is trivial over $S^{\pm}(M, g)$. Choosing a smooth unit section to $\mathcal{W}$ defines $\phi(P)$ and shows that ϕ can be chosen to vary smoothly. This completes the proof of Assertion (2).

We follow the argument given by Tricerri and Vanhecke [152, Theorem 12.7] to prove Assertion (3); we modify their argument slightly to extend it from the Riemannian to the pseudo-Riemannian setting. Since λ_0 and $\lambda_1 \neq 0$ are constants and since $\nabla R_{\text{Id}} = 0$, we have $\nabla R = \lambda_1 \nabla R_\phi$. Let $\phi_{;a} := \nabla_a \phi$ and $\phi^2 := \varepsilon \,\text{Id}$. Let $\{a, b, c, d\}$ be arbitrary elements of V. The second Bianchi identity then yields:

$$\begin{aligned}
0 =&\lambda_1^{-1}\{\nabla_a R(b,c)d + \nabla_b R(c,a)d + \nabla_c R(a,b)d\}\\
=&(\phi_{;a}c, d)\phi b + (\phi_{;b}a, d)\phi c + (\phi_{;c}b, d)\phi a\\
&+(\phi c, d)\phi_{;a}b + (\phi a, d)\phi_{;b}c + (\phi b, d)\phi_{;c}a\\
&-(\phi_{;a}b, d)\phi c - (\phi_{;b}c, d)\phi a - (\phi_{;c}a, d)\phi b\\
&-(\phi b, d)\phi_{;a}c - (\phi c, d)\phi_{;b}a - (\phi a, d)\phi_{;c}b\\
&-2(\phi_{;a}b, c)\phi d - 2(\phi_{;b}c, a)\phi d - 2(\phi_{;c}a, b)\phi d\\
&-2(\phi b, c)\phi_{;a}d - 2(\phi c, a)\phi_{;b}d - 2(\phi a, b)\phi_{;c}d.
\end{aligned} \tag{1.15.1.b}$$

Let $x \in S^{\pm}(T_PM)$ and let $\pi_1 := \{x, \phi x\}$ be a non-degenerate 2 plane. Since ϕ is skew-adjoint and since π is ϕ invariant, ϕ preserves $\pi^\perp$. Since $m \geq 4$, we can choose a non-degenerate vector w in $\pi^\perp$. Then $\{x, \phi x, w, \phi w\}$ forms an orthogonal basis for a non-degenerate 4 plane. We set $a = w$, $b = x$, $c = \phi w$, $d = x$ in Equation (1.15.1.b) to compute:

$$\begin{aligned}
0 =&(\phi_{;w}\phi w, x)\phi x + (\phi_{;x}w, x)\phi\phi w + (\phi_{;\phi w}x, x)\phi w\\
&+(\phi\phi w, x)\phi_{;w}x + (\phi w, x)\phi_{;x}\phi w + (\phi x, x)\phi_{;\phi w}w\\
&-(\phi_{;w}x, x)\phi\phi w - (\phi_{;x}\phi w, x)\phi w - (\phi_{;\phi w}w, x)\phi x\\
&-(\phi x, x)\phi_{;w}\phi w - (\phi\phi w, x)\phi_{;x}w - (\phi w, x)\phi_{;\phi w}x\\
&-2(\phi_{;w}x, \phi w)\phi x - 2(\phi_{;x}\phi w, w)\phi x - 2(\phi_{;\phi w}w, x)\phi x\\
&-2(\phi x, \phi w)\phi_{;w}x - 2(\phi\phi w, w)\phi_{;x}x - 2(\phi w, x)\phi_{;\phi w}x.
\end{aligned} \tag{1.15.1.c}$$

The operators ϕ and $\nabla\phi$ are skew-adjoint. As $\phi^2 = \varepsilon\,\mathrm{Id}$, $\phi(\nabla\phi)+(\nabla\phi)\phi = 0$ so $\phi\nabla\phi$ is skew-adjoint. Thus:

$$\begin{aligned}0 =&(\phi_{;u}v, v) = (\phi_{;u}\phi v, v) = (\phi_{;u}v, \phi v)\\ =&(\phi_{;u}v, w) + (v, \phi_{;u}w) \text{ for all } u, v, w \in V.\end{aligned} \tag{1.15.1.d}$$

We use Equation (1.15.1.d) and the fact that $\{x, \phi x, w, \phi w\}$ is an orthogonal set to simplify Equation (1.15.1.c) by eliminating terms which vanish and by combining other terms to see:

$$\begin{aligned}0 =&3(\phi_{;w}\phi w, x)\phi x + (\phi_{;x}w, x)\phi\phi w\\ &-(\phi_{;x}\phi w, x)\phi w - 3(\phi_{;\phi w}w, x)\phi x - 2(\phi\phi w, w)\phi_{;x}x.\end{aligned} \tag{1.15.1.e}$$

We take the inner product of Equation (1.15.1.e) with the vector w to see

$$0 = \varepsilon(\phi_{;x}w, x)(w, w) - 2\varepsilon(w, w)(\phi_{;x}x, w) = -3\varepsilon(w, w)(\phi_{;x}x, w).$$

Since w is non-degenerate, $(w, w) \neq 0$ and thus $(\phi_{;x}x, w) = 0$. Since w was an arbitrary non-degenerate vector of $\pi^\perp$, this implies that

$$\phi_{;x}x \in \pi.$$

ByEquation (1.15.1.d), $(\phi_{;x}x, x) = (\phi_{;x}x, \phi x) = 0$. Thus, we conclude therefore $\phi_{;x}x = 0$ for every non-degenerate vector $x \in V$. Continuity then implies $\phi_{;x}x = 0$ for all $x \in V$. Since $\phi(\nabla\phi) + (\nabla\phi)\phi = 0$, we also have $\phi_{;x}\phi x = 0$ for all x. We polarize to see

$$\phi_{;a}b + \phi_{;b}a = \phi_{;a}\phi a = \phi_{;a}a = 0 \text{ for all } a \in V. \tag{1.15.1.f}$$

Next, we set $a = w$, $b = d = x$ and $c = \phi x$ in Equation (1.15.1.b) to see:

$$\begin{aligned}0 =&(\phi_{;w}\phi x, x)\phi x + (\phi_{;x}w, x)\phi\phi x + (\phi_{;\phi x}x, x)\phi w\\ &+(\phi\phi x, x)\phi_{;w}x + (\phi w, x)\phi_{;x}\phi x + (\phi x, x)\phi_{;\phi x}w\\ &-(\phi_{;w}x, x)\phi\phi x - (\phi_{;x}\phi x, x)\phi w - (\phi_{;\phi x}w, x)\phi x\\ &-(\phi x, x)\phi_{;w}\phi x - (\phi\phi x, x)\phi_{;x}w - (\phi w, x)\phi_{;\phi x}x\\ &-2(\phi_{;w}x, \phi x)\phi x - 2(\phi_{;x}\phi x, w)\phi x - 2(\phi_{;\phi x}w, x)\phi x\\ &-2(\phi x, \phi x)\phi_{;w}x - 2(\phi\phi x, w)\phi_{;x}x - 2(\phi w, x)\phi_{;\phi x}x\end{aligned}$$

We use Equation (1.15.1.d), we use Equation (1.15.1.f), and we use the fact that $\{x, \phi x, w, \phi w\}$ is an orthogonal set to simplify this equation by eliminating terms which vanish and by combining other terms to see:

$$0 = -4(\phi\phi x, x)\phi_{;x}w.$$

Since $(\phi^2 x, x) \neq 0$, we see that $\phi_{;x}w = 0$ for any vector $w \in \pi^\perp$ which is not degenerate. Continuity then implies $\phi_{;x}w = 0$ for all $w \in \pi^\perp$. Since $\phi_{;x}$ vanishes on π, this implies that $\phi_{;x} = 0$ for any non degenerate vector in V. By continuity, we then have $\phi_{;x} = 0$ for all $x \in V$ and thus $\nabla\phi = 0$.

As above, we let $\{x, \phi x, w, \phi w\}$ be an orthonormal set. We compute

$$\begin{array}{ll} R_{\mathrm{Id}}(x,w)x = -(x,x)w, & R_\phi(x,w)x = 0, \\ R_{\mathrm{Id}}(x,w)\phi x = 0, & R_\phi(x,w)\phi x = \varepsilon(x,x)\phi w, \\ R_{\mathrm{Id}}(x,w)w = (w,w)x, & R_\phi(x,w)w = 0, \\ R_{\mathrm{Id}}(x,w)\phi w = 0, & R_\phi(x,w)\phi w = -\varepsilon(w,w)\phi x. \end{array}$$

Consequently, we have that

$$\begin{array}{ll} {}^gR(x,w)x = -\lambda_0(x,x)w, & {}^gR(x,w)\phi x = \quad \lambda_1\varepsilon(x,x)\phi w, \\ {}^gR(x,w)w = \quad \lambda_0(w,w)x, & {}^gR(x,w)\phi w = -\lambda_1\varepsilon(w,w)\phi x. \end{array}$$

Since $\nabla\phi = 0$, we have $\nabla {}^gR = 0$ so $\nabla^2 {}^gR = 0$. We apply the identity

$$0 = {}^gR(a,b,c,d;e,f) - {}^gR(a,b,c,d;f,e)$$

with $a = x$, $b = \phi x$, $c = \phi w$, $d = x$, $e = x$, and $f = w$ to see:

$$\begin{aligned} 0 =& {}^gR({}^gR(x,w)x, \phi x, \phi w, x) + {}^gR(x, {}^gR(x,w)\phi x, \phi w, x) \\ &+ {}^gR(x, \phi x, {}^gR(x,w)\phi w, x) + {}^gR(x, \phi x, \phi w, {}^gR(x,w)x) \\ =& - \lambda_0(x,x){}^gR(w, \phi x, \phi w, x) + \lambda_1\varepsilon(x,x){}^gR(x, \phi w, \phi w, x) \\ &- \lambda_1\varepsilon(w,w){}^gR(x, \phi x, \phi x, x) - \lambda_0(x,x){}^gR(x, \phi x, \phi w, w). \end{aligned} \tag{1.15.1.g}$$

We compute that:

$$\begin{array}{ll} R_{\mathrm{Id}}(w,\phi x,\phi w,x) = 0, & R_\phi(w,\phi x,\phi w,x) = (x,x)(w,w), \\ R_{\mathrm{Id}}(x,\phi w,\phi w,x) = -\varepsilon(x,x)(w,w), & R_\phi(x,\phi w,\phi w,x) = 0, \\ R_{\mathrm{Id}}(x,\phi x,\phi x,x) = -\varepsilon(x,x)^2, & R_\phi(x,\phi x,\phi x,x) = 3(x,x)^2, \\ R_{\mathrm{Id}}(x,\phi x,\phi w,w) = 0, & R_\phi(x,\phi x,\phi w,w) = 2(w,w)(x,x). \end{array}$$

We may now complete the proof by using Equation (1.15.1.g) and these relations to compute:

$$\begin{aligned} 0 =& \{-\lambda_0\lambda_1 - \lambda_0\lambda_1 + \lambda_1(\lambda_0 - 3\varepsilon\lambda_1) - 2\lambda_0\lambda_1\}(x,x)^2(w,w) \\ =& \lambda_1(-3\varepsilon\lambda_1 - 3\lambda_0)(x,x)^2(w,w). \quad \square \end{aligned}$$

The condition $\nabla\phi = 0$ has very strong implications. Let the associated eigenbundles of ϕ be given by:

$$T^{\pm}(M) := \begin{cases} \{X \in TM \otimes_{\mathbb{R}} \mathbb{C} : \phi X = \pm\sqrt{-1}X\} & \text{if } \varepsilon = -1, \\ \{X \in TM : \phi X = \pm X\} & \text{if } \varepsilon = +1. \end{cases}$$

Let $[\cdot,\cdot]$ denote the Lie bracket. We say that ϕ is an *integrable almost (para-)complex structure* if the bundles $T^{\pm}(M)$ are integrable in the sense of Frobenius; this means that

$$X_1, X_2 \in C^\infty(T^\pm(M)) \text{ implies } [X_1, X_2] \in C^\infty(T^\pm(M)).$$

Let $\delta = \pm 1$. Let X_i be complex vector fields on M so that $\phi X_i = \delta\sqrt{\varepsilon}X_i$ for $i = 1, 2$. Since $\nabla\phi = 0$, $\nabla_X\phi = \phi\nabla_X$. We show $T^\delta(M)$ is integrable by computing:

$$\begin{aligned}&\phi([X_1, X_2]) = \phi(\nabla_{X_1}X_2 - \nabla_{X_2}X_1)\\ =&\nabla_{X_1}\phi X_2 - \nabla_{X_2}\phi X_1 = \nabla_{X_1}\delta\sqrt{\varepsilon}X_2 - \nabla_{X_2}\delta\sqrt{\varepsilon}X_1\\ =&\delta\sqrt{\varepsilon}\{\nabla_{X_1}X_2 - \nabla_{X_2}X_1\} = \delta\sqrt{\varepsilon}[X_1, X_2].\end{aligned}$$

Suppose that $\varepsilon = -1$. We use the Nirenberg–Newlander theorem [122] to show that there exist local holomorphic coordinates $z_j = x_j + \sqrt{-1}y_j$ so

$$\phi\partial_j^x = \partial_j^y \text{ and } \phi\partial_j^y = -\partial_j^x.$$

Thus, if $\nabla\phi = 0$, then (M, g) has an underlying holomorphic structure.

The *Kähler form* is (up to a normalizing constant which plays no role in our development) defined by $\Omega(X, Y) = g(X, \phi Y)$. Let $\{e_i\}$ be a local frame field for TM and let $\{e^i\}$ be the dual coframe field for the cotangent bundle T^*M. Let $\text{ext}(e^i)$ be left *exterior multiplication* by the covector e^i and let $\text{int}(e^i)$ denote the dual map, *interior multiplication.* We then have:

$$\begin{aligned}&(\nabla_Z\Omega)(X, Y) = (\nabla_Z g)(X, \phi Y) + g(X, (\nabla_Z\phi)Y) = 0 \text{ so}\\ &d\Omega = \textstyle\sum_{e_i} \text{ext}(e^i)\nabla_{e_i}\Omega = 0 \text{ and } \delta\Omega = \textstyle\sum_{e_i} \text{int}(e^i)\nabla_{e_i}\Omega = 0.\end{aligned}$$

This shows that the Kähler form is closed, coclosed, and parallel. Thus, the metric in question is, by definition, a (para-)Kähler metric.

Assertion (2) of Lemma 1.15.1 can fail if (M, g) is not simply connected. Let $\mathbb{CP}^{2n-1}$ be complex projective space equipped with the Fubini-Study metric (see Section 3.6 for details). Let

$$\Theta(z_1, ..., z_{2n}) := (-\bar{z}_2, \bar{z}_1, -\bar{z}_4, \bar{z}_3, ..., -\bar{z}_{2n}, \bar{z}_{2n-1})$$

define a conjugate linear isometry of $\mathbb{C}^{2n}$. Since $\Theta(\lambda\vec{z}) = \bar{\lambda}\vec{z}$, Θ preserves complex lines and induces a fixed point free isometry $[\Theta]$ of $\mathbb{CP}^{2n-1}$ with $\Theta^2 = \text{Id}$. Let $M := \mathbb{CP}^{2n-1}/\mathbb{Z}_2$ be the associated quotient space; we have a normal covering

$$\mathbb{Z}_2 \to \mathbb{CP}^{2n-1} \to M.$$

Since the arithmetic genus of $\mathbb{CP}^{2n-1}$ is 1, (M,g) is not a holomorphic manifold and hence ϕ is not globally defined on (M,g). On the other hand, the Fubini-Study metric on $\mathbb{CP}^{2n-1}$ is invariant under this $\mathbb{Z}_2$ action and induces a metric on (M,g) relative to which (M,g) is a complex space form.

Let $t \to \sigma(t)$ be a curve in a pseudo-Riemannian manifold (M,g). If Y_1 and Y_2 are parallel vector fields along σ, then

$$\partial_t(Y_1,Y_2) = (Y_1',Y_2) + (Y_1,Y_2') = 0 \tag{1.15.1.h}$$

so (Y_1,Y_2) is constant. In particular, if Y is a parallel vector field, then (Y,Y) is constant.

We now study generalized (para-)complex spaceforms defined by Clifford modules.

1.15.2 Lemma. *Let (M,g) and $(\tilde M,\tilde g)$ be pseudo-Riemannian manifolds of the same signature (p,q). Assume that there exist Clifford families of skew-adjoint endomorphisms $\{\phi_i\}_{1\le i\le\ell}$ and $\{\tilde\phi_i\}_{1\le i\le\ell}$ of TM and $T\tilde M$ and constants $\varepsilon_i=\pm1$ so that*

$$\begin{aligned}
&\phi_i^2=\varepsilon_i\,\mathrm{Id}_M \text{ for } 1\le i\le\ell,\\
&\tilde\phi_i^2=\varepsilon_i\,\mathrm{Id}_{\tilde M} \text{ for } 1\le i\le\ell,\\
&\phi_i\phi_j+\phi_j\phi_i=0 \text{ for } 1\le i<j\le\ell, \text{ and}\\
&\tilde\phi_i\tilde\phi_j+\tilde\phi_j\tilde\phi_i=0 \text{ for } 1\le i<j\le\ell.
\end{aligned}$$

Assume there exist constants $\lambda_0,\dots,\lambda_\ell$ so

$$\begin{aligned}
&{}^gR_P=\lambda_0R_{\mathrm{Id}}+\textstyle\sum_{1\le i\le\ell}\lambda_iR_{\phi_i}, \text{ and}\\
&{}^{\tilde g}R_{\tilde P}=\lambda_0R_{\mathrm{Id}}+\textstyle\sum_{1\le i\le\ell}\lambda_iR_{\tilde\phi_i}.
\end{aligned}$$

Assume $\nabla\phi_i=0$ and $\tilde\nabla\tilde\phi_i=0$ for $1\le i\le\ell$. Choose points $P\in M$ and $\tilde P\in\tilde M$. Suppose there exists an isometry $\Psi: T_PM\to T_{\tilde P}\tilde M$ so that $\tilde\phi_i(\tilde P)\Psi=\Psi\phi_i(P)$ for $1\le i\le\ell$. Then there exists a local isometry ψ from (M,g) to $(\tilde M,\tilde g)$ so that $\psi(P)=\tilde P$ and so that $\psi_*(P)=\Psi$.

Proof. We use the same strategy as that used to prove Lemma 1.14.2 and define a local diffeomorphism between a neighborhood of P in M and of $\tilde P$ in $\tilde M$ by setting:

$$\psi=\exp_{\tilde P}\Psi\exp_P^{-1}.$$

Let v be a spacelike vector in T_PM. We use Lemma 1.4.5 to see that $\{\phi_1v,\dots,\phi_\ell v\}$ is an orthonormal set with $v\perp\phi_iv$. Thus, we may choose an orthonormal basis $\{w_1,\dots,w_{m-1}\}$ for $v^\perp$ so that $\phi_iv=w_i$ for $1\le i\le\ell$.

Let e_i be parallel vector fields along the geodesic $\sigma := \exp_P(tv)$ so that $e_i(0) = w_i$. Since $\nabla\phi_i = 0$, we have

$$\nabla_{\sigma'}\{\phi_i\sigma'\} = \phi_i\nabla_{\sigma'}\sigma' = 0.$$

Thus, $e_i = \phi_i\sigma'$ for $1 \le i \le \ell$. Let $c^2 = (v,v)$. We may then compute:

$$\mathcal{J}_R(\sigma')e_i = \begin{cases} c^2(\lambda_0 - 3\varepsilon_i\lambda_i)e_i & \text{if } 1 \le i \le \ell, \\ c^2\lambda_0 e_i & \text{if } \ell < i \le m-1. \end{cases}$$

We replace κ by $\lambda_0 - 3\varepsilon_i\lambda_i$ or by λ_0 in Equations (1.14.2.b) and (1.14.2.c) to construct a suitable family of Jacobi vector fields Z_i and $\tilde{Z}_i$ such that $\psi_* Z_i = \tilde{Z}_i$. We apply the same argument as that used to prove Lemma 1.14.2 to complete the proof. □

There can be inequivalent Clifford module structures on $\mathbb{R}^{(p,q)}$ if $\ell \ge 3$; this is not the case if $\ell = 1$ so this case is particularly simple. In the following Lemma we show that (para-)complex space forms are determined up to local isometry by (p,q,λ_1) if $m \ne 2,4$.

1.15.3 Lemma.

(1) *Let V_1 and V_2 be vector spaces of the same signature (p,q). Fix $\varepsilon = \pm 1$. Assume given $\phi_i \in \mathfrak{so}(V_i)$ with $\phi_i^2 = \varepsilon\,\mathrm{Id}$ for $i = 1,2$. Then there exists an isometry $\Psi : V_1 \to V_2$ such that $\phi_2\Psi = \Psi\phi_1$.*

(2) *Assume either that (M_1,g_1) and (M_2,g_2) are complex space forms of the same signature (p,q) and dimension $m = p+q$ or that (M_1,g_1) and (M_2,g_2) are para-complex space forms of the same dimension m. Assume that $m \ne 2,4$ and that $\lambda_1(M_1,g_1) = \lambda_1(M_2,g_2)$. Choose points $P_1 \in M_1$ and $P_2 \in M_2$. Let Ψ be any linear isometry from $T_{P_1}M_1$ to $T_{P_2}M_2$ which intertwines the (para-)complex structures; such maps Ψ exist by Assertion (1). Then there exists a local isometry ψ from (M_1,g_1) to (M_2,g_2) such that $\psi(P_1) = P_2$, such that $\psi_*(P_1) = \Psi$, and such that ψ_* intertwines the (para-)complex structures on each tangent space.*

Proof. Let $\phi_i \in \mathfrak{so}(V_i)$ with $\phi_i^2 = \varepsilon\,\mathrm{Id}$ for $i = 1,2$. We suppose first $\varepsilon = +1$. Then $(p,q) = (2r,2s)$ and we can choose orthonormal bases for V_1 and V_2 of the form

$$\mathcal{B}_1 = \{e_1^-, \phi_1 e_1^-, ..., e_r^-, \phi_1 e_r^-, e_1^+, \phi_1 e_1^+, ..., e_s^+, \phi_1 e_s^+\} \text{ and}$$
$$\mathcal{B}_2 = \{f_1^-, \phi_2 f_1^-, ..., f_r^-, \phi_2 f_r^-, f_1^+, \phi_2 f_1^+, ..., f_s^+, \phi_2 f_s^+\}.$$

The vectors with the superscript '+' are spacelike and the vectors with the superscript '−' are timelike. The desired map Ψ which intertwines ϕ_1 and ϕ_2 is then defined by setting:

$$\Psi e_i^{\pm} = f_i^{\pm} \text{ and } \Psi\phi_1 e_i^{\pm} = \phi_2 f_i^{\pm}.$$

If $\varepsilon = +1$, then $p = q$ and the maps ϕ_i are a para-isometries which interchange the roles of timelike and spacelike vectors. Thus, we can choose orthonormal bases for V_1 and V_2 of the form

$$\mathcal{B}_1 = \{e_1^+, \phi e_1^+, ..., e_r^+, \phi e_r^+\} \text{ and}$$
$$\mathcal{B}_2 = \{f_1^+, \phi f_1^+, ..., f_r^+, \phi f_r^+\}.$$

The vectors $\{e_1^+, ..., e_r^+\}$ and $\{f_1^+, ..., f_r^+\}$ are spacelike and the vectors $\{\phi_1 e_1^+, ..., \phi_1 e_r^+\}$ and $\{\phi_2 f_1^+, ..., \phi_2 f_r^+\}$ are timelike. We set $\Psi(e_i^+) = f_i^+$ and $\Psi(\phi_1 e_i^+) = \phi_2 f_i^+$ to complete the proof of Assertion (1). We apply Lemma 1.15.1 to see $\nabla\phi = 0$ and $\tilde{\nabla}\tilde{\phi} = 0$. Assertion (2) now follows from Lemma 1.15.2. □

Chapter 2

The Skew-Symmetric Curvature Operator

2.1 Introduction

In Chapter 2, we study the geometry of the skew-symmetric curvature operator. Let V be a vector space with an inner product of signature (p, q). Let $\mathrm{Gr}^+_{2,0}(V)$, $\mathrm{Gr}^+_{1,1}(V)$, and $\mathrm{Gr}^+_{0,2}(V)$ be the *Grassmannian manifolds* of all timelike, mixed, and spacelike oriented 2 planes. If the metric is positive definite (i.e. $p = 0$), then we shall let $\mathrm{Gr}^+(V) := \mathrm{Gr}^+_{0,2}(V)$.

Let R be an algebraic curvature tensor on V. If $\{x, y\}$ is an oriented orthonormal basis for a non-degenerate 2 plane π, then we have defined the skew-symmetric curvature operator in Equation (1.5.5.a) by setting:

$$\mathcal{R}(\pi) := |(x,x)(y,y) - (x,y)^2|^{-1/2}\mathcal{R}(x,y).$$

We use Lemma 1.9.1 to see that $\mathcal{R}(\pi)$ is independent of the particular oriented basis which is chosen. Since $R(x,y,z,w) = -R(x,y,w,z)$, $\mathcal{R}(\pi)$ is a skew-symmetric operator. We say that R has *constant timelike rank* r, *constant mixed rank* r, or *constant spacelike rank* r if the $\mathrm{rank}(\mathcal{R}(\cdot)) = r$ on $\mathrm{Gr}^+_{2,0}(V)$, on $\mathrm{Gr}^+_{1,1}(V)$, or on $\mathrm{Gr}^+_{0,2}(V)$, respectively. Similarly, we say that R is *timelike IP*, *mixed IP*, or *spacelike IP* if the eigenvalues of $\mathcal{R}(\pi)$ are constant on $\mathrm{Gr}^+_{2,0}(V)$, on $\mathrm{Gr}^+_{1,1}(V)$ or on $\mathrm{Gr}^+_{0,2}(V)$, respectively.

Recall that two linear transformations T_1 and T_2 of V are said to be *Jordan equivalent* if they have the same Jordan normal form or, equivalently, if they are *conjugate* which means that there exists an invertible linear transformation ψ of V so that $T_1 = \psi T_2 \psi^{-1}$.

We say that R is *timelike Jordan IP*, *mixed Jordan IP*, or *spacelike Jordan IP* if the Jordan normal form of R is constant on $\operatorname{Gr}_{2,0}^+(V)$, on $\operatorname{Gr}_{1,1}^+(V)$, or on $\operatorname{Gr}_{0,2}^+(V)$, respectively.

In Section 2.2, we will explore relationships between these various notions. In Lemma 2.2.2, we will exhibit algebraic curvature tensors which have constant spacelike, but neither constant timelike nor mixed rank. It is immediate that if R is timelike, mixed, or spacelike Jordan IP, then R is, respectively, timelike, mixed, or spacelike IP and furthermore, R has constant timelike, mixed, or spacelike rank. In Lemma 1.10.5, we showed that spacelike IP, timelike IP, and mixed IP are equivalent notions; we therefore say that such an algebraic curvature tensor is IP. In Lemma 2.2.5, we construct algebraic curvature tensors which are IP but which are neither spacelike, timelike, nor mixed Jordan IP. We also show that there exist algebraic curvature tensors which are spacelike Jordan IP but which are neither timelike, nor mixed Jordan IP.

Let V have signature (p,q) and let R be an algebraic curvature tensor which has constant spacelike, timelike, and mixed rank r. In Chapter 4, we will use topological methods to prove the following result (see the discussion following Theorem 4.1.4 for details). In the Riemannian setting, this result is due to Gilkey, Leahy, and Sadofsky [83]; in the higher signature setting it is due to Zhang [163].

2.1.1 Theorem. *Let V be a vector space of signature (p,q) and let R be an algebraic curvature tensor of constant timelike, mixed, and spacelike rank $r > 0$ on V.*

(1) *Let $p = 0$, let $q \geq 5$, and let $q \neq 7, 8$. Then $r = 2$.*

(2) *Let $p = 1$ and let $q \geq 9$. Then $r = 2$.*

(3) *Let $p = 2$ and let $q \geq 10$. Then $r \leq 4$. Assume that neither q nor $q + 2$ are powers of 2. Then $r = 2$.*

Stavrov [146] has recently extended this result:

2.1.2 Theorem. *Let V be a vector space of signature (p,q) and let R be an algebraic curvature tensor of constant spacelike rank $r > 0$ on V. If $p \leq \frac{q-6}{4}$ and if the set $\{q, q+1, ..., q+p\}$ does not contain a power of 2, then $r \leq 2$.*

Theorems 2.1.1 and 2.1.2 focus attention on the case $r = 2$. In Theorem 2.2.3, we will exhibit algebraic curvature tensors which have constant spacelike rank 4 if $p \geq q$; thus some restrictions on p and q are necessary in trying to generalize Theorem 2.1.1 to higher signatures.

Dimensions 4, 7, and 8 are exceptional. In Section 2.8, we will construct IP algebraic curvature tensors of rank 4 on $\mathbb{R}^4$ and on $\mathbb{R}^{(2,2)}$. In Section 2.9, we shall exhibit a 4 tensor which satisfies the first two curvature identities given in Equations (1.5.1.e) and (1.5.1.f), which has constant rank 6, and which is IP, but which does not satisfy the first Bianchi identity of Equation (1.5.1.g). In Section 2.10, we will study the 8 dimensional setting and prove:

2.1.3 Theorem. *Let R be an IP algebraic curvature tensor on $\mathbb{R}^8$. Then* $\operatorname{rank}(R) \leq 2$.

Let $m \neq 4, 8$. Kowalski, Sekizawa, and Vlasek [111] showed that the tangent sphere bundle of a Riemannian manifold of dimension m can not have strictly positive curvature. Crucial to their argument was a technical result that if R is an algebraic curvature tensor, then there exists a 2 plane π so that $\operatorname{rank} R(\pi) < m$. This result follows from Theorem 2.1.1 (1); it also follows from work of Yampolsky [162]. What is a bit surprising, however, is that there is a stronger result available. In Section 4.3, we will use a result of Adams [1] concerning $\widetilde{KO}(\mathbb{RP}^n)$ to derive the following result from a slightly more general result (see Lemma 4.3.22 for details):

2.1.4 Theorem. *Let R be an algebraic curvature tensor on a Riemannian vector space V of dimension $m \geq 3$, where $m \neq 4, 8$. Fix a non-zero vector $v \in V$. Then there exists a 2 plane π so that $R(\pi)v = 0$.*

Let V have signature (p, q). We suppose $q \geq 5$. In Sections 2.3, 2.5, and 2.7, we obtain various classification results. These results were first established in the Riemannian setting by Gilkey, Leahy, and Sadofsky [83] and were subsequently extended to the pseudo-Riemannian setting by Gilkey and Zhang [88] and by Zhang [163].

Let ϕ be a self-adjoint linear transformation of V. We showed in Lemma 1.8.1 that

$$R_\phi(x, y)z := (\phi y, z)\phi x - (\phi x, z)\phi y$$

is an algebraic curvature tensor on V. Furthermore, by Lemma 1.8.6, ϕ is determined by R_ϕ, modulo a possible sign ambiguity, if $\ker \phi$ contains no spacelike vectors. In Section 2.3, we prove the following classification result:

2.1.5 Theorem. *Let R be an algebraic curvature tensor on a vector space V of signature (p, q), where $q \geq 5$. The following assertions are equivalent:*

(1) *R has constant spacelike rank* 2.
(2) *There exists a self-adjoint map ϕ so that* $\ker \phi$ *contains no spacelike vectors such that $R = \pm R_\phi$.*

In Section 2.4, we show that every spacelike rank 2 algebraic curvature tensor is geometrically realizable by the germ of a spacelike rank 2 pseudo-Riemannian hypersurface in flat space. We also establish a smooth analogue of Theorem 2.1.5.

In Section 2.5, we establish the following algebraic classification result:

2.1.6 Theorem. *Let R be an algebraic curvature tensor on a vector space V of signature (p, q), where $q \geq 5$. The following conditions are equivalent:*

(1) *R is a spacelike rank 2 Jordan IP algebraic curvature tensor.*

(2) *There exists a non-zero constant C and a self-adjoint map ϕ of V so that $R = CR_\phi$, where ϕ satisfies exactly one of the following three conditions:*

  2a) *ϕ is an isometry of V.*

  2b) *ϕ is a para-isometry of V.*

  2c) *$\phi^2 = 0$, and $\ker\phi$ contains no spacelike vectors.*

Let $R = CR_\phi$, where $C \neq 0$ and where $\phi^2 = \pm\,\mathrm{Id}$. Then R is also timelike and mixed rank 2 Jordan IP. If $\phi^2 = 0$, then $R_\phi(x, y)^2 = 0$ for all (x, y) so R_ϕ is *nilpotent.* Such a tensor is timelike rank 2 Jordan IP if and only if $\ker\phi$ contains no timelike vectors; it does not have constant mixed rank.

We say that a pseudo-Riemannian manifold is spacelike, mixed, or timelike *rank* 2 *Jordan IP* if the associated curvature tensor ${}^gR$ is a spacelike, mixed, or timelike rank 2 Jordan IP algebraic curvature tensor at each point of the manifold. In Section 2.6, we construct three families of such manifolds. The pseudo-spheres discussed in Lemma 2.6.1 and the warped products discussed in Lemma 2.6.3 are spacelike, mixed, and timelike rank 2 Jordan IP manifolds. The manifolds discussed in Lemma 2.6.5 are nilpotent spacelike and timelike rank 2 Jordan IP manifolds; they are not mixed rank 2 Jordan IP manifolds.

In Section 2.7, we prove the following geometric classification result:

2.1.7 Theorem. *Let (M, g) be a connected rank 2 spacelike Jordan IP pseudo-Riemannian manifold of signature (p, q), where $q \geq 5$. Assume that ${}^gR$ is not nilpotent for at least one point P of M.*

(1) *For each point $P \in M$, we have ${}^gR_P = C(P)R_{\phi(P)}$, where ϕ is self-adjoint map of T_PM so that $\phi^2 = \mathrm{Id}$ and $C \neq 0$. In particular, therefore, ${}^gR_P$ is never nilpotent.*

(2) *If $\phi = \pm\,\mathrm{Id}$, then (M, g) has constant sectional curvature.*

(3) *If $\phi \neq \pm\,\mathrm{Id}$, then (M, g) is locally isometric to one of the warped product manifolds given in* Definition 2.6.2.

It is natural to wonder if every spacelike Jordan IP algebraic curvature tensor can be geometrically realized by the germ of a spacelike Jordan IP pseudo-Riemannian manifold. Let V be a Riemannian vector space of dimension $m \geq 6$. Let ϕ be a self-adjoint map of V such that $\phi^2 = \mathrm{Id}$.

If R_ϕ has constant sectional curvature, then $\operatorname{Tr}\phi = \pm m$. If R_ϕ arises from a warped product manifold, as discussed in Definition 2.6.2, then $\operatorname{Tr}\phi = \pm(m-2)$. Thus, by taking ϕ so $\operatorname{Tr}\phi$ is not one of these values, we may combine Theorems 2.1.6 and 2.1.7 to see, in contrast to the situation for rank 2 algebraic curvature tensors, that not every rank 2 spacelike Jordan IP algebraic curvature tensor is geometrically realizable by a rank 2 spacelike Jordan IP pseudo-Riemannian manifold.

As noted previously, the 4, 7, and 8 dimensional geometries are exceptional; the topological result bounding the rank does not hold in those dimensions. In Section 2.8, we discuss a rank 4 algebraic curvature tensor on $\mathbb{R}^4$ which was discovered by Ivanov and Petrova [97]; Zhang has also constructed a similar tensor on $\mathbb{R}^{(2,2)}$. We discuss these tensors in Theorem 2.8.1.

In Section 2.9, we discuss some partial results in the Riemannian setting if $m = 7$; these are closely related to the exceptional group G_2. In Section 2.10, we prove Theorem 2.1.3 and show any Riemannian IP algebraic curvature tensor in dimension 8 must have rank 2. The classification results cited above can then be applied.

Let J be a pseudo-Hermitian structure on a vector space V of signature (p,q). Recall that an algebraic curvature tensor R is said to be almost complex if $JR(x, Jx) = R(x, Jx)J$ for all x; we refer to Lemma 1.6.2 for other equivalent conditions. We assume that R is almost complex. This implies that $R(\pi)$ is complex linear if π is a non-degenerate complex line.

We consider the Riemannian setting for the moment. Let $\mathbb{CP}(V)$ be the projective space of complex lines in V. Suppose that R is an almost complex IP algebraic curvature tensor on V; as the inner product is positive definite, $R(\pi)$ is diagonalizable as a complex map and therefore the eigenvalue structure determines the Jordan form. We postpone until Chapter 4 the proof of the following result (see the discussion following Theorem 4.1.3):

2.1.8 Theorem. *Let R be an almost complex IP algebraic curvature tensor on a Riemannian vector space V of dimension m. Let $\{\lambda_s, \mu_s\}$ be the eigenvalues and multiplicities of the complex operator $R(\pi)$ for $\pi \in \mathbb{CP}(V)$. We order the multiplicities μ_s so that $\mu_0 \geq ... \geq \mu_\ell$. Suppose that $\ell \geq 1$. If $m \equiv 2$ mod 4, then $\ell = 1$ and $\mu_1 = 1$. If $m \equiv 0$ mod 4, then either $\ell = 1$ and $\mu_1 \leq 2$ or $\ell = 2$ and $\mu_1 = \mu_2 = 1$.*

In Section 2.11, we discuss the skew-symmetric curvature operator in the complex setting. We present results of Gilkey and Ivanova [78] constructing almost complex Jordan IP algebraic curvature tensors which, in the Riemannian setting, realize the maximal eigenvalue structures permitted by Theorem 2.1.8.

We conclude our study of the skew-symmetric curvature operator in Section 2.12 by discussing a higher order analogue defined by Stanilov [141].

2.2 Examples

Let V be a vector space of signature (p,q). Let ϕ be a self-adjoint map of V. In Equation (1.1.1.b), we defined

$$R_\phi(x,y)z = (\phi y, z)\phi x - (\phi x, z)\phi y.$$

We showed in Lemma 1.8.1 that R_ϕ is an algebraic curvature tensor. Let π be a non-degenerate 2 plane. If $\ker\phi\cap\pi\neq\{0\}$, then we can choose a basis $\{x,y\}$ for π so $\phi x = 0$. This implies that $\operatorname{rank} R_\phi(\pi) = 0$. On the other hand, if $\ker\phi\cap\pi = \{0\}$, then $\{\phi x, \phi y\}$ is a linearly independent set for any basis $\{x,y\}$ for π. We may use Lemma 1.2.2 to see that $\operatorname{rank}(R_\phi(\pi)) = 2$. Consequently, we have:

$$\text{(2.2.1.a)} \qquad \operatorname{rank}\{R_\phi(\pi)\} = \begin{cases} 2 & \text{if } \ker\phi\cap\pi = \{0\}, \\ 0 & \text{if } \ker\phi\cap\pi \neq \{0\}. \end{cases}$$

We use this observation to show that the notions of constant spacelike, timelike, and mixed rank are distinct.

2.2.2 Lemma. *Let V be a vectorspace of signature (p,q), where $p\geq 2$ and where $q\geq 2$.*

(1) *There exists an algebraic curvature tensor R on V which has constant spacelike rank 2, but which has neither constant timelike rank nor constant mixed rank.*

(2) *There exists an algebraic curvature tensor R on V which has constant timelike rank 2, but which has neither constant spacelike rank nor constant mixed rank.*

(3) *There exists an algebraic curvature tensor R on V which has constant spacelike rank 2 and constant timelike rank 2, but which does not have constant mixed rank.*

Proof. Let $\{e_1^-,...,e_p^-,e_1^+,...,e_q^+\}$ be a normalized orthonormal basis for V. Define

$$\phi(e_i^-) = 0 \quad \text{for } 1\leq i\leq p, \text{ and}$$

$$\phi(e_i^+) = e_i^+ \quad \text{for } 1\leq i\leq q.$$

We apply Display (2.2.1.a) to relate $\operatorname{rank} R_\phi(\pi)$ to $\pi\cap\ker\phi$.

It is immediate that $\ker\phi = \operatorname{span}\{e_1^-, ..., e_p^-\}$ contains no spacelike vectors. Thus, $\operatorname{rank} R_\phi(\pi) = 2$ for any spacelike 2 plane. We show that R_ϕ has neither constant timelike rank nor constant spacelike rank and complete the proof of Assertion (1) by noting:

$$\begin{aligned}
&\ker\phi \cap \operatorname{span}\{2e_1^- + e_1^+, 2e_2^- + e_2^+\} = \{0\},\\
&\ker\phi \cap \operatorname{span}\{e_1^-, e_2^-\} \neq \{0\},\\
&\ker\phi \cap \operatorname{span}\{2e_1^- + e_1^+, e_2^+\} = \{0\}, \text{ and}\\
&\ker\phi \cap \operatorname{span}\{e_1^-, e_2^+\} \neq \{0\}.
\end{aligned}$$

We interchange the roles of spacelike (+) and timelike (−) indices to prove Assertion (2) similarly.

To prove Assertion (3), we define

$$\begin{aligned}
\phi(e_1^-) &:= e_1^- + e_1^+, \ \phi(e_i^-) := e_i^- \text{ for } i > 1,\\
\phi(e_1^+) &:= -e_1^- - e_1^+, \ \phi(e_j^+) := e_j^+ \text{ for } j > 1.
\end{aligned}$$

Since $\ker\phi = \operatorname{span}\{e_1^- + e_1^+\}$ is totally isotropic, $\ker\phi$ contains neither spacelike nor timelike vectors so R_ϕ has constant spacelike rank 2 and constant timelike rank 2. We show R_ϕ does not have constant mixed rank 2 and complete the proof of Assertion (3) by computing:

$$\begin{aligned}
&\ker\phi \cap \operatorname{span}\{e_2^-, e_2^+\} = \{0\}, \text{ and}\\
&\ker\phi \cap \operatorname{span}\{e_1^-, e_1^+\} \neq \{0\}. \quad \square
\end{aligned}$$

Let R be an algebraic curvature tensor of constant spacelike rank r on a vector space V of signature (p, q). In Theorem 2.1.1, we showed that $r = 2$ in many cases if $p = 0$, $p = 1$, or $p = 2$. There are, however, examples where $r = 4$.

2.2.3 Theorem. *Let V be a vector space of signature (p, q). If $p \geq q \geq 2$, then there exists an algebraic curvature tensor R on V so that:*

1. (1) *R has constant spacelike rank 4.*
2. (2) *R does not have constant mixed rank 4.*
3. (3) *R has constant timelike rank 4 if $p = q$.*
4. (4) *R does not have constant timelike rank if $p > q$.*

Proof. Let $\{e_1^-, ..., e_p^-, e_1^+, ..., e_q^+\}$ be a normalized orthonormal basis for V. We define a self adjoint map ϕ of V by setting:

$$\phi(e_i^\pm) = \begin{cases} \pm e_i^\mp & \text{if } 1 \leq i \leq q,\\ 0 & \text{if } i > q. \end{cases}$$

Let $v \in V$. We expand v relative to the basis given above in the form $v = \sum_i c_i e_i^- + \sum_j d_j e_j^+$ and compute that:

$$\begin{aligned}(v,v) &= -\textstyle\sum_{1\le i\le p} c_i^2 + \sum_{1\le j\le q} d_j^2, \text{ and}\\ (\phi v,\phi v) &= \textstyle\sum_{1\le i\le q} c_i^2 - \sum_{1\le j\le q} d_j^2.\end{aligned} \tag{2.2.3.a}$$

Let $R := R_{\text{Id}} + R_\phi$. Then

$$R(x,y)z = (y,z)x - (x,z)y + (\phi y,z)\phi x - (\phi x,z)\phi y.$$

If $\{x,y\}$ spans a spacelike 2 plane π, then we use Display (2.2.3.a) to see that $\{\phi x,\phi y\}$ spans a timelike 2 plane $\phi\pi$. Thus, $\{x,y,\phi x,\phi y\}$ is a set of linearly independent vectors. We may then use Lemma 1.2.2 to see that $\operatorname{rank} R(\pi) = 4$ and, consequently, R has constant spacelike rank 4.

If $p = q$, then a similar argument shows that R has constant timelike rank 4. If $p > q$, then we may demonstrate that R does not have constant timelike rank:

$$\operatorname{rank} R(\operatorname{span}\{e_1^-,e_p^-\}) = 2 \text{ and } \operatorname{rank} R(\operatorname{span}\{e_1^-,e_2^-\}) = 4.$$

We show that R never has mixed rank 4 by noting:

$$\operatorname{rank} R(\operatorname{span}\{e_1^-,e_1^+\}) = 2 \text{ and } \operatorname{rank} R(\operatorname{span}\{e_1^-,e_2^+\}) = 4. \quad \square$$

The case $q = 4$ is also of interest.

2.2.4 Lemma. *Let V be a vector space of signature $(p,4)$. Then there exists an algebraic curvature tensor R on V of constant spacelike rank 4. If $p \ge 1$, then R does not have constant mixed rank; if $p \ge 2$, then R does not have constant timelike rank.*

Proof. Let $a_2 + 2a_1 = 0$. We define an alternating 4 tensor on $\mathbb{R}^4$ whose non-zero entries, up to the curvature symmetries given in Equation (1.5.1.e), are given by:

$$\begin{array}{llll} R_{1212} = a_1, & R_{1234} = a_2, & R_{1313} = a_2, & R_{1324} = -a_1,\\ R_{1414} = a_2, & R_{1423} = a_1, & R_{2323} = a_2, & R_{2314} = a_1,\\ R_{2424} = a_2, & R_{2413} = -a_1, & R_{3434} = a_1, & R_{3412} = a_2. \end{array}$$

In Theorem 2.8.1, we will show that R is an IP algebraic curvature tensor of rank 4. Let $\{e_1^-,...,e_p^-,e_1^+,...,e_4^+\}$ be a normalized orthonormal basis for V. Let $\rho(e_i^-) = 0$ and $\rho(e_i^+) = e_i^+$ define the natural projection from V to

$$V^+ := \operatorname{span}\{e_1^+,e_2^+,e_3^+,e_4^+\}.$$

Let $R_V := \rho^* R$, i.e.

$$R_V(x, y, z, w) = R(\rho x, \rho y, \rho z, \rho w).$$

If $\{x, y\}$ spans a spacelike 2 plane in V, then $\{\rho x, \rho y\}$ spans a spacelike 2 plane in V^+ so $R_V(x, y) = R(\rho x, \rho y)$ has rank 4. We show that R does not have constant mixed rank by computing:

$$\begin{aligned} &\operatorname{rank} R_V\{\operatorname{span}\{e_1^-, e_1^+\}\} = 0, \\ &\operatorname{rank} R_V\{\operatorname{span}\{2e_1^- + e_1^+, e_2^+\}\} = 4. \end{aligned}$$

If $p \geq 2$, then we show that R does not have constant timelike rank by computing:

$$\begin{aligned} &\operatorname{rank} R_V\{\operatorname{span}\{e_1^-, e_2^-\}\} = 0, \\ &\operatorname{rank} R_V\{\operatorname{span}\{2e_1^- + e_1^+, 2e_2^- + e_2^+\}\} = 4. \quad \square \end{aligned}$$

We can now show that the eigenvalue structure does not determine the Jordan normal form and, furthermore, that spacelike Jordan IP, timelike Jordan IP, and mixed Jordan IP are not equivalent concepts:

2.2.5 Lemma. *Let V be a vector space of signature (p, q).*

(1) *If $p > 2$ and if $q > 2$, then there exists an algebraic curvature tensor R on V which is IP, but which is neither spacelike Jordan IP, nor timelike Jordan IP, nor mixed Jordan IP.*

(2) *If $p = q \geq 2$, then there exists an algebraic curvature tensor R on V which is spacelike and timelike Jordan IP, but which is not mixed Jordan IP.*

(3) *If $p > q \geq 2$, then there exists an algebraic curvature tensor R on V which is spacelike Jordan IP, but which is neither timelike Jordan IP nor mixed Jordan IP.*

Proof. Let $\{e_1^-, ..., e_p^-, e_1^+, ..., e_q^+\}$ be a normalized orthonormal basis for V. Let $1 \leq a \leq \min(p, q)$. As in the proof of Theorem 3.3.2, define:

$$\Phi_a(e_k^\pm) = \begin{cases} \pm(e_k^+ + e_k^-) & \text{if } k \leq a, \\ 0 & \text{if } k > a. \end{cases}$$

Then Φ_a is self-adjoint and $\Phi_a^2 = 0$. Let $R_a := R_{\Phi_a}$. Thus, by Lemma 1.8.1, $R_a(x, y)^2 = 0$. Since $R_a(\pi)$ is nilpotent, we apply Lemma 1.2.10 to see that

0 is the only eigenvalue of R_Φ so R is IP. We use Display (2.2.1.a) to see $\operatorname{rank} R(\pi) = 0$ if and only if $\ker \Phi_a \cap \pi \neq \{0\}$.

We take $a = 2$ to complete the proof of Assertion (1) by computing:

$$\begin{aligned}
&\ker \Phi_2 \cap \operatorname{span}\{e_1^-, e_2^-\} = \{0\},\ \ker \Phi_2 \cap \operatorname{span}\{e_1^-, e_3^-\} \neq \{0\},\\
&\ker \Phi_2 \cap \operatorname{span}\{e_1^+, e_2^+\} = \{0\},\ \ker \Phi_2 \cap \operatorname{span}\{e_1^+, e_3^+\} \neq \{0\},\\
&\ker \Phi_2 \cap \operatorname{span}\{e_1^-, e_2^+\} = \{0\},\ \ker \Phi_2 \cap \operatorname{span}\{e_1^-, e_3^+\} \neq \{0\}.
\end{aligned}$$

To prove Assertions (2) and (3), we set $a = q$. We have:

$$\ker \Phi_q = \operatorname{span}\{e_1^- + e_1^+, ..., e_q^- + e_q^+, e_{q+1}^-, ..., e_p^-\}.$$

Since the metric on $\ker \Phi_q$ is negative-semi definite, $\ker \Phi_q$ contains no spacelike vectors and thus R_q is spacelike Jordan Osserman. If $p = q$, the same argument shows R_q is timelike Jordan Osserman. If $p > q$, then we may demonstrate that R_q is not timelike Jordan Osserman by computing:

$$\ker \Phi_a \cap \operatorname{span}\{e_1^-, e_2^-\} = \{0\},\ \ker \Phi_a \cap \operatorname{span}\{e_1^-, e_p^-\} \neq \{0\}.$$

We show that R_q is never mixed Jordan Osserman and complete the proof by computing:

$$\ker \Phi_a \cap \operatorname{span}\{e_1^-, e_2^+\} = \{0\}, \text{ and } \ker \Phi_a \cap \operatorname{span}\{e_1^-, e_1^+\} \neq \{0\}. \quad \square$$

2.3 Rank 2 algebraic curvature tensors

Let V be a vector space of signature (p, q). If ϕ is a self-adjoint linear map from V to V, then we have defined:

$$\begin{aligned}
&R_\phi(x, y)z := (\phi y, z)\phi x - (\phi x, z)\phi y, \text{ and}\\
&R_\phi(x, y, z, w) := (\phi y, z)(\phi x, w) - (\phi x, z)(\phi y, w);
\end{aligned}$$

by Lemma 1.8.1, R_ϕ is an algebraic curvature tensor. In this section, we shall prove Theorem 2.1.5. This means we must show that if R is any spacelike rank 2 algebraic curvature tensor and if $q \geq 5$, then there exists a self-adjoint map ϕ of V, where $\ker \phi$ contains no spacelike vector, such that $R = \pm R_\phi$.

For the moment, we concentrate on curvature symmetry

$$R(x, y, z, w) = -R(y, x, z, w) = -R(x, y, z, w)$$

given in Equation (1.5.1.e). We shall decouple the domain from the range and establish a slightly more general classification result which we will then use to derive Theorem 2.1.5.

Let A and B be vector spaces of signatures (p_A, q_A) and (p_B, q_B), respectively. We say $T \in \otimes^2 A^* \otimes^2 B^*$ is an *alternating mixed tensor* if

$$T(a_1, a_2, b_1, b_2) = -T(a_2, a_1, b_1, b_2) = -T(a_1, a_2, b_2, b_1). \tag{2.3.1.a}$$

We associate an *alternating bilinear map* from $A \otimes A$ to $\mathfrak{so}(B)$ by:

$$(T(a_1, a_2)b_1, b_2) = T(a_1, a_2, b_1, b_2).$$

We extend T to a map from $\widetilde{\mathrm{Gr}}_2^+(V)$ to $\mathfrak{so}(B)$ by setting:

$$T(\mathrm{span}\{a_1, a_2\}) := \frac{T(a_1, a_2)}{|(a_1, a_1)(a_2, a_2) - (a_1, a_2)^2|^{\frac{1}{2}}}.$$

By Lemma 1.9.1, $T(\pi)$ is independent of the oriented basis $\{a_1, a_2\}$ chosen for π. We say that T has *constant spacelike rank* r if $\mathrm{rank}(T(\cdot)) = r$ on $\mathrm{Gr}_{0,2}^+(A)$.

Let ϕ be a linear map from A to B. We generalize the tensors R_ϕ defined in Equation (1.1.1.b) by setting:

$$\begin{aligned} &T_\phi(a_1, a_2)b := (\phi(a_2), b)\phi(a_1) - (\phi(a_1), b)\phi(a_2), \text{ and} \\ &T_\phi(a_1, a_2, b_1, b_2) := (\phi(a_2), b_1)(\phi(a_1), b_2) \\ &\qquad - (\phi(a_1), b_1)(\phi(a_2), b_2). \end{aligned} \tag{2.3.1.b}$$

There is another family we must consider in this more general setting. If $\xi \in B$ and if χ is an alternating bilinear map from $A \otimes A$ to B, then define:

$$\begin{aligned} &T_{\chi,\xi}(a_1, a_2)b := (\chi(a_1, a_2), b)\xi - (\xi, b)\chi(a_1, a_2), \text{ and} \\ &T_{\chi,\xi}(a_1, a_2, b_1, b_2) := (\chi(a_1, a_2), b_1)(\xi, b_2) \\ &\qquad - (\xi, b_1)(\chi(a_1, a_2), b_2). \end{aligned} \tag{2.3.1.c}$$

It is immediate that T_ϕ and $T_{\chi,\xi}$ are alternating mixed tensors. We classify the alternating mixed tensors of constant spacelike rank 2 as follows:

2.3.2 Theorem. *Let A and B be vector spaces of signatures (p_A, q_A) and (p_B, q_B), respectively, where $q_A \geq 5$.*

(1) *Let ϕ be a linear map from A to B. The tensor T_ϕ has constant spacelike rank 2 if and only if $\ker(\phi)$ contains no spacelike vectors.*

(2) *Let* χ *be an alternating bilinear map from* $A\otimes A$ *to* B *and let* $\xi\in B$. *The tensor* $T_{\chi,\xi}$ *has constant spacelike rank* 2 *if and only if* $\chi(a_1,a_2)$ *and* ξ *are linearly independent vectors if* $\operatorname{span}\{a_1,a_2\}\in\operatorname{Gr}^+_{0,2}(V)$.

(3) *If* $T\in\otimes^2A^*\otimes^2B^*$ *is an alternating mixed tensor which has constant spacelike rank* 2*, then either* $T=\pm T_\phi$ *for suitably chosen* ϕ *or* $T=(\chi,\xi)$ *for suitably chosen* (χ,ξ).

Before beginning the proof of Theorems 2.1.5 and 2.3.2, we must establish some technical preliminaries. Let A^+ be the set of spacelike vectors in A. If $0\neq a\in A$, let $[a]=a\cdot\mathbb{R}\in\mathbb{P}(A)$. Let $\Phi:\mathbb{P}(A^+)\to\mathbb{P}(B)$ be a map from the set of spacelike lines in A to the set of all lines in B. A linear map ϕ from A to B is said to *linearize the projective map* Φ if $\ker(\phi)$ contains no spacelike vectors and if $[\phi(a)]=\Phi[a]$ for all $[a]\in\mathbb{P}(A^+)$.

Let T be an alternating mixed tensor of rank 2. Our primary task in the proof of Theorem 2.3.2 is to construct χ or ϕ from T. Let $a\in A^+$. We generalize the map Φ defined in the proof of Lemma 1.8.6 setting:

(2.3.2.a) $$\Phi([a]):=\cap_{\{\pi:a\in\pi\in\operatorname{Gr}^+_{0,2}(A)\}}\operatorname{range}(T(\pi)).$$

Here is a brief outline to the strategy we will employ. We first consider the case $p_A=p_B=0$. We will show in Lemma 2.3.3 that $\dim\Phi([a])=1$ so that Φ is a projective map from $\mathbb{P}(A^+)$ to $\mathbb{P}(B)$. We will also show Φ is either injective or constant. We will show in Lemma 2.3.4 that $T=T_{\chi,\xi}$ if Φ is constant. We will also show that if Φ is non-constant, then Φ is linearizable; by Lemma 1.2.11 the linearization is unique up to scale. We will then show that $T=CT_\phi$ for some constant C. In Lemma 2.3.7, we remove the hypothesis that $p_A=0$. We then use Lemma 1.2.13 to change the signature of the metric on B and complete the proof of Theorem 2.3.2. We then use Theorem 2.3.2 to prove Theorem 2.1.5. Thus, the map Φ will play a crucial role in our discussion.

If $a_1,...,a_k$ are vectors of A, then $\{a_1,...,a_k\}$ is a set of linearly independent vectors if and only if $a_1\wedge...\wedge a_k\neq0$. Let $T\in\otimes^2A^*\otimes^2B^*$ be an alternating mixed tensor. We suppose $p_A=p_B=0$ for the moment and work in the Riemannian setting. We show that $\Phi([a])$ is one dimensional and hence determines a map from $\mathbb{P}(A)$ to $\mathbb{P}(B)$:

2.3.3 Lemma. *Let* A *and* B *be Riemannian vector spaces of dimensions* q_A *and* q_B*, respectively. Let* $T\in\otimes^2A^*\otimes^2B^*$ *be an alternating mixed tensor which has constant spacelike rank* 2*. Let* Φ *be defined by Equation* (2.3.2.a). *Let* $a_1\wedge a_2\wedge a_3\neq0$.

(1) *If* $q_A\geq3$*, then* $\dim\{\operatorname{range}\{T(a_1,a_2)\}\cap\operatorname{range}\{T(a_1,a_3)\}\}=1$.

(2) *If* $q_A \geq 5$, *then* $\dim \Phi([a_1]) = 1$.
(3) *If* $q_A \geq 5$, *then* $\Phi([a_1]) = \text{range}\{T(a_1, a_2)\} \cap \text{range}\{T(a_1, a_3)\}$.
(4) *If* $q_A \geq 5$, *then either* Φ *is injective or* Φ *is constant.*

Proof. Let $u := \dim\{\text{range}\{T(a_1, a_2)\} + \text{range}\{T(a_1, a_3)\}\}$. We have:

$$\begin{aligned} &\dim\{\text{range}\{T(a_1, a_2)\}\} + \dim\{\text{range}\{T(a_1, a_3)\}\} \\ =&\dim\{\text{range}\{T(a_1, a_2)\} + \text{range}\{T(a_1, a_3)\}\} \\ +&\dim\{\text{range}\{T(a_1, a_2)\} \cap \text{range}\{T(a_1, a_3)\}\}. \end{aligned}$$

Consequently, $4 - u = \dim\{\text{range}\{T(a_1, a_2)\} \cap \text{range}\{T(a_1, a_3)\}\}$ so

$$4 \geq u \geq \dim \text{range}(T(a_1, a_2)) = 2.$$

To prove Assertion (1) we must show that $u = 3$, i.e. that

$$u \neq 2 \text{ and } u \neq 4.$$

First suppose that $u = 2$. Then $\text{range}(T(a_1, a_2)) = \text{range}(T(a_1, a_3))$ is 2 dimensional so $T(a_1, a_2) = cT(a_1, a_3)$ by Lemma 1.2.15. Thus,

$$T(a_1, a_2 - ca_3) = 0$$

which contradicts the assumption T has constant spacelike rank 2. Thus,

$$u \neq 2.$$

Suppose $u = 4$. We argue for a contradiction. Let $\{b_1, ..., b_{\dim B}\}$ be an orthonormal basis for B so

$$\begin{aligned} &\text{range}(T(a_1, a_2)) = \text{span}\{b_1, b_2\}, \\ &T(a_1, a_3)b_1 \in \text{span}\{b_1, b_2, b_3\}, \quad \text{and} \\ &\text{range}(T(a_1, a_2)) + \text{range}(T(a_1, a_3)) = \text{span}\{b_1, b_2, b_3, b_4\}. \end{aligned} \tag{2.3.3.a}$$

Define $\mathcal{T}_{ij} \in \mathfrak{so}(B)$ so that $\omega(\mathcal{T}_{ij}) = b_i^* \wedge b_j^*$ by setting:

$$\begin{aligned} &\mathcal{T}_{ij} : b_i \to b_j, \\ &\mathcal{T}_{ij} : b_j \to -b_i, \text{ and} \\ &\mathcal{T}_{ij} : b_k \to 0 \text{ for } k \neq i, j. \end{aligned}$$

The set $\{\mathcal{T}_{ij}\}_{i<j}$ is a basis for the vector space $\mathfrak{so}(B)$ which is orthonormal relative to the normalized Killing form on $\mathfrak{so}(B)$ defined by:

$$\langle T_1, T_2 \rangle := -\tfrac{1}{2}\operatorname{Tr}(T_1 T_2).$$

By rescaling T, we may assume $T(a_1, a_2) = \mathcal{T}_{12}$. We use the properties given in Display (2.3.3.a) to expand:

$$\begin{aligned} &T(a_1, a_2) = \mathcal{T}_{12}, \text{ and} \\ &T(a_1, a_3) = c_{12}\mathcal{T}_{12} + c_{13}\mathcal{T}_{13} + c_{23}\mathcal{T}_{23} + c_{24}\mathcal{T}_{24} + c_{34}\mathcal{T}_{34}. \end{aligned}$$

As $T(a_1, a_2)b_1 \in \operatorname{span}\{b_1, b_2, b_3\}$, $T(a_1, a_2)$ does not involve $\mathcal{T}_{14}$. We have:

$$\begin{aligned} \omega(T(a_1, a_2)) &= b_1^* \wedge b_2^*, \qquad \text{and} \\ \omega(T(a_1, a_3)) &= c_{12}b_1^* \wedge b_2^* + c_{13}b_1^* \wedge b_3^* + c_{23}b_2^* \wedge b_3^* \\ &\quad + c_{24}b_2^* \wedge b_4^* + c_{34}b_3^* \wedge b_4^*. \end{aligned}$$

Let $\mathcal{T}(\varepsilon) := T(a_1, a_2 + \varepsilon a_3)$. Since $\operatorname{rank}(\mathcal{T}(\varepsilon)) = 2$ for any value of the parameter ε, we use Lemma 1.2.14 to see that:

$$\begin{aligned} 0 =&\omega(\mathcal{T}(\varepsilon)) \wedge \omega(\mathcal{T}(\varepsilon)) \\ =&2\{\varepsilon c_{34} + \varepsilon^2(c_{12}c_{34} - c_{13}c_{24})\}b_1^* \wedge b_2^* \wedge b_3^* \wedge b_4^*. \end{aligned} \tag{2.3.3.b}$$

Since Equation (2.3.3.b) holds for all $\varepsilon \in \mathbb{R}$, we may conclude that

$$c_{34} = 0 \text{ and } c_{13}c_{24} = 0.$$

We complete the proof of Assertion (1) by considering two cases:

(1) If $c_{24} = 0$, then $T(a_1, a_3) = c_{12}\mathcal{T}_{12} + c_{13}\mathcal{T}_{13} + c_{23}\mathcal{T}_{23}$ so

$$\operatorname{range}(T(a_1, a_3)) \subseteq \operatorname{span}\{b_1, b_2, b_3\}.$$

This shows that $u \leq 3$ which contradicts the assumption that $u = 4$.

(2) If $c_{13} = 0$, then $T(a_1, a_3) = c_{12}\mathcal{T}_{12} + c_{23}\mathcal{T}_{23} + c_{24}\mathcal{T}_{24}$ so

$$\operatorname{range}(T(a_1, a_3)) \subseteq \operatorname{span}\{b_1, b_2, c_{23}b_3 + c_{24}b_4\}.$$

Thus, again $u \leq 3$, which contradicts the assumption that $u = 4$.

By Assertion (1), $\dim \Phi[a] \leq 1$ for any non-zero vector a. We suppose that Assertion (2) is false and argue for a contradiction. Let $0 \neq a_1$ be such that $\Phi([a_1]) = \{0\}$. Let a_2, and a_3 be vectors so that $a_1 \wedge a_2 \wedge a_3 \neq 0$. Then by Assertion (1), we see that

$$\dim\{\text{range}\{T(a_1, a_2)\} \cap \text{range}\{T(a_1, a_3)\}\} = 1.$$

Since $\Phi([a_1]) = \{0\}$, there exists a vector a_4 so that $a_1 \wedge a_4 \neq 0$ and so that:

$$\text{(2.3.3.c)} \quad \text{range}\{T(a_1, a_1)\} \cap \text{range}\{T(a_1, a_3)\} \cap \text{range}\{T(a_1, a_4)\} = \{0\}.$$

We must then have that

$$\text{range}\{T(a_1, a_i)\} \neq \text{range}\{T(a_1, a_j)\} \text{ for } 2 \leq i < j \leq 4.$$

Consequently, $a_1 \wedge a_i \wedge a_j \neq 0$ for $2 \leq i < j \leq 4$. We use Assertion (1) to define lines L_{ij} for $2 \leq i < j \leq 4$ and to define a linear subspace E:

$$\begin{aligned} &L_{ij} := \text{range}\{T(a_1, a_i)\} \cap \text{range}\{T(a_1, a_j)\} \in \mathbb{P}(B), \text{ and} \\ &E := L_{23} + L_{24} + L_{34}. \end{aligned}$$

Let $\{i, j, k\}$ be a permutation of $\{2, 3, 4\}$. Then

$$\begin{aligned} \{0\} &= \text{range}\{T(a_1, a_j)\} \cap \text{range}\{T(a_1, a_i)\} \cap \text{range}\{T(a_1, a_k)\} \\ &= L_{ij} \cap L_{ik}. \end{aligned}$$

Thus, L_{ij} and L_{ik} are distinct lines contained in $\text{range}\{T(a_1, a_i)\}$ so

$$\text{range}\{T(a_1, a_i)\} = L_{ij} \oplus L_{ik} \text{ so } \text{range}\{T(a_1, a_i)\} \subseteq E.$$

Let $0 \neq a_5 \perp a_1$. We shall show that $\text{range}\{T(a_1, a_5)\} \subseteq E$. If we have $a_5 \in \text{span}\{a_1, a_i\}$ for some value of i with $2 \leq i \leq 4$, then

$$T(a_1, a_5) = cT(a_1, a_i) \text{ so } \text{range}\{T(a_1, a_5)\} \subseteq E.$$

Thus, we suppose that $a_1 \wedge a_i \wedge a_5 \neq 0$ for $2 \leq i \leq 4$. We use Assertion (1) to define the lines

$$L_{i5} := \text{range}\{T(a_1, a_i)\} \cap \text{range}\{T(a_1, a_5)\} \in \mathbb{P}(B) \text{ for } 2 \leq i \leq 4.$$

If $L_{25} = L_{35} = L_{45}$, then

$$L_{25} = L_{i5} \subseteq \text{range}\{T(a_1, a_i)\} \text{ for } 2 \leq i \leq 4$$

which contradicts Equation (2.3.3.c). Thus, at least two of the lines L_{i5} are distinct lines which are contained in range$\{T(a_1,a_5)\}$. Consequently, the lines $\{L_{25},L_{35},L_{45}\}$ span range$\{T(a_1,a_5)\}$ and

$$\begin{aligned}&\text{range}\{T(a_1,a_5)\}\\ &\subseteq \text{range}\{T(a_1,a_2)\}+\text{range}\{T(a_1,a_3)\}+\text{range}\{T(a_1,a_4)\}\\ &\subset E.\end{aligned}$$

Consequently, the map $\psi: a \to \text{range}\{T(a_1,a)\}$ is a well defined map from the projective space $\mathbb{P}(a_1^\perp)$ to $\text{Gr}_2(E)$. Suppose that $[a]$ and $[\tilde{a}]$ are distinct lines in $\mathbb{P}(a_1^\perp)$. As $a_1\wedge a\wedge\tilde{a}\neq 0$, we may use Assertion (1) to see

$$\text{range}\{T(a_1,a)\}\neq\text{range}\{T(a_1,\tilde{a})\}\text{ so }\psi([a])\neq\psi([\tilde{a}]).$$

Therefore, the map ψ is injective. As $\dim(E)\leq 3$, $\dim\text{Gr}_2(E)\leq 2$. We use Lemma 4.2.7 to see that ψ is continuous. Thus, by Theorem 4.3.2,

$$\dim(\mathbb{P}(a_1^\perp)\}\leq 2.$$

Since $\dim\mathbb{P}(a_1^\perp)=q_A-2\geq 3$, this contradiction completes the proof of Assertion (2). Example 2.3.8 shows that Assertion (2) can fail if $q_A=4$.

Assertion (3) follows directly from Assertions (1) and (2).

To prove Assertion (4), we shall suppose that $\Phi:\mathbb{P}(A)\to\mathbb{P}(B)$ is not injective. We must show that Φ is constant. Choose distinct lines $[a_i]$ so that $\Phi([a_1])=\Phi([a_2])$. If

$$[a_3]\in\mathbb{P}(A)-\mathbb{P}(\text{span}\{a_1,a_2\}),$$

then the vectors a_1, a_2, and a_3 are linearly independent. Thus,

$$\begin{aligned}\Phi([a_1])=\Phi([a_2])&\subseteq\text{range}\{T(a_3,a_1)\}\cap\text{range}\{T(a_3,a_2)\}\\ &=\Phi([a_3]).\end{aligned}$$

Thus, Φ is constant on $\mathbb{P}(A)-\mathbb{P}(\text{span}\{a_1,a_2\})$. We use Assertion (2) to see Φ is continuous. Since $\mathbb{P}(A)-\mathbb{P}(\text{span}\{a_1,a_2\})$ is an open dense subset of $\mathbb{P}(A)$, Φ is constant on all of $\mathbb{P}(A)$. □

We can now classify the mixed alternating tensors of rank 2 in the Riemannian setting.

2.3.4 Lemma. *Let A and B be Riemannian vector spaces of dimensions q_A and q_B, respectively, where $q_A \geq 5$. Let $T \in \otimes^2 A^* \otimes^2 B^*$ be an alternating mixed tensor which has constant spacelike rank 2. Let Φ be the associated map from $\mathbb{P}(A)$ to $\mathbb{P}(B)$.*

(1) *Let Φ be constant. Fix a unit vector ξ belonging to the line Φ. Set $\chi(a_1, a_2) := -T(a_1, a_2)\xi$. Then $T = T_{\chi,\xi}$.*
(2) *Let Φ be non-constant. Then there exists a linearization ϕ of Φ so $T = \pm T_\phi$.*

Proof. Suppose that $\Phi : \mathbb{P}(A) \to \mathbb{P}(B)$ is constant. Choose a unit vector $\xi \in \Phi$. Then $\xi \in \Phi[a]$ for all a and we can define an alternating bilinear map χ from $A \otimes A$ to $\xi^\perp \subseteq B$ by setting:

$$\chi(a_1, a_2) := -T(a_1, a_2)\xi \in B.$$

Let $a_1, a_2 \in A$. Suppose that $a_1 \wedge a_2 \neq 0$. We use Lemma 1.2.15 to see:

$$\text{range}(T(a_1, a_2)) = \text{span}\{\chi(a_1, a_2), \xi\} = \text{range}(T_{\chi,\xi}(a_1, a_2)).$$

Thus, by Lemma 1.2.15, there exists a constant $c = c(a_1, a_2)$ such that

$$T(a_1, a_2) = cT_{\chi,\xi}(a_1, a_2).$$

Since $T(a_1, a_2)\xi = -\chi(a_1, a_2) = T_{\chi,\xi}(a_1, a_2)\xi$, $c = 1$ so

$$T(a_1, a_2) = T_{\chi,\xi}(a_1, a_2).$$

On the other hand, if $a_1 \wedge a_2 = 0$, then

$$T(a_1, a_2) = 0 \text{ and } T_{\chi,\xi}(a_1, a_2) = 0$$

since T and $T_{\chi,\xi}$ alternating. Thus, $T = T_{\chi,\xi}$ and Assertion (1) follows.

To prove Assertion (2), we suppose that Φ is not constant. We may then apply Lemma 2.3.3 to see that Φ is injective. The spheres $S(A)$ and $S(B)$ are the universal covers of the corresponding projective spaces $\mathbb{P}(A)$ and $\mathbb{P}(B)$. We lift Φ to define a map

$$\bar{\Phi} : S(A) \to S(B) \text{ so } [\bar{\Phi}(a)] = \Phi([a]).$$

Let $\{e_1, ..., e_{q_A}\}$ be a basis for A and let $\{e^1, ..., e^{q_A}\}$ be the corresponding dual basis for A^*. Let $i < j$. Since $\Phi([e_i])$ and $\Phi([e_j])$ are distinct lines which are contained in $T(e_i, e_j)$,

$$\text{range}(T(e_i, e_j)) = \text{span}\{\bar{\Phi}(e_i), \bar{\Phi}(e_j)\}.$$

Let $a_1, a_2 \in A$. Since T is bilinear, we have

$$\begin{aligned} T(a_1, a_2) &= \textstyle\sum_{i,j} e^i(a_1) e^j(a_2) T(e_i, e_j), \\ \operatorname{range}(T(a_1, a_2)) &\subseteq \textstyle\sum_{i,j} \operatorname{range}(T(e_i, e_j)) \\ &\subseteq \operatorname{span}\{\bar{\Phi}(e_1), ..., \bar{\Phi}(e_{q_A})\}. \end{aligned}$$

Thus, by replacing B by $\operatorname{span}\{\bar{\Phi}(e_1), ..., \bar{\Phi}(e_{q_A})\}$, we may assume without loss of generality that:

$$\dim(B) \leq \dim(A).$$

Since $\mathbb{P}(A)$ is compact, $\operatorname{range}(\Phi)$ is compact and hence closed. We have that Φ is injective and that $\dim(\mathbb{P}(B)) \leq \dim(\mathbb{P}(A))$. We use Lemma 4.2.7 to see that Φ is continuous. Thus, by Theorem 4.3.2,

$$\dim(\mathbb{P}(B)) = \dim(\mathbb{P}(A))$$

and hence Φ is an open map. Thus, $\operatorname{range}(\Phi)$ is also an open subset of $\mathbb{P}(B)$. Since $\mathbb{P}(B)$ is connected and the range of Φ is non-empty, we may conclude that Φ is a surjective map. Since $\mathbb{P}(A)$ is compact and since $\mathbb{P}(B)$ is Hausdorff, we can use Theorem 4.3.1 to see that Φ is a homeomorphism from $\mathbb{P}(A)$ onto $\mathbb{P}(B)$. It now follows that the lift $\bar{\Phi}$ is a homeomorphism from $S(A)$ onto $S(B)$. Since Φ induces an isomorphism from the fundamental group of $\mathbb{P}(A)$ to the fundamental group of $\mathbb{P}(B)$, we have:

$$\bar{\Phi}(-a) = -\bar{\Phi}(a).$$

We now show that Φ is linearizable; such a linearization is unique modulo rescaling by Lemma 1.2.11. Let $a_1, a_2 \in A$ be a pair of vectors which are linearly independent. We must first show that

$$\Phi([a_1 + a_2]) \subseteq \Phi([a_1]) + \Phi([a_2]).$$

Since $\dim A = \dim B \geq 5$ and since $\bar{\Phi}$ is surjective, we choose unit vectors a_3 and a_4 in A so that

$$\begin{aligned} &\bar{\Phi}(a_3) \perp \{\bar{\Phi}(a_1), \bar{\Phi}(a_2), \bar{\Phi}(a_1 + a_2)\}, \text{ and} \\ &\bar{\Phi}(a_4) \perp \{\bar{\Phi}(a_1), \bar{\Phi}(a_2), \bar{\Phi}(a_1 + a_2), \bar{\Phi}(a_3)\}. \end{aligned} \tag{2.3.4.a}$$

We then have that $a_i \wedge a_1 \neq 0$, that $a_i \wedge a_2 \neq 0$, and that $a_i \wedge (a_1 + a_2) \neq 0$ for $i = 3, 4$. Since T is bilinear,

$$\begin{aligned} \operatorname{range}(T(a_3, a_1 + a_2)) &\subseteq \operatorname{range}(T(a_3, a_1)) + \operatorname{range}(T(a_3, a_2)), \text{ and} \\ \operatorname{range}(T(a_4, a_1 + a_2)) &\subseteq \operatorname{range}(T(a_4, a_1)) + \operatorname{range}(T(a_4, a_2)). \end{aligned}$$

We apply Lemma 2.3.3 to see that:

$$\begin{aligned}[\bar{\Phi}(a_1+a_2)] &= \cap_{i=3,4}\operatorname{range}\{T(a_1+a_2,a_i)\}\\ &\subseteq \cap_{i=3,4}\big\{\operatorname{range}\{T(a_1,a_i)\}+\operatorname{range}\{T(a_2,a_i)\}\big\}\\ &\subseteq \cap_{i=3,4}\operatorname{span}\{\bar{\Phi}(a_1),\bar{\Phi}(a_2),\bar{\Phi}(a_i)\}\\ &= \operatorname{span}\{\bar{\Phi}(a_1),\bar{\Phi}(a_2)\} = \Phi([a_1])+\Phi([a_2]).\end{aligned}$$

Consequently, we have the desired inclusion:

$$\Phi([a_1+a_2]) \subseteq \Phi([a_1])+\Phi([a_2]). \tag{2.3.4.b}$$

If $0 \neq a \in A$, then let:

$$\mathcal{A}([a]) := \big\{a_1 \in A-\{0\} : \Phi([a_1]) \perp \Phi([a])\big\} \cup \big\{0\big\}.$$

We argue as follows to show that $\mathcal{A}([a])$ is a linear subspace of A. Since $\Phi[ca_1] = \Phi[a_1]$ for $a_1 \neq 0$, $\mathcal{A}([a])$ is closed under scalar multiplication. Let a_i be elements of $\mathcal{A}([a])$. We wish to show that $a_1+a_2 \in \mathcal{A}([a])$. This is immediate if a_1 and a_2 are linearly dependent. If a_1 and a_2 are linearly independent, then we use Equation (2.3.4.b) to see $a_1+a_2 \in \mathcal{A}([a])$. Thus, in fact, $\mathcal{A}([a])$ is a linear subspace of A.

Since Φ is a homeomorphism from $\mathbb{P}(\mathcal{A}([a]))$ to $\mathbb{P}(\bar{\Phi}(a)^\perp)$, we use Theorem 4.3.2 to show that $\mathcal{A}([a]) \in \mathrm{Gr}_{\dim A-1}(A)$ by computing:

$$\dim\mathbb{P}(\mathcal{A}([a])) = \dim\mathbb{P}(\bar{\Phi}(a)^\perp) \text{ so } \dim(\mathcal{A}([a])) = \dim A-1.$$

We will apply Lemma 1.2.12 to the family of hyperplanes and associated linearizations:

$$\begin{aligned}&\mathcal{F} := \big\{\mathcal{A}([a]) : a \in A-\{0\}\big\}, \text{ where}\\ &\phi_a(a_1) := T(a,a_1)\bar{\Phi}(a) \text{ for } a_1 \in \mathcal{A}([a]).\end{aligned}$$

Since T is bilinear, ϕ_a is linear. If $0 \neq a_1 \in \mathcal{A}([a])$, then $\Phi([a_1]) \perp \Phi([a])$ so $0 \neq a \wedge a_1$. We use Lemma 1.2.15 to see that $\{\bar{\Phi}(a), \phi_a(a_1)\}$ is an orthogonal basis for $\operatorname{range}(T(a,a_1))$. Since

$$\Phi([a]) \perp \Phi([a_1]) \text{ we have } \phi_a(a_1) \in \Phi([a_1]).$$

Thus, ϕ_a is a linearization of Φ on $\mathcal{A}([a])$. Let $a_1, a_2 \in A$ be non-zero. Since $\dim A \geq 3$ and since Φ is bijective, we may choose $a \in A$ so

$$\Phi([a]) \perp \Phi([a_1]) \text{ and } \Phi([a]) \perp \Phi([a_2]).$$

Thus, $a_1, a_2 \in \mathcal{A}([a])$. This shows that every 2 plane of A is contained in some element of $\mathcal{F}$. Thus, by Lemma 1.2.12, there exists a linearization ϕ of Φ on all of A.

We use Equation (2.3.1.b) to define

$$T_\phi(a_1, a_2)b := (\phi a_2, b)\phi a_1 - (\phi a_1, b)\phi a_2.$$

Let a_1 and a_2 be linearly independent vectors in A. Since $T_\phi(a_1, a_2)$ and $T(a_1, a_2)$ belong to $\mathfrak{so}(B)$ and since

$$\text{range}(T_\phi(a_1, a_2)) = \text{range}(T(a_1, a_2)) = \text{span}\{\phi(a_1), \phi(a_2)\},$$

we may apply Lemma 1.2.15 to see there exists a constant $\mu(a_1, a_2)$ so that

$$T_\phi(a_1, a_2) = \mu(a_1, a_2)T(a_1, a_2).$$

Let $a_1, a_2, a_3 \in A$ with $a_1 \wedge a_2 \wedge a_3 \neq 0$. We compute:

$$\begin{aligned}
&\mu(a_1, a_2)T(a_1, a_2) + \mu(a_1, a_3)T(a_1, a_3)\\
=&T_\phi(a_1, a_2) + T_\phi(a_1, a_3)\\
=&T_\phi(a_1, a_2 + a_3)\\
=&\mu(a_1, a_2 + a_3)T(a_1, a_2 + a_3)\\
=&\mu(a_1, a_2 + a_3)T(a_1, a_2) + \mu(a_1, a_2 + a_3)T(a_1, a_3).
\end{aligned}$$

We apply Lemma 2.3.3 to see that $T(a_1, a_2)$ and $T(a_1, a_3)$ are linearly independent maps. Thus, we may conclude that:

$$\mu(a_1, a_2) = \mu(a_1, a_2 + a_3) = \mu(a_1, a_3).$$

Consequently, $\mu(a_1, a_2)$ is independent of a_2; since $\mu(a_1, a_2) = \mu(a_2, a_1)$, we also have that $\mu(a_1, a_2)$ is independent of a_2. Denote this common value by μ and express

$$T(a_1, a_2) = \mu T_\phi(a_1, a_2) \text{ if } a_1 \wedge a_2 \neq 0. \tag{2.3.4.c}$$

If a_1 and a_2 are linearly dependent, then $T(a_1, a_2) = 0$ and $T_\phi(a_1, a_2) = 0$ since both are alternating. Thus, Equation (2.3.4.c) holds for all (a_1, a_2). We replace ϕ by $\sqrt{|\mu|}\phi$ to complete the proof. □

We will apply Lemma 2.3.4 to the restriction of T to a maximal spacelike subspace S. This will yield a linearization of Φ on every such subspace S; we will then have to patch these linearizations together to complete the proof of Theorem 2.3.2. We begin this process with a technical Lemma. Let A be a vector space of signature (p_A, q_A), where $q_A \geq 5$. Let

$$\mathcal{S}_r := \cup_{s \geq r} \operatorname{Gr}_{0,s}(A)$$

be the set of all spacelike subspaces of A which have dimension at least r. We have the following inclusions:

$$\mathcal{S}_{q_A} \subsetneq ... \subsetneq \mathcal{S}_3.$$

Let $S \in \mathcal{S}_3$ and $\tilde{S} \in \mathcal{S}_3$. We say an ordered collection $\{S_1, ..., S_\ell\}$ is a *chain linking S and $\tilde{S}$* if

(1) $S = S_1$ and $\tilde{S} = S_\ell$,
(2) $S \cap \tilde{S} \subset S_i \cap S_{i+1}$ for all i, and
(3) $S_i \cap S_{i+1} \in \mathcal{S}_3$ for all i.

2.3.5 Lemma. *Let A be a vector space of signature (p_A, q_A). If $q_A \geq 5$, then any two elements of $\mathcal{S}_3$ can be linked by a chain.*

Proof. Let $S, \tilde{S} \in \mathcal{S}_3$. Choose $U, \tilde{U} \in \operatorname{Gr}_{0,q_A}(A)$ so $S \subset U$ and $\tilde{S} \subset \tilde{U}$. Then

$$S \cap \tilde{S} \subset U \cap \tilde{U}.$$

If $\{U = U_1, ..., U_\ell = \tilde{U}\}$ is a chain linking U and $\tilde{U}$, then the augmented collection $\{S, U_1, ..., U_\ell, \tilde{S}\}$ is a chain linking S and $\tilde{S}$. Thus, without loss of generality, we may assume that S and $\tilde{S}$ are maximal spacelike subspaces.

Let $\{s_1, ..., s_{q_A}\}$ and $\{\tilde{s}_1, ..., \tilde{s}_{q_A}\}$ be orthonormal bases for S and $\tilde{S}$ so that $s_1 = \tilde{s}_1$, ..., $s_k = \tilde{s}_k$ for $k = \dim(S \cap \tilde{S})$. If $k = q_A$, then $S = \tilde{S}$ and Lemma 2.3.5 is immediate. We therefore proceed by reverse induction on k. Let

$$C := \operatorname{span}\{s_1, ..., s_k, s_{k+2}, ..., s_{q_A}\} \in \mathcal{S}_{q_A - 1} \subset \mathcal{S}_3.$$

Since $\dim(C^\perp) = p_A + 1$, we have that $\dim(C^\perp \cap \tilde{S}) \geq 1$. Thus, we can assume the basis for $\tilde{S}$ is chosen so that $\tilde{s}_{k+1} \perp C$. We then define

$$S_1 := \{s_1, ..., s_k, \tilde{s}_{k+1}, s_{k+2}, ..., s_{q_A}\} \in \mathcal{S}_{q_A}.$$

Since $S_1 \cap \tilde{S} = k + 1$, by induction we can find a chain

$$\{S_1, S_2, ..., S_\nu = \tilde{S}\}$$

linking S_1 and $\tilde{S}$. We augment the chain to define the collection

$$\{S, S_1, ..., S_2, ..., S_\nu = \tilde{S}\}.$$

Since $S \cap \tilde{S} = S_1 \cap \tilde{S}$, this collection is a chain linking S to $\tilde{S}$. □

We continue to assume that the metric on B is positive definite; we drop this assumption on A and allow A to have arbitrary signature, where $q_A \geq 5$. We suppose given an alternating mixed tensor $T : \otimes^2 A \to \mathfrak{so}(B)$ of constant spacelike rank 2. If $S \in \mathcal{S}_3$, let T_S be the restriction of T to $S \otimes S$ and let

$$\Phi_S := \Phi_{T_S}.$$

If $\dim S \geq 5$, then Lemma 2.3.3 implies $\dim \Phi_S = 1$. However, if $S \in \mathcal{S}_3$, then we can always choose $S_1 \in \mathcal{S}_5$ so $S \subset S_1$. Since $\Phi_{S_1}([a]) \subset \Phi_S([a])$, we can use Lemma 2.3.3 to see equality holds and thus in fact

$$\dim \Phi_S([a]) = 1 \text{ for any } S \in \mathcal{S}_3.$$

This defines a family of maps

$$\Phi_S : \mathbb{P}(S) \to \mathbb{P}(B) \text{ such that}$$
$$\Phi_S = \Phi_{\tilde{S}} \text{ if } S \subset \tilde{S} \text{ and } S \in \mathcal{S}_3. \tag{2.3.5.a}$$

We study this family as follows:

2.3.6 Lemma. *Let A and B be vector spaces of signatures (p_A, q_A) and (p_B, q_B), respectively, where $p_B = 0$ and $q_A \geq 5$. Let $T \in \otimes^2 A^* \otimes^2 B^*$ be an alternating mixed tensor of spacelike rank 2.*

(1) *There exists $\Phi : \mathbb{P}(A^+) \to \mathbb{P}(B)$ so $\Phi|_{\mathbb{P}(S)} = \Phi_S$ for all $S \in \mathcal{S}_3$.*

(2) *Either the map Φ is constant or all the maps Φ_S are injective.*

Proof. Let $[a] \in \mathbb{P}(A^+)$ be a spacelike line. Let S and $\tilde{S}$ be any two spaces in $\mathcal{S}_3$ which contain $[a]$. We use Lemma 2.3.5 to link S and $\tilde{S}$ by a chain $\{S_i\}$. We use Equation (2.3.5.a) to see that

$$\Phi_{S_i}([a]) = \Phi_{S_i \cap S_{i+1}}([a]) = \Phi_{S_{i+1}}([a]) \text{ for all } i.$$

Consequently, $\Phi_S([a]) = \Phi_{\tilde{S}}([a])$. We may therefore set

$$\Phi([a]) := \phi_S([a]) \text{ for any } S \in \mathcal{S}_3 \text{ with } a \in S;$$

this is well defined and independent of the particular S chosen. This proves Assertion (1).

We use Lemma 2.3.3 to prove Assertion (2). Suppose there exists $S \in \mathcal{S}_3$ so that Φ_S is not injective. Choose $S_1 \in \mathcal{S}_{q_A}$ so $S \subset S_1$. Then Φ_{S_1} is not injective so Φ_{S_1} and hence Φ_S is constant by Lemma 2.3.3. Use Lemma

2.3.6 to link S to any other element $\tilde{S} \in \mathcal{S}_3$ by a chain $\{S_i\}$. If Φ_{S_i} is not injective, then Φ_{S_i} is constant. Thus, $\Phi_{S_i \cap S_{i+1}}$ is constant. Since

$$\dim S_i \cap S_{i+1} \geq 3,$$

$\Phi_{S_{i+1}}$ is not injective and hence $\Phi_{S_{i+1}}$ is constant and takes the same values. Thus, if Φ_S is not injective, then $\Phi_{\tilde{S}}$ and Φ_S are both constant and take the same value. □

We can now generalize Lemma 2.3.4 to the case $p_A \neq 0$:

2.3.7 Lemma. *Let A and B be vector spaces of signatures (p_A, q_A) and (p_B, q_B), respectively, where $p_B = 0$ and $q_A \geq 5$. Let $T \in \otimes^2 A^* \otimes^2 B^*$ be an alternating mixed tensor of spacelike rank 2. Let $\Phi : \mathbb{P}(A^+) \to \mathbb{P}(B)$ be the associated map.*

(1) *Suppose that Φ is constant. Set $\chi(a_1, a_2) := -T(a_1, a_2)\xi$, where $\xi \in S(\Phi)$. Then $T = T_{\chi,\xi}$.*
(2) *Suppose that Φ is not constant. Then there exists a linearization ϕ of Φ so $T = \pm T_\phi$.*

Proof. Suppose first that the map Φ is constant. If a_1 and a_2 span a spacelike 2 plane, then Lemma 2.3.4 shows that $T(a_1, a_2) - T_{\chi,\xi}(a_1, a_2) = 0$. We then use Lemma 1.2.5 to see that $T - T_{\chi,\xi} = 0$ in general; this proves Assertion (1).

Suppose that Φ is non-constant. We use Lemma 2.3.6 to see that the map Φ_S is injective for all $S \in \mathcal{S}_3$. As the sphere of unit spacelike vectors $S^+(A)$ is the universal cover of the projective space $\mathbb{P}(A^+)$, we lift Φ to define a continuous map

$$\begin{aligned} &\bar{\Phi} : S^+(A) \to S^+(B) \text{ so that} \\ &[\bar{\Phi}(a)] = \Phi([a]) \ \forall\, a \in S^+(A). \end{aligned}$$

Let $S \in \mathcal{S}_3$. Let $T_S := T|_S$. Choose $\bar{S} \in \mathcal{S}_{q_A}$ so $S \subset \bar{S}$. Since $\Phi_{\bar{S}}$ is injective, we apply Lemma 2.3.4 (2) to the alternating mixed tensor T_S to find $\phi_{\bar{S}}$ so $T_{\bar{S}} = \pm T_{\phi_{\bar{S}}}$; we let $\phi_S = \phi_{\bar{S}}|_S$. Then $T_S = T_{\phi_S}$. By Lemma 1.8.6, ϕ_S is determined up to sign. We use $\bar{\Phi}$ to normalize the choice of sign by requiring

$$(\bar{\Phi}(a), \phi_S(a)) > 0 \text{ for all } a \in S - \{0\}.$$

If $S_1 \in \mathcal{S}_3$, if $S_2 \in \mathcal{S}_3$, and if $S_1 \subset S_2$, then $\phi_{S_2}|_{S_1} = \phi_{S_1}$. Thus, we may use Lemma 2.3.5 to see that if $a \in S_1 \cap S_2$ and if $S_i \in \mathcal{S}_3$, then $\phi_{S_1}(a) = \phi_{S_2}(a)$. Let ϕ denote this common value;

$$\phi : A^+ \to B \text{ so } \phi|_S = \phi_S \text{ for all } S \in \mathcal{S}_3.$$

We now extend ϕ to all of A. Let a be an arbitrary element of A. Choose unit spacelike vectors a_1 and a_2 which are perpendicular to a. Choose a third unit spacelike vector a_3 which is perpendicular to a, to a_1, and to a_2. Choose large real numbers ε_i so that

$$|\varepsilon_i^2| > |(a,a)| \text{ for } 1 \le i \le 3.$$

As $a + \varepsilon_i a_i \in A^+$, $\phi(a+\varepsilon_i a_i) - \phi(\varepsilon_i a_i)$ is well defined. Note that

$$\begin{aligned}
&\operatorname{span}\{a+\varepsilon_1 a_1, \varepsilon_3 a_3\} \in \operatorname{Gr}_{0,2}(A),\\
&\operatorname{span}\{\varepsilon_1 a_1, \varepsilon_3 a_3\} \qquad \in \operatorname{Gr}_{0,2}(A), \text{ and}\\
&\operatorname{span}\{a+\varepsilon_3 a_3, \varepsilon_1 a_1\} \in \operatorname{Gr}_{0,2}(A).
\end{aligned}$$

Any spacelike 2 plane is contained in a spacelike 3 plane. Thus, ϕ is linear on spacelike planes and we may compute:

$$\begin{aligned}
&\phi(a+\varepsilon_1 a_1) - \phi(\varepsilon_1 a_1)\\
&=\phi(a+\varepsilon_1 a_1+\varepsilon_3 a_3) - \phi(\varepsilon_3 a_3) - \phi(\varepsilon_1 a_1+\varepsilon_3 a_3) + \phi(\varepsilon_3 a_3)\\
&=\phi(a+\varepsilon_1 a_1+\varepsilon_3 a_3) - \phi(\varepsilon_1 a_1) - \phi(\varepsilon_1 a_1+\varepsilon_3 a_3) + \phi(\varepsilon_1 a_1)\\
&=\phi(a+\varepsilon_3 a_3) - \phi(\varepsilon_3 a_3).
\end{aligned}$$

As the roles of a_1 and a_2 are symmetric, this shows that

$$\begin{aligned}
\phi(a+\varepsilon_1 a_1) - \phi(\varepsilon_1 a_1) =& \phi(a+\varepsilon_3 a_3) - \phi(\varepsilon_3 a_3)\\
=& \phi(a+\varepsilon_2 a_2) - \phi(\varepsilon_2 a_2)
\end{aligned}$$

is independent of the choice of (a_1, ε_1). If $a \in A^+$, then $\operatorname{span}\{a, a_1\}$ is spacelike so this difference yields $\phi(a)$. Consequently, we may extend ϕ from A^+ to A by defining

$$\phi(a) := \phi(a+\varepsilon_1 a_1) - \phi(\varepsilon_1 a_1).$$

It is immediate that $\phi(\varrho a) = \varrho\phi(a)$. To check ϕ is linear, we must show that ϕ is additive. If $\{a, \tilde{a}\}$ are given, choose $a_1 \in S^+(A)$ and $\tilde{a}_1 \in S^+(A)$ so

$$\{a, a_1, \tilde{a}_1\} \text{ and } \{\tilde{a}, a_1, \tilde{a}_1\}$$

are orthonormal sets. Since $\{a+\tilde{a}, a_1+\tilde{a}_1\}$ is an orthogonal set and since $a_1 + \tilde{a}_1$ is spacelike, we may compute:

$$\begin{aligned}
\phi(a+\tilde{a}) =& \phi(a+\tilde{a}+\varepsilon(a_1+\tilde{a}_1)) - \varepsilon\phi(a_1+\tilde{a}_1)\\
=& \phi(a+\varepsilon a_1) + \phi(\tilde{a}+\tilde{\varepsilon} a_1) - \varepsilon\phi(a_1) - \varepsilon\phi(\tilde{a}_1)\\
=& \phi(a) + \phi(\tilde{a}).
\end{aligned}$$

This shows that ϕ is a linear map from A to B. Furthermore, there exists a choice of sign $\varepsilon(a_1, a_2)$ so

$$T(a_1, a_2) - \varepsilon(a_1, a_2)T_\phi(a_1, a_2) = 0$$

if $\operatorname{span}\{a_1, a_2\} \in \operatorname{Gr}_{0,2}^+(V)$. As $\operatorname{Gr}_{0,2}^+(V)$ is connected, $\varepsilon(a_1, a_2) = \varepsilon$ is universal and independent of (a_1, a_2). By Lemma 1.2.5, $T - \varepsilon T_\phi = 0$ in general. □

Proof of Theorem 2.3.2. Let A and B be vector spaces of signatures (p_A, q_A) and (p_B, q_B), respectively, where $q_A \geq 5$.

Suppose first that we are given a linear map ϕ from A to B or that we are given an alternating bilinear map χ from $A \otimes A$ to B and an element $\xi \in B$. It is clear that the tensors T_ϕ and $T_{\chi,\xi}$ are alternating mixed tensors. Furthermore, we have:

$$\operatorname{rank}(T_\phi) \leq 2 \text{ and } \operatorname{rank}(T_{\chi,\xi}) \leq 2.$$

We apply Lemma 1.2.2 to see that T_ϕ has constant spacelike rank 2 if and only if $\{\phi(a_1), \phi(a_2)\}$ are linearly independent vectors whenever $\{a_1, a_2\}$ spans a spacelike 2 plane. This is equivalent to the condition that $\ker(\phi)$ contains no spacelike vector. Similarly, $T_{\chi,\xi}$ has constant spacelike rank 2 if and only if $\chi(a_1, a_2)$ and ξ are linearly independent whenever $\{a_1, a_2\}$ spans a spacelike 2 plane. This establishes Assertions (1) and (2).

Suppose that $T \in \otimes^2 A^* \otimes^2 B^*$ is an alternating mixed tensor which has constant spacelike rank 2. We must show that $T = \pm T_\phi$ or that $T = T_{\chi,\xi}$. This follows from Lemma 2.3.7 if the metric on B is positive definite. We argue as follows to complete the proof in the higher signature setting. We use Lemma 1.2.13 to find ψ which relates $(\cdot,\cdot)$ to a positive definite metric $(\cdot,\cdot)_+$. We have the associated map

$$T : \Lambda^2 A \to \mathfrak{so}(B);$$

by assumption $T(\pi)$ has rank 2 on $\operatorname{Gr}_{0,2}^+(A)$. If we change the metric, then the associated map becomes

$$\psi T : \Lambda^2 A \to \mathfrak{so}(B, (\cdot,\cdot)_+).$$

The rank, however, is unchanged. We may therefore apply Lemma 2.3.7 to express:

$$\text{(2.3.7.a)} \quad \begin{aligned} &\psi T(a_1, a_2)b = (\hat{\chi}(a_1, a_2), b)_+ \cdot \hat{\xi} - (\hat{\xi}, b)_+ \cdot \hat{\chi}(a_1, a_2) \text{ or} \\ &\psi T(a_1, a_2)b = \pm\{(\hat{\phi}(a_2), b)_+ \cdot \hat{\phi}(a_1) - (\hat{\phi}(a_1), b)_+ \cdot \hat{\phi}(a_2)\}. \end{aligned}$$

Since $\psi^2 = \text{Id}$, we may apply ψ to Equation (2.3.7.a) to see:

$$T(a_1,a_2)b = (\psi\hat{\chi}(a_1,a_2),b)\psi\hat{\xi} - (\psi\hat{\xi},b)\psi\hat{\chi}(a_1,a_2) \text{ or}$$
$$T(a_1,a_2)b = \pm\{(\psi\hat{\phi}(a_2),b)\psi\hat{\phi}(a_1) - (\psi\hat{\phi}(a_1),b)\psi\hat{\phi}(a_2)\}.$$

We complete the proof by setting $\chi = \psi\hat{\chi}$ and $\xi = \psi\hat{\xi}$, or $\phi := \psi\hat{\phi}$ as appropriate. □

Proof of Theorem 2.1.5. Suppose that $\phi = \phi^*$ and that $\ker(\phi)$ contains no spacelike vectors. By Theorem 2.3.2, R_ϕ has constant spacelike rank 2. By Lemma 1.8.1, R_ϕ is an algebraic curvature tensor.

Conversely, let R be an algebraic curvature tensor which has constant spacelike rank 2. We apply Theorem 2.3.2. Suppose first that $R = R_{\chi,\xi}$. Choose $\{x,y\}$ a spacelike orthonormal set so $\{\xi,x,y\}$ is an orthogonal set. Choose z so $(\xi,z) \neq 0$. We apply the first Bianchi identity to see:

$$\begin{aligned} 0 =& \{(\chi(x,y),z) + (\chi(y,z)x) + (\chi(z,x),y)\}\xi \\ &- (\xi,z)\chi(x,y) - (\xi,x)\chi(y,z) - (\xi,y)\chi(z,x). \end{aligned}$$

Since $(\xi,z) \neq 0$ but $(\xi,x) = (\xi,y) = 0$, $\chi(x,y)$ is a multiple of ξ so

$$\dim\{\text{range}(R(x,y))\} \leq 1$$

which is false. Thus, $R = \pm R_\phi$, where $\ker(\phi) \cap V^+ = \emptyset$. By replacing R by $-R$ we may suppose $R = R_\phi$. Since R_ϕ is an algebraic curvature tensor, we may apply Lemma 1.8.6 to see $\phi = \phi^*$. □

2.3.8 Example. We conclude this section by showing that Theorem 2.3.2 is false if $q_A = 4$. Let $\{e_1^-, ..., e_p^-, e_1^+, ..., e_q^+\}$ be a normalized orthonormal basis for a vector space A of signature (p_A, q_A), where $q_A = 4$. Let

$$\omega := e_1^+ \wedge e_2^+ \wedge e_3^+ \wedge e_4^+.$$

Extend the inner product on V to an inner product on $\Lambda^4(V)$ and define:

$$T : V \otimes V \to \mathfrak{so}(V) \text{ by}$$
$$(T(v_1,v_2)v_3,v_4) = (v_1 \wedge v_2 \wedge v_3 \wedge v_4, \omega).$$

Let $\{f_1, f_2\}$ be an orthonormal basis for a spacelike 2 plane π. We may decompose $f_i = f_i^- + f_i^+$, where $f_i^+ \in W := \text{span}\{e_1^+, e_2^+, e_3^+, e_4^+\}$ and where $f_i^- \in W^\perp$. Let $\bar{\pi}$ be the projection of π to W;

$$\pi_W = \text{span}\{f_1^+, f_2^+\}$$

is also a spacelike 2 plane. There exists a non-zero constant so

$$T(\pi) = T(f_1^+, f_2^+) = cT(\bar{\pi}).$$

Thus, to show T has constant spacelike rank 2, it suffices to show that $\operatorname{rank} T(\bar{\pi}) = 2$ for every spacelike 2 plane $\bar{\pi}$ contained in W.

Let $\{f_1, f_2\}$ be an oriented orthonormal basis for a 2 plane $\pi \subset W$. We extend $\{f_1, f_2\}$ to an oriented orthonormal basis $\{f_1, f_2, f_3, f_4\}$ for W and compute:

$$T(f_1, f_2)f_1 = 0, \qquad T(f_1, f_2)f_2 = 0,$$
$$T(f_1, f_2)f_3 = f_4, \text{ and } T(f_1, f_2)f_4 = -f_3.$$

Thus, T has constant spacelike rank 2. Since $\operatorname{range}(T(\pi)) = \pi^\perp \cap W$, we have that

$$\begin{aligned}
&\Phi([e_1^+]) \\
&\subset \operatorname{span}\{e_1^+, e_2^+\}^\perp \cap \operatorname{span}\{e_1^+, e_3^+\}^\perp \cap \operatorname{span}\{e_1^+, e_4^+\}^\perp \cap W \\
&= \operatorname{span}\{e_3^+, e_4^+\} \cap \operatorname{span}\{e_2^+, e_3^+\} \cap \operatorname{span}\{e_2^+, e_4^+\} \cap W \\
&= \{0\}.
\end{aligned}$$

Thus, $T \neq T_\phi$ for any ϕ and $T \neq T_{\chi,\xi}$ for any (χ, ξ).

2.4 Geometric realizations of rank 2 tensors

Let R be an algebraic curvature tensor on a vector space V of signature (p, q). We say that a pseudo-Riemannian manifold (M, g) is a *geometric realization* of R at a point $P \in M$ if there is an isomorphism ψ from T_PM to V so that

$$\psi^*(\cdot, \cdot) = (\cdot, \cdot)_P \text{ and } \psi^*R = {}^gR_P.$$

We say that a pseudo-Riemannian manifold (M, g) has *spacelike rank* 2 if the associated algebraic curvature tensor ${}^gR$ has spacelike rank 2 at every point of the manifold. By Theorem 1.12.2, every algebraic curvature tensor is geometrically realizable. We now show that every spacelike rank 2 algebraic curvature tensor is geometrically realizable by the germ of a spacelike rank 2 pseudo-Riemannian manifold.

2.4.1 Theorem. *Let R be a spacelike rank 2 algebraic curvature tensor which is defined on a vector space V of signature (p, q), where $q \geq 5$. Then there exists the germ of hypersurface (M, g) in flat space which is a spacelike rank 2 pseudo-Riemannian manifold and which is a geometric realization of*

R at a point of M. If R is also timelike and/or mixed rank 2, then (M, g) can be chosen to have timelike and/or mixed rank 2 as well.

Proof. We apply Theorem 2.1.5 to see that $R = \varepsilon R_\phi$, where $\varepsilon = \pm 1$, where ϕ is self-adjoint, and where $\ker \phi$ contains no spacelike vectors.

Let $m = p + q = \dim V$. As in the proof of Theorem 1.12.3, we let

$$W := V \oplus e_{m+1}\mathbb{R}$$

and extend the metric from V to W so that

$$V \perp \operatorname{span} e_{m+1} \qquad \text{and} \qquad (e_{m+1}, e_{m+1}) = \varepsilon.$$

Let $L_\phi(v, w) := (\phi v, w)$ be the associated symmetric bilinear form. We define an embedding of V into W by setting:

$$F(v) := v \oplus \tfrac{\varepsilon}{2} L_\phi(v, v) e_{m+1}.$$

The induced pseudo-Riemannian metric on V agrees with the original inner product at $v = 0$ and thus is non-degenerate in some small neighborhood of $v = 0$. We restrict to this neighborhood henceforth. Let ν be the unit normal vector field which agrees with e_{m+1} at the origin; ν is defined by the relations:

$$(dF, \nu) = 0, \ (\nu, \nu) = \varepsilon, \text{ and } \nu(0) = e_{m+1}.$$

The second fundamental form L_F of the embedding is given by:

$$L_F(v_1, v_2) = (v_1 v_2 F, \nu) = \varepsilon(\nu, e_{m+1}) L_\phi(v_1, v_2).$$

Thus, $L_\phi = L_F$ when $v = 0$. Let S_F be the associated shape operator. At the origin, the shape operator is given by the defining map ϕ. We use Lemma 1.12.3 to see that

$${}^g R(0) = \varepsilon R_{S[0]} = \varepsilon R_\phi = R.$$

Thus, this hypersurface is a geometric realization of R at the origin. To complete the proof, we must check that ${}^g R$ has constant spacelike rank on a neighborhood of the origin. We identify $T_v V$ with V at any point $v \in V$. Since $(\nu, e_{m+1}) \neq 0$, we compute:

$$\begin{aligned}
\ker S_F[v] :=& \{v_1 \in V : (S_F[v]v_1, v_2) = 0 \ \forall \ v_2 \in V\} \\
=& \{v_1 \in V : L_F[v](v_1, v_2) = 0 \ \forall \ v_2 \in V\} \\
=& \{v_1 \in V : \varepsilon(\nu, e_{m+1}) L_\phi(v_1, v_2) = 0 \ \forall \ v_2 \in V\} \\
=& \ker \phi.
\end{aligned}$$

This shows that the kernel of S_F is independent of v. Choose a basis $\{e_1, ..., e_m\}$ for V so that $\{e_1, ..., e_\ell\}$ is a basis for $\ker\phi$ and let $\{e^1, ..., e^m\}$ be the dual basis for V^*. Let $y^i := e^i(v)$ be the associated coordinates on V. Then:

$$\begin{aligned} L_\phi(v,v) &= \textstyle\sum_{i,j} L_\phi(y^i e_i, y^j e_j) = \sum_{i,j} y^i y^j (\phi e_i, e_j) \\ &= \textstyle\sum_{i,j>\ell} y^i y^j (\phi e_i, e_j), \\ F(v) &= \textstyle\sum_i y^i e_i \oplus \frac{\varepsilon}{2} \sum_{i,j>\ell} y^i y^j (\phi e_i, e_j) e_{m+1}, \\ F_*(\partial_i) &= e_i \oplus 0 \text{ for } i \leq \ell, \\ F_*(\partial_j) &= e_j \oplus \varepsilon \textstyle\sum_{k>\ell} y^k (\phi e_j, e_k) e_{m+1}, \text{ and} \\ g_{ij}(v) &= (e_i, e_j) \text{ for } i \leq \ell \text{ and arbitrary } j. \end{aligned}$$

Thus, the identification of $\ker S_F = \ker\phi$ defined above is an isometry as

$$g_v(v_1, v_2) = (v_1, v_2) \text{ if } v_1 \in \operatorname{span}\{e_1, ..., e_\ell\} = \ker\phi.$$

Let $S_F[v]$ be the shape operator at a point v of the manifold. The argument given above shows that if $\ker\phi$ contains no spacelike vectors, then $\ker S_F[v]$ contains no spacelike vectors. Similarly, if $\ker\phi$ contains no timelike vectors or is trivial, then $\ker S_F[v]$ contains no timelike vectors or is trivial. □

We wish to extend Theorem 2.1.5 from the algebraic to the geometric category. We begin with a technical result concerning smoothness. Let $\mathcal{O}$ be a small open ball around the origin in some Euclidean space. Let A and B be vector spaces of signatures (p_A, q_A) and (p_B, q_B), respectively, where $q_A \geq 5$. We suppose given a smooth map

$$T : \mathcal{O} \times A \times A \to \mathfrak{so}(B).$$

We let $T_P(a_1, a_2) = T(P, a_1, a_2)$. We suppose that T_P is an alternating mixed tensor of constant spacelike rank 2 for any $P \in \mathcal{O}$. We use Equation (2.3.2.a) to define:

$$\Phi_P([a]) := \cap_{\{\pi \in \mathrm{Gr}^+_{0,2}(A) : a \in \pi\}} \operatorname{range}(T_P(\pi)).$$

Let $a_1 \in S^+(A)$. Choose vectors $a_2, a_3 \in A$ so that $\operatorname{span}\{a_1, a_2, a_3\}$ is a spacelike 3 plane which is contained in a spacelike 5 plane. We may then use Lemma 2.3.6 to see that:

$$\begin{aligned} &\dim \Phi_P([a_1]) = 1, \text{ and} \\ &\Phi_P([a_1]) := \operatorname{range}(T_P(a_1, a_2)) \cap \operatorname{range}(T_P(a_1, a_3)). \end{aligned}$$

Then, by Theorem 4.2.7, we have that:

$$\Phi : \mathcal{O} \times \mathbb{P}(A^+) \to \mathbb{P}(B)$$

is smooth. Thus, in particular if Φ_0 is not constant, then Φ_P is not constant for P close to 0.

Let $\operatorname{Hom}(A, B)$ be the space of linear maps from A to B.

2.4.2 Lemma. *Let $\mathcal{O}$ be a small open ball around the origin 0 in $\mathbb{R}^m$. Let A and B be vector spaces of signatures (p_A, q_A) and (p_B, q_B), respectively, where $q_A \geq 5$. Let $T : \mathcal{O} \times A \times A \to \mathfrak{so}(B)$ be smooth. We suppose that T_P is a spacelike rank 2 alternating tensor for each $P \in \mathcal{O}$. Suppose that Φ_P is not-constant for any $P \in \mathcal{O}$. Then there exists a smooth map $\phi_P : \mathcal{O} \to \mathrm{Hom}(A, B)$ and a choice of sign $\varepsilon = \pm 1$ so that $\ker \phi_P$ contains no spacelike vector, and so that $T_P = \varepsilon T_{\phi_P}$ for all $P \in \mathcal{O}$.*

Proof. As noted in the proof of Theorem 2.3.2, by replacing T by ψT, we may assume without loss of generality that the metric on B is positive definite. We lift the map Φ to define a smooth map on the universal cover

$$\bar{\phi} : \mathcal{O} \times S^+(A) \to S(B) \text{ so that } [\bar{\phi}_P a] = \Phi_P([a]).$$

By Lemma 2.3.4, for each P, there exist a linearization ϕ_P of Φ_P so that

$$T_P = \varepsilon_P T_{\phi_P} \text{ for } \varepsilon_P = \pm 1.$$

The map ϕ_P is uniquely determined modulo sign. Fix $a_1 \in S^+(A)$. We normalize the choice of ϕ_P by requiring that

$$(\phi_P(a_1), \bar{\phi}_P(a_1)) > 0.$$

Since $[\phi_P(a)] = [\bar{\phi}_P(a)] = \Phi_P[a]$ for $a \in S^+(A)$, there exists a function $c : \mathcal{O} \times S^+(A) \to \mathbb{R} - \{0\}$ so that

$$\phi_P(a) = c(P, a)\bar{\phi}_P(a).$$

For fixed P, the function $a \to c(P, a)$ is smooth. Since $c(P, a_1) > 0$,

$$c(P, a) > 0 \text{ for all } P \in \mathcal{O} \text{ and } a \in S^+(A).$$

Thus, the choice of a_1 plays no role in the construction; only the $\mathbb{Z}_2$ normalization inherent in the choice of the lift $\bar{\phi}$ is relevant.

Let $a_1 \in S^+(A)$. Choose vectors a_2 and a_3 so that $\{a_1, a_2, a_3\}$ is an orthnormal spacelike subset of A. For $1 \leq i < j \leq 3$, define

$$\begin{aligned}
\mathcal{T}_{ij}(P) :=& (T_P(a_i, a_j)\bar{\phi}_P a_j, \bar{\phi}_P a_i),\\
\mathcal{D}_{ij}(P) :=& (\bar{\phi}_P a_i, \bar{\phi}_P a_i)(\bar{\phi}_P a_j, \bar{\phi}_P a_j) - (\bar{\phi}_P a_i, \bar{\phi}_P a_j)(\bar{\phi}_P a_i, \bar{\phi}_P a_j),\\
\mathcal{E}_{ij}(P) :=& (\bar{\phi}_P a_i, \phi_P a_i)(\bar{\phi}_P a_j, \phi_P a_j) - (\bar{\phi}_P a_i, \phi_P a_j)(\bar{\phi}_P a_i, \phi_P a_j)\\
=& c(P, a_i)c(P, a_j)\mathcal{D}_{ij}(P) = \varepsilon_P \mathcal{T}_{ij}(P).
\end{aligned}$$

Since $\bar{\phi}$ and T are smooth, $\mathcal{D}_{ij}$ and $\mathcal{T}_{ij}$ are smooth functions of P. Since $\{\bar{\phi}_P a_i, \bar{\phi}_P a_j\}$ are linearly independent vectors, we may use the Cauchy–Schwarz inequality to see $\mathcal{D}_{ij}(P) > 0$. Since $c(P, a_i) > 0$,

$$\varepsilon_P = \text{sign}(\mathcal{T}_{ij}(P))$$

is a smooth function of P and hence constant. By replacing T by $-T$ if necessary, we may therefore assume $\varepsilon_P = +1$. Thus, we have

$$c(P, a_i)c(P, a_j) = \tfrac{\mathcal{T}_{ij}(P)}{\mathcal{D}_{ij}(P)} \text{ for } i < j.$$

We therefore conclude that

$$c(P, a_1)^2 = \tfrac{c(P,a_1)c(P,a_2)c(P,a_1)c(P,a_2)}{c(P,a_2)c(P,a_3)}$$

is a smooth function of P. Since $c(P, a_1) > 0$, we may take the square root to see the map

$$P \to c(P, a_1) \text{ is smooth in } P \text{ for any } a_1 \in S^+(V).$$

Since we can choose a basis for A consisting of spacelike unit vectors and since ϕ_P is linear, we conclude finally that $P \to \phi_P$ is smooth as desired. □

The following is now an immediate consequence of Theorem 2.1.5 and Lemma 2.4.2:

2.4.3 Theorem. *Let (M, g) be a pseudo-Riemannian manifold. Assume that ${}^gR$ defines a spacelike rank 2 algebraic curvature tensor on a contractible open set $\mathcal{O}$. Then there exists a choice of sign $\varepsilon = \pm 1$ and there exists a smooth section to the bundle of endomorphisms of the tangent bundle defined over $\mathcal{O}$ so that for all points $P \in \mathcal{O}$ we have ϕ_P is self-adjoint,* $\ker \phi_P$ *contains no spacelike vectors, and ${}^gR_P = \varepsilon R_{\phi_P}$.*

2.5 IP algebraic curvature tensors

Let V be a vector space of signature (p, q), where $q \geq 5$. An algebraic curvature tensor R on V is said to be *timelike Jordan IP*, *mixed Jodan IP*, or *spacelike Jordan IP* if the Jordan normal form of $R(\cdot)$ is constant on the Grassmannians $\text{Gr}^+_{2,0}(V)$, $\text{Gr}^+_{1,1}(V)$, or $\text{Gr}^+_{0,2}(V)$ of oriented timelike, mixed, or spacelike 2 planes, respectively.

In this section, we establish the basic classification result of spacelike rank 2 Jordan IP algebraic curvature tensors which is given in Theorem 2.1.6. We will show that a spacelike rank 2 algebraic curvature tensor R is spacelike Jordan IP if and only if there exists a non-zero constant C and a self-adjoint map ϕ so that $R = \pm C R_\phi$, where ϕ satisfies one of the following conditions:

(1) The map ϕ is an isometry of V, i.e. $\phi^2 = \mathrm{Id}$.
(2) The map ϕ is a para-isometry of V, i.e. $\phi^2 = -\mathrm{Id}$.
(3) The map ϕ satisfies $\phi^2 = 0$ and $\ker\phi$ contains no spacelike vectors.

Note that if possibility (2) holds, then $p = q$ and that if possibility (3) holds, then $p \geq q$.

Proof of Theorem 2.1.6. Let R be a spacelike rank 2 algebraic curvature tensor. We use Theorem 2.1.5 to find a self-adjoint map ϕ so that $\ker\phi$ contains no spacelike vectors and so that

$$R = \pm R_\phi.$$

We shall have to renormalize ϕ suitably in order to construct the map described in Theorem 2.1.6. Let v_1 be a unit spacelike vector. Choose a spacelike subspace $S \in \mathcal{S}_3$ so that $v_1 \in S$. Choose a unit spacelike vector $v_2 \in S$ so that

$$v_2 \perp v_1 \text{ and } \phi v_2 \perp \phi v_1; \tag{2.5.1.a}$$

Equation (2.5.1.a) imposes at most 2 linear conditions on v_2, these can be satisfied since $\dim S \geq 3$. Let

$$\pi := \operatorname{span}\{v_1, v_2\} \text{ and } \sigma := \operatorname{range}(R_\phi(\pi)) = \phi\pi.$$

Since

$$R_\phi(x, y)z = (\phi y, z)\phi x - (\phi x, z)\phi y,$$

we use Equation (2.5.1.a) to see that:

$$\begin{aligned} R_\phi(\pi) &: \phi v_1 \to -(\phi v_1, \phi v_1)\phi v_2, \text{ and} \\ R_\phi(\pi) &: \phi v_2 \to \quad (\phi v_2, \phi v_2)\phi v_1. \end{aligned}$$

Let $(\cdot,\cdot)_\sigma$ be the restriction of $(\cdot,\cdot)$ to σ. Since $\phi v_1 \perp \phi v_2$, $(\cdot,\cdot)_\sigma$ is determined by the two inner products

$$(\phi v_1, \phi v_1) \text{ and } (\phi v_2, \phi v_2).$$

We can relate the Jordan structure of $R_\phi(\pi)$ to the induced innerproduct $(\cdot,\cdot)_\sigma$ on σ. There are four cases:

(1) The inner product $(\cdot,\cdot)_\sigma$ is positive or negative definite, i.e.

$$(\phi v_1, \phi v_1) \cdot (\phi v_2, \phi v_2) > 0.$$

Then $R_\phi(\pi)$ has two non-zero purely imaginary eigenvalues of the form $\pm\sqrt{-1}\lambda$, where

$$\lambda := \sqrt{(\phi v_1, \phi v_1) \cdot (\phi v_2, \phi v_2)}.$$

(2) The inner product $(\cdot,\cdot)_\sigma$ is indefinite and non-degenerate, i.e.

$$(\phi v_1, \phi v_1) \cdot (\phi v_2, \phi v_2) < 0.$$

Then $R_\phi(\pi)$ has two non-zero real eigenvalues of the form $\pm\lambda$, where

$$\lambda = \sqrt{-(\phi v_1, \phi v_1) \cdot (\phi v_2, \phi v_2)}.$$

(3) The inner product $(\cdot,\cdot)_\sigma$ is degenerate but non-trivial. Then $R_\phi^2 \neq 0$ and $R_\phi^3 = 0$.

(4) The inner product $(\cdot,\cdot)_\sigma$ is trivial, i.e. σ is totally isotropic. Then $R_\phi \neq 0$ and $R_\phi^2 = 0$.

We can now prove one implication of Theorem 2.1.6. Suppose that we can express $R = CR_\phi$, where ϕ is an isometry or a para-isometry. Then case (1) holds and $\lambda(\pi) = 1$ is constant. Furthermore, the discussion given above shows the Jordan normal form of R is constant. Suppose that $R = CR_\phi$, where $\phi^2 = 0$. Then the range of ϕ is totally isotropic so case (4) holds and again R has constant Jordan form.

To prove the other implication of Theorem 2.1.6, we suppose that R_ϕ is spacelike rank 2 Jordan IP. We must first eliminate the Jordan normal forms described by cases (2) and (3) above.

Let v_1 be a unit spacelike vector. Since $q \geq 5$, we may choose a spacelike plane $S \in \mathcal{S}_5$ containing v_1. Since $\dim(S) \geq 5$, we can impose 2 linear conditions and choose a unit vector $v_2 \in S$ so that:

$$(v_1, v_2) = 0 \text{ and } (\phi v_1, \phi v_2) = 0.$$

Since $\dim(S) \geq 5$, we can impose 4 linear conditions and choose a unit vector $v_3 \in S$ which satisfies the relations:

$$(v_1, v_3) = 0, \ (v_2, v_3) = 0, \ (\phi v_1, \phi(v_3)) = 0, \text{ and } (\phi v_2, \phi(v_3)) = 0.$$

Since R_ϕ has spacelike rank 2, $\ker\phi \cap S = \{0\}$. Consequently, the vectors ϕv_1, ϕv_2, and $\phi(v_3)$ are linearly independent. Let

$$\pi_{ij} := \operatorname{span}\{v_i, v_j\} \text{ for } 1 \leq i < j \leq 3.$$

Suppose case (2) holds, i.e. $R_\phi(\pi_{ij})$ has two non-zero real eigenvalues. Then we have

$$\begin{aligned}&(\phi v_1, \phi v_1) \cdot (\phi v_2, \phi v_2) < 0,\\&(\phi v_1, \phi v_1) \cdot (\phi(v_3), \phi(v_3)) < 0, \text{ and}\\&(\phi v_2, \phi v_2) \cdot (\phi(v_3), \phi(v_3)) < 0.\end{aligned}$$

We multiply these three equations together to derive a contradiction by computing:

$$(\phi v_1, \phi v_1)^2 \cdot (\phi v_2, \phi v_2)^2 \cdot (\phi(v_3), \phi(v_3))^2 < 0.$$

Next, we suppose that case (3) holds. Then the set of inner products

$$\{(\phi(v_i), \phi(v_i)), (\phi(v_j), \phi(v_j))\}$$

consists of one zero number and one non-zero number for $1 \leq i < j \leq 3$. Again, this is not possible as there are 3 such pairs.

We complete the proof by considering the remaining eigenvalue structures. We suppose first that $R_\phi(\pi)$ has two non-zero purely imaginary eigenvalues $\pm\sqrt{-1}\lambda$ for any $\pi \in \operatorname{Gr}^+_{0,2}(V)$. Let v_1 and v_2 be arbitrary unit spacelike vectors in V. Let S be any maximal spacelike subspace; we do not assume that v_1 or v_2 belong to S. Since $q \geq 5$, we can impose 4 linear conditions and choose a unit vector v_3 in S which satisfies the equations:

$$(v_1, v_3) = 0,\ (v_2, v_3) = 0,\ (\phi v_1, \phi(v_3)) = 0, \text{ and } (\phi v_2, \phi(v_3)) = 0.$$

Let $\pi_i := \operatorname{span}\{v_i, v_3\}$ for $i = 1, 2$. We compute:

$$\begin{aligned}\lambda^2 =&(\phi v_1, \phi v_1) \cdot (\phi(v_3), \phi(v_3))\\=&(\phi v_2, \phi v_2) \cdot (\phi(v_3), \phi(v_3)).\end{aligned} \tag{2.5.1.b}$$

Since $\lambda \neq 0$, we can divide the identity given in Equation (2.5.1.b) by $(\phi(v_3), \phi(v_3))$ to see that

$$(\phi v_1, \phi v_1) = (\phi v_2, \phi v_2) \text{ if } v_1, v_2 \in S^+(V). \tag{2.5.1.c}$$

Let C be this common non-zero value. We rescale Equation (2.5.1.c) to see:

$$(\phi(v), \phi(v)) = C|v|^2 = C(v, v) \text{ if } v \in V^+. \tag{2.5.1.d}$$

Now let v be arbitrary. Choose v_1 spacelike. Then $v+\varepsilon v_1$ is spacelike for large ε and hence Display (2.5.1.d) implies that

$$(2.5.1.e)\qquad (\phi(v+\varepsilon v_1),\phi(v+\varepsilon v_1)) = C(v+\varepsilon v_1, v+\varepsilon v_1) \text{ for } \varepsilon > \varepsilon_0.$$

Since Equation (2.5.1.e) is quadratic in the parameter ε, it must hold when $\varepsilon = 0$. Thus, we have that

$$(\phi(v),\phi(v)) = C(v,v) \text{ for all } v \in V.$$

We renormalize ϕ to complete the proof:

(1) if $C > 0$, then $C^{-\frac{1}{2}}\phi$ is an isometry.

(2) If $C < 0$, then $|C|^{-\frac{1}{2}}\phi$ is a para-isometry.

Second, we suppose that $R_\phi(\pi)^2 = 0$ for any spacelike 2 plane π. Then range$(R_\phi(\pi))$ is totally isotropic. Let v_1 and v_2 be arbitrary (not necessarily distinct) vectors in V. Choose unit spacelike vectors w_1 and w_2 so $w_1 \perp w_2$. Let

$$v_i(\varepsilon) := v_i + \varepsilon w_i.$$

For large ε, span$\{v_1(\varepsilon), v_2(\varepsilon)\}$ is spacelike so

$$(2.5.1.f)\qquad (\phi(v_1(\varepsilon)),\phi(v_2(\varepsilon))) = 0 \text{ for } \varepsilon \geq \varepsilon_0.$$

Since Equation (2.5.1.f) is a quadratic identity which holds for large values of the parameter ε, Equation (2.5.1.f) continues to hold for $\varepsilon = 0$ and thus

$$(\phi v_1, \phi v_2) = 0$$

for all $v_i \in V$. This shows that range(ϕ) is totally isotropic. Since R_ϕ is spacelike rank 2, the kernel of ϕ contains no spacelike vectors. □

We can use Theorem 2.1.6 to obtain the following Corollary:

2.5.2 Corollary. *Let V be a vector space of signature (p,q), where $q \geq 5$. Let R be a spacelike rank 2 algebraic curvature tensor on V. We may decompose $R = CR_\phi$, where ϕ satisfies one of the following 3 conditions:*

(1) $\phi^2 = \mathrm{Id}$. *Then ϕ is an isometry of V. If $\pi \in \mathrm{Gr}^+_{0,2}(V)$, then the range of $R_\phi(\pi)$ is spacelike. R is also timelike and mixed Jordan IP.*

(2) *The map ϕ is a para-isometry of V. If $\pi \in \mathrm{Gr}^+_{0,2}(V)$, then the range of $R_\phi(\pi)$ is timelike. R is also timelike and mixed Jordan IP.*

(3) *The map ϕ satisfies $\phi^2 = 0$ and $\ker\phi$ contains no spacelike vectors. If $\pi \in \mathrm{Gr}^+_{0,2}(V)$, then the range of $R_\phi(\pi)$ is totally isotropic. R is timelike Jordan IP if and only if* range $\phi = \ker\phi$. *R is not mixed Jordan IP.*

Proof. Suppose first either that ϕ is an isometry of V or that ϕ is a para-isometry of V. Then $\ker\phi = \{0\}$ so by Lemma 1.2.2,

$$\operatorname{rank}(R_\phi(x,y)) = 2 \text{ if } x \wedge y \neq 0.$$

Suppose that $\pi \in \operatorname{Gr}^+_{2,0}(V)$ or that $\pi \in \operatorname{Gr}^+_{0,2}(V)$. Let $\{v_1, v_2\}$ be an orthonormal basis for π. Then $\{\phi v_1, \phi v_2\}$ is an orthonormal basis for $\phi\pi = \operatorname{range} R_\phi(\pi)$; the roles of spacelike and timelike are preserved if ϕ is an isometry and reversed if ϕ is a para-isometry. Set $\varepsilon = +1$ if $\phi\pi$ is spacelike and $\varepsilon = -1$ if $\phi\pi$ is timelike. We show that R_ϕ is timelike and spacelike Jordan IP by observing that $R_\phi(\pi)$ is conjugate to a transformation which has a standard form:

$$\begin{aligned}
&R_\phi(v_1,v_2)\phi v_2 = \varepsilon\phi v_1,\\
&R_\phi(v_1,v_2)\phi v_1 = -\varepsilon\phi v_2, \text{ and}\\
&R_\phi(v_1,v_2)v = 0 \text{ if } v \perp \operatorname{span}\{\phi v_1, \phi v_2\}.
\end{aligned}$$

Let $\{v_1, v_2\}$ be an orthonormal basis for $\pi \in \operatorname{Gr}^+_{1,1}(V)$. Then $\{\phi v_1, \phi v_2\}$ is an orthonormal basis for $\operatorname{range} R_\phi(\pi) = \phi\pi \in \operatorname{Gr}^+_{1,1}(V)$. We may choose a basis so that ϕv_1 is timelike and so that ϕv_2 is spacelike. We check that R_ϕ is mixed Jordan IP by showing that $R_\phi(\pi)$ is conjugate to a transformation which has a standard form:

$$\begin{aligned}
&R_\phi(v_1,v_2)\phi v_1 = \phi v_2,\\
&R_\phi(v_1,v_2)\phi v_2 = \phi v_1, \text{ and}\\
&R_\phi(v_1,v_2)v = 0 \text{ if } v \perp \operatorname{span}\{\phi v_1, \phi v_2\}.
\end{aligned}$$

Finally, suppose that $\phi^2 = 0$. Then the range of ϕ is totally isotropic. This implies that $R_\phi^2 = 0$. Since $\operatorname{range}\phi \subset \ker\phi$ contains at least one mixed 2 plane but does not intersect every mixed 2 plane, R_ϕ is not mixed Jordan IP. Suppose that R_ϕ is spacelike and timelike Jordan IP. Then $\ker\phi$ contains no spacelike and no timelike vectors. Let $V^\pm$ be maximal spacelike and timelike subspaces of V. Then

$$\begin{aligned}
&\dim\{\ker\phi \cap V^+\} = 0 \text{ so } \dim(\ker\phi) \leq m - \dim V^+ = p,\\
&\dim\{\ker\phi \cap V^-\} = 0 \text{ so } \dim(\ker\phi) \leq m - \dim V^- = q, \text{ and}\\
&\dim\ker\phi \leq \min(p,q).
\end{aligned} \tag{2.5.2.a}$$

On the other hand, because $\operatorname{range}\phi \subset \ker\phi$, we may estimate:

$$\dim(\operatorname{range}\phi) \leq \dim(\ker\phi).$$

Therefore, we may use Equation (2.5.2.a) to conclude that:

$$
\begin{aligned}
\text{(2.5.2.b)} \qquad p+q = &\dim(\operatorname{range}\phi) + \dim(\ker\phi) \le 2\dim(\ker\phi) \\
\le &2\min(p,q) \le p+q.
\end{aligned}
$$

Thus, all the inequalities in Equation (2.5.2.b) must have been equalities. This implies $p = q$ and $\operatorname{range}\phi = \ker\phi$. Conversely, of course, if we suppose that $\operatorname{range}\phi = \ker\phi$, then we may conclude that $\ker\phi$ is totally isotropic and thus contains no spacelike or timelike vectors. Corollary 2.5.2 now follows from this analysis and from Theorem 2.1.6. □

2.6 Examples of IP manifolds

Ivanov and Petrova [97] constructed two families of IP Riemannian manifolds; these families can be generalized to the pseudo-Riemannian setting. In Lemma 2.6.1, we discuss manifolds of constant sectional curvature, and in Lemma 2.6.3, we discuss warped products of an interval with a metric of constant sectional curvature, where the warping function is a suitably chosen quadratic function. These examples are spacelike, timelike, and mixed rank 2 Jordan IP pseudo-Riemannian manifolds that can be defined in any signature. The final example does not have a Riemannian analogue. If $p = q$, then we construct in Lemma 2.6.5 a nilpotent timelike and spacelike rank 2 Jordan IP pseudo-Riemannian hypersurface in flat space.

A pseudo-Riemannian manifold (M, g) is said to be a *homogeneous space* if the isometries act transitively on M; (M, g) is said to be a 2 *point homogeneous space* if the isometries act transitively on the unit sphere bundles $S^{\pm}(M, g)$; (M, g) is said to be a *symmetric space* if, for every point P of M, there is an isometry ψ fixing P so that $\psi_* = -\mathrm{Id}$ on T_PM.

We begin by studying the geometry of the pseudo-spheres:

2.6.1 Lemma. *Let W be a vector space of signature (r, s). Let $0 \neq K \in \mathbb{R}$. Let $S := \{v \in W : (v, v) = \frac{1}{K}\}$ have the metric induced from the inclusion $S \subset W$. Then:*

(1) *S has signature $(r, s-1)$ if $K > 0$ and signature $(r-1, s)$ if $K < 0$.*

(2) *S has constant sectional curvature K.*

(3) *S is spacelike, timelike, and mixed rank 2 Jordan IP.*

(4) *S is both a 2 point homogeneous space and a symmetric space.*

Proof. Let ε be the sign of K and let $P \in S$. We set $\nu = |K|^{\frac{1}{2}} \cdot P$ and write S as a graph over $P^{\perp}$ by defining:

$$F(y) := y \oplus f(y) \cdot \nu \text{ where } f(y) := \sqrt{\varepsilon\{\tfrac{1}{K} - (y, y)\}}.$$

Note that $F(0) = P$. Since $df(0) = 0$, the identification of $P^\perp$ with T_PS is an isometry. Thus, the metric on S has the signature given in Assertion (1). Furthermore, since ν is a unit normal at P, the second fundamental form and shape operators at P are given by:

$$L_P(v_1, v_2) = \varepsilon(\partial_i\partial_j f)(0) = -\varepsilon\sqrt{|K|}\cdot(v_1, v_2), \text{ and}$$
$$S_P = -\varepsilon\sqrt{|K|}\cdot \mathrm{Id}.$$

We may therefore use Lemma 1.12.3 to see that:

$$^gR_P = \varepsilon(-\varepsilon\sqrt{|K|})^2 R_{\mathrm{Id}} = KR_{\mathrm{Id}}.$$

Thus, S has constant sectional curvature K; Assertion (2) now follows. We use Theorem 2.1.6 and Assertion (2) to derive Assertion (3).

Let $O(W)$ be the Lie group of inner product preserving linear transformations of W. This group preserves the structures in question and defines a transitive isometric action on S. Thus, S is a homogeneous space. The isotropy subgroup of all elements of $O(W)$ which fixes a point $P \in S$ may be identified with $O(T_PS)$. This group acts transitively on the pseudo-spheres $S^\pm(T_PS)$. Since S is homogeneous, we may conclude S is a 2 point homogeneous space. We define the map $\psi \in O(W)$ by requiring that:

$$\psi(P) = P \text{ and } \psi|_{P^\perp} = -\mathrm{Id}.$$

This map is an isometry of S which acts as -1 on T_PS; thus S is a symmetric space. □

In Lemma 1.14.2, we showed that any two manifolds of constant sectional curvature K and signature (p, q) are locally isometric; thus the pseudo-spheres are local models for these spaces.

We can now construct warped product manifolds which are spacelike, mixed, and timelike rank 2 Jordan IP manifolds.

2.6.2 Definition. Let $S = \{v \in W : (v, v) = \frac{1}{K}\}$ be the pseudo-sphere discussed in Lemma 2.6.1; if $K = 0$, then let S be flat space. Let $\varepsilon := \pm 1$ and let

$$f(t) := \varepsilon K t^2 + At + B \text{ where } A^2 - 4\varepsilon KB \neq 0.$$

Choose a connected open interval $I \subset \mathbb{R}$ so that $f(t) \neq 0$ on I. Let

$$M := I \times S \text{ and } ds_M^2 := \varepsilon dt^2 + f(t)ds_S^2.$$

Suppose S has signature (r, s). The signature of M is (p, q), where:

$$(p,q) = \begin{cases} (r+1, s) & \text{if } f(t) > 0 \text{ and } \varepsilon < 0, \\ (r, s+1) & \text{if } f(t) > 0 \text{ and } \varepsilon > 0, \\ (s+1, r) & \text{if } f(t) < 0 \text{ and } \varepsilon < 0, \\ (s, r+1) & \text{if } f(t) < 0 \text{ and } \varepsilon > 0. \end{cases}$$

2.6.3 Lemma. *Let M be the pseudo-Riemannian manifold of* Definition 2.6.2. *Then M is a spacelike, mixed, and timelike rank* 2 *Jordan IP pseudo-Riemannian manifold which does not have constant sectional curvature.*

Proof. Let $m := \dim M$. Fix $P \in S$. Choose local coordinates

$$x = (x_1, ..., x_{m-1})$$

for S which are centered at P. We let $x_0 = t \in I$. We also define local coordinates $(x_0, ..., x_{m-1})$ on M. We let indices a, b, c, and d range from 1 to $m-1$ and index the local coordinate frame $\{\partial_1, ..., \partial_{m-1}\}$ for S. We let indices i, j, k, and ℓ range from 0 to $m-1$ and index the full local coordinate frame $\{\partial_0, ..., \partial_{m-1}\}$ for M.

Let g_{ij} denote the components of the metric tensor on M and let $\tilde{g}_{ab}$ denote the components of the metric tensor on S relative to these local coordinate frames. We normalize the coordinates on S so that $\partial_a \tilde{g}_{bc}(P) = 0$. We have

$$g_{ab} = f\tilde{g}_{ab},\ g_{0a} = 0, \text{ and } g_{00} = \varepsilon.$$

Let $f_t := \partial_t f$ and $f_{tt} = \partial_t^2 f$. Let Γ and $\tilde{\Gamma}$ be the Christoffel symbols of the Levi-Civita connections defined by the metrics on M and on S, respectively. The Christoffel symbols where 0 appears twice or three times vanish; the non-zero components are given by:

$$\begin{aligned}
&\Gamma_{ijk} = \tfrac{1}{2}\{\partial_j g_{ik} + \partial_i g_{jk} - \partial_k g_{ij}\}, && \tilde{\Gamma}_{abc} = \tfrac{1}{2}\{\partial_b \tilde{g}_{ac} + \partial_a \tilde{g}_{bc} - \partial_c \tilde{g}_{ab}\},\\
&\Gamma_{0ab} = \Gamma_{a0b} = \tfrac{1}{2} f_t \tilde{g}_{ab}, && \Gamma_{ab0} = -\tfrac{1}{2} f_t \tilde{g}_{ab},\quad \Gamma_{abc} = f\tilde{\Gamma}_{abc},\\
&\Gamma_{0a}{}^{b} = \Gamma_{a0}{}^{b} = \tfrac{1}{2} f^{-1} f_t \tilde{g}_a{}^{b}, && \Gamma_{ab}{}^{0} = -\tfrac{1}{2}\varepsilon f_t \tilde{g}_{ab},\ \Gamma_{ab}{}^{c} = \tilde{\Gamma}_{ab}{}^{c}.
\end{aligned}$$

Therefore, the components of the curvature tensor R on M are:

$$\begin{aligned}
R_{ijk}{}^{l} &= \partial_i \Gamma_{jk}{}^{l} - \partial_j \Gamma_{ik}{}^{l} + \textstyle\sum_n \Gamma_{in}{}^{l}\Gamma_{jk}{}^{n} - \sum_n \Gamma_{jn}{}^{l}\Gamma_{ik}{}^{n},\\
R_{abc}{}^{d} &= \{\partial_a \tilde{\Gamma}_{bc}{}^{d} - \partial_b \tilde{\Gamma}_{ac}{}^{d}\} + \Gamma_{a0}{}^{d}\Gamma_{bc}{}^{0} - \Gamma_{b0}{}^{d}\Gamma_{ac}{}^{0} + O(|x|)\\
&= \tilde{R}_{abc}{}^{d} - \tfrac{1}{4}\varepsilon f_t^2 f^{-1}\{\tilde{g}_a{}^{d}\tilde{g}_{bc} - \tilde{g}_{ac}\tilde{g}_b{}^{d}\} + O(|x|),\\
R_{abcd} &= (fK - \tfrac{1}{4}\varepsilon f_t^2)\{\tilde{g}_{ad}\tilde{g}_{bc} - \tilde{g}_{ac}\tilde{g}_{bd}\} + O(|x|)\\
&= f^{-2}(fK - \tfrac{1}{4}\varepsilon f_t^2)\ \{g_{ad}g_{bc} - g_{ac}g_{bd}\} + O(|x|).
\end{aligned}$$

We assumed that $A^2 - 4\varepsilon KB \neq 0$. Thus, the metric is not flat since

$$\begin{aligned}
&fK - \tfrac{1}{4}\varepsilon f_t^2\\
=&(\varepsilon K t^2 + At + B)K - \tfrac{\varepsilon}{4}(2\varepsilon K t + A)^2\\
=&KB - \tfrac{\varepsilon}{4}A^2 \neq 0.
\end{aligned}$$

Furthermore, the sectional curvatures of 2 planes tangent to S are given by:

$$\frac{fK - \frac{\varepsilon}{4}f_t^2}{f^2} = \frac{4KB - \varepsilon A^2}{4f^2}.$$

These exhibit non trivial dependence on the parameter t so M does not have constant sectional curvature.

By Lemma 2.6.1, the pseudo-sphere S is a symmetric space. Thus, there is an isometry of S fixing P which acts as multiplication by -1 on T_PS. We extend this isometry to an isometry of M which is trivial on I to see that

$$R_{0abc} = 0 \text{ for any } a, b, c.$$

Let $a \neq b$. The isotropy group of isometries of S fixing P acts on T_PM is $O(T_PS)$. Thus, we can find an isometry of S which sends $e_a \to e_a$ and $e_b \to -e_b$. This implies

$$R_{0ab0} = 0 \text{ if } a \neq b.$$

We compute the remaining curvature:

$$\begin{aligned} R_{0aa}{}^0 &= \partial_0 \Gamma_{aa}{}^0 - \Gamma_{ab}{}^0 \Gamma_{0a}{}^b + O(|x|) \\ &= -\tfrac{1}{2}\varepsilon f_{tt}\tilde{g}_{aa} + \tfrac{1}{4}\varepsilon f^{-1} f_t^2 \tilde{g}_{ab}\tilde{g}_a{}^b + O(|x|), \text{ and} \\ R_{0aa0} &= f^{-2}\varepsilon\{-\tfrac{1}{2}ff_{tt} + \tfrac{1}{4}f_t^2\}\{g_{aa}g_{00} - g_{0a}g_{a0}\} + O(|x|). \end{aligned}$$

Define:

$$\phi(\partial_t) = \partial_t \text{ and } \phi(\partial_a) = -\partial_a.$$

Since $f(t) = \varepsilon K t^2 + At + B$, $K = \frac{1}{2}\varepsilon f_{tt}$ and hence

$$(fK - \tfrac{1}{4}\varepsilon f_t^2) = -\varepsilon\{-\tfrac{1}{2}ff_{tt} + \tfrac{1}{4}f_t^2\}.$$

Thus, $R = CR_\phi$ and the desired conclusion follows from Theorem 2.1.6. □

We use the methods developed in the proof of Theorem 2.4.1 to construct a family of nilpotent rank 2 Jordan IP pseudo-Riemannian manifold such that if $\{x, y\}$ spans a spacelike or timelike 2 plane, then

$${}^gR(x,y)^2 = 0 \text{ and } \operatorname{rank}({}^gR(x,y)) = 2.$$

2.6.4 Definition. Let $\{e_i, \tilde{e}_i\}_{1\leq i\leq p}$ be a basis for a vector space V. We give V the 'hyperbolic' inner product defined by the following relations:

$$(e_i, \tilde{e}_j) = \delta_{ij}, \ (e_i, e_j) = (\tilde{e}_i, \tilde{e}_j) = 0.$$

The inner product has signature (p, p). Let $\{x_i, \tilde{x}_i\}_{1\leq i\leq p}$ be the corresponding coordinates on V^*. Let $f(\tilde{x})$ be a real valued function of the coordinates $\{\tilde{x}_1, ..., \tilde{x}_p\}$ which is independent of the remaining coordinates. We let

$$\tilde{H}(f)_{ij} := \partial_i^{\tilde{x}} \partial_j^{\tilde{x}} f \text{ for } 1 \leq i, j \leq p$$

be the Hessian with respect to $\tilde{x}$. We suppose that

$$f(0) = 0, \ df(0) = 0, \text{ and } \det\{\tilde{H}(f)_{ij}\}(0) \neq 0.$$

We set $W := V \oplus \mathbb{R} \cdot \nu$; the metric on W is defined by requiring that this is an orthogonal direct sum and that $(\nu, \nu) = \pm 1$. We define an embedding $\mathrm{Id} \oplus f$ of V into W by setting

$$F(x, \tilde{x}) = \textstyle\sum_{1\leq i\leq p}\{x_i e_i + \tilde{x}_i \tilde{e}_i\} + f(\tilde{x})\nu.$$

Let $M_f := \mathrm{range}(F)$. Since $df(0) = 0$, the identification $T_0 M_f = V$ is an isometry. Therefore, M_f is the germ of a pseudo-Riemannian manifold of signature (p, p).

2.6.5 Lemma. *Let M_f be the germ of the pseudo-Riemannian manifold given in* Definition 2.6.4. *Then M_f is a timelike and spacelike rank 2 nilpotent Jordan IP manifold; M_f does not have constant mixed rank.*

Proof. Let L_P and S_P be the associated second fundamental form and shape operator at P. We apply Lemma 1.12.3 to see that

$${}^{g}R = \pm R_{S_P}.$$

Since $\partial_i^x F = 0$, $L_P(\partial_i^x, *) = 0$ for any point of the manifold. Consequently, $S_P(\partial_i^x) = 0$. This implies that:

$$\mathrm{span}\{\partial_i^x\} \subset \ker(S_P).$$

Since $\det \tilde{H}(f)(0) \neq 0$, S_0 is invertible on $\mathrm{span}\{\partial_i^{\tilde{x}}\}$. Thus, the kernel of S_0 has dimension p. By shrinking the neighborhood $\mathcal{O}$, if necessary, we may suppose $\dim\ker(S_P) \leq p$ for $P \in \mathcal{O}$. It now follows that

$$\ker(S_P) = \mathrm{span}\{\partial_i^x\} \text{ for all } P \in \mathcal{O}.$$

Since $F_*(\partial_i^x) = e_i$, we have $g(\partial_i^x, \partial_j^x) = 0$ for all i, j. This shows that the kernel of S_P is totally isotropic. Thus, in particular $\ker(S_P)$ contains no spacelike or timelike vectors. We have

$$\dim \ker(S_P)^\perp = \dim V - \dim \ker(S_P) = \dim \ker(S_P).$$

Since $\ker(S_P)$ is totally isotropic, $\ker(S_P) \subset \ker(S_P)^\perp$. This implies that

$$\ker(S_P) = \ker(S_P)^\perp.$$

As S_P is self-adjoint, $\operatorname{range}(S_P) \perp \ker(S_P)$. Thus, $\operatorname{range}(S_P) \subset \ker(S_P)$. As

$$\dim \operatorname{range}(S_P) = \dim V - \dim \ker(S_P) = \dim \ker(S_P),$$

we have $\operatorname{range}(S_P) = \ker(S_P)$ so $S_P^2 = 0$. The desired conclusion now follows from Corollary 2.5.2. □

2.7 Classification of IP manifolds

In this section, we complete the proof of Theorem 2.1.7. Throughout this section, let (M, g) be a connected pseudo-Riemannian manifold of signature (p, q) with $q \geq 5$ and dimension $m := p + q$. We assume that M is rank 2 spacelike Jordan IP. We use Theorem 2.1.6 to express

$$^gR_P = C(P)R_{\phi(P)} \text{ for } \phi = \phi^* \text{ and } C \neq 0.$$

We suppose for the moment that ^gR is never nilpotent so that we can normalize ϕ so $\phi^2 = \pm \operatorname{Id}$. We use Lemma 2.4.2 to see that the maps $P \to C(P)$ and $P \to \phi(P)$ can be chosen to be smooth, at least locally. To have a unified notation, we complexify the tangent bundle and extend the map ϕ, the innerproduct $(\cdot, \cdot)$, and the curvature tensor ^gR to be complex multi-linear. Let:

$$\tilde{\phi} := \begin{cases} \phi & \text{if } \phi^2 = \operatorname{Id}, \\ \sqrt{-1}\phi & \text{if } \phi^2 = -\operatorname{Id}. \end{cases}$$

Since $\tilde{\phi}^2 = \operatorname{Id}$, the eigenvalues of $\tilde{\phi}$ are ± 1 and $\tilde{\phi}$ is diagonalizable. Let $\mathcal{F}^\pm$ be the associated eigendistributions and $\nu^\pm$ the associated dimensions:

$$\mathcal{F}^\pm := \{X \in TM \otimes \mathbb{C} : \tilde{\phi}X = \pm X\} \text{ and } \nu^\pm := \dim \mathcal{F}^\pm.$$

If $\nu^+ = 0$ or $\nu^- = 0$, then $\tilde{\phi} = \pm \operatorname{Id}$. In this setting, ϕ will be a multiple of the identity. As ϕ is real, we can not have $\phi^2 = -\operatorname{Id}$ and thus we have $\phi^2 = \operatorname{Id}$. For $\nu^+ = 0$ and $\nu^- = 0$, this then implies that (M, g) has constant sectional curvature. Consequently, we shall assume, at least implicitly, that $\nu^\pm > 0$ for much of our development.

Let $u^\pm \in \mathcal{F}^\pm$. Since

$$(u^+, u^-) = (\tilde\phi u^+, u^-) = (u^+, \tilde\phi u^-) = -(u^+, u^-),$$

$\mathcal{F}^-$ and $\mathcal{F}^+$ are non-degenerate orthogonal complex distributions. Since we are working over $\mathbb{C}$, we can find a local orthonormal frame $\mathcal{B} := \{e_1, ..., e_m\}$ for $TM \otimes \mathbb{C}$ so

$$\begin{aligned}\mathcal{F}^- &= \operatorname{span}\{e_1, ..., e_{\nu^-}\},\\ \mathcal{F}^+ &= \operatorname{span}\{e_{\nu^-+1}, ..., e_m\}, \text{ and}\\ (e_i, e_j) &= \delta_{ij}.\end{aligned}$$

Let Roman indices a, b, etc. range from 1 to ν^- and index the frame for $\mathcal{F}^-$, let Greek indices α, β, etc. range from $\nu^- + 1$ to m and index the frame for $\mathcal{F}^+$, and let Roman indices i, j, etc. range from 1 to m and index the full frame for $TM \otimes \mathbb{C}$. Let $\tilde\phi_{ij;k}$ be the components of the covariant derivative $\nabla\tilde\phi$ relative to such a normalized basis. We begin our analysis by studying this tensor:

Lemma 2.7.1. *Let (M, g) be a connected pseudo-Riemannian manifold of signature (p, q), where $q \geq 5$. Assume that we have ${}^gR_P = C(P)R_{\phi(P)}$, where $\phi^2 = \pm\,\mathrm{Id}$, where $\phi = \phi^*$, and where $C \neq 0$.*

1. *We have $\tilde\phi_{ij;k} = \tilde\phi_{ji;k}$, $\tilde\phi_{ab;k} = 0$, $\tilde\phi_{\alpha\beta;k} = 0$, and $\tilde\phi_{a\alpha;i} = -2\Gamma_{ia\alpha}$.*
2. *If i, j, and k are distinct indices, then $\tilde\phi_{ij;k} = \tilde\phi_{ik;j}$.*
3. *The only possibly non-zero components of $\nabla\tilde\phi$ are given by $\tilde\phi_{a\alpha;a} = \tilde\phi_{\alpha a;a}$ and by $\tilde\phi_{a\alpha;\alpha} = \tilde\phi_{\alpha a;\alpha}$.*

Proof. We adapt arguments of Gilkey, Leahy, and Sadofsky [83] and Zhang [163] for the proof. As the frame is orthonormal, $\Gamma_{kij} = -\Gamma_{kji}$. Let

$$\varepsilon_i := (\phi e_i, e_i) = \pm 1;\ \varepsilon_a = -1 \text{ and } \varepsilon_\alpha = +1.$$

We prove Assertion (1) by computing:

$$\begin{aligned}\tilde\phi_{ij;k} =& (\nabla_{e_k}\tilde\phi e_i - \tilde\phi\nabla_{e_k}e_i, e_j) = (\nabla_{e_k}\tilde\phi e_i, e_j) - (\nabla_{e_k}e_i, \tilde\phi e_j)\\ =& (\varepsilon_i - \varepsilon_j)(\nabla_{e_k}e_i, e_j) = (\varepsilon_i - \varepsilon_j)\Gamma_{kij}.\end{aligned}$$

Let ${}^gR_{ijkl;n}$ be the components of the covariant derivative $\nabla{}^gR$. Let $\varepsilon = \pm 1$ be such that $\phi^2 = \varepsilon\,\mathrm{Id}$. We may then express:

$$\begin{aligned}&{}^gR_{ijkl} = \varepsilon C\{\tilde\phi_{il}\tilde\phi_{jk} - \tilde\phi_{ik}\tilde\phi_{jl}\} \text{ and,}\\ (2.7.1.a)\qquad &{}^gR_{ijkl;n} = \varepsilon C(\tilde\phi_{il;n}\tilde\phi_{jk} + \tilde\phi_{il}\tilde\phi_{jk;n} - \tilde\phi_{ik;n}\tilde\phi_{jl} - \tilde\phi_{ik}\tilde\phi_{jl;n})\\ &\qquad + \varepsilon C_{;n}(\tilde\phi_{il}\tilde\phi_{jk} - \tilde\phi_{ik}\tilde\phi_{jl}).\end{aligned}$$

We now use the *second Bianchi identity* established in Theorem 1.12.1 (2c):

$$0 = {}^gR_{ijkl;n} + {}^gR_{ijln;k} + {}^gR_{ijnk;l}.$$

Fix distinct indices i, j, and k. As $q \geq 5$, we may choose an index l which is distinct from i, j, and k. We have $\tilde{\phi}_{ij} = 0$ for $i \neq j$. Since $C \neq 0$, we may prove Assertion (2) by using Equation (2.7.1.a) to compute:

$$\begin{aligned} 0 =& {}^gR_{illj;k} + {}^gR_{iljk;l} + {}^gR_{ilkl;j} \\ =& \varepsilon C \varepsilon_\ell \tilde{\phi}_{ij;k} + 0 - \varepsilon C \varepsilon_\ell \tilde{\phi}_{ik;j}. \end{aligned}$$

If i, j, and k are distinct indices, then

$$\tilde{\phi}_{ij;k} = \tilde{\phi}_{ik;j} = \tilde{\phi}_{jk;i}.$$

Since at least two of the indices must refer to elements from the same distribution, $\tilde{\phi}_{ij;k} = 0$ by Assertion (1). Thus, if $\tilde{\phi}_{ij;k} \neq 0$, then we must have that (i, j, k) is a permutation of (a, a, α) or (a, α, α). Assertion (3) follows as

$$\tilde{\phi}_{aa;\alpha} = \tilde{\phi}_{\alpha\alpha;a} = 0 \text{ and } \tilde{\phi}_{a\alpha;i} = \tilde{\phi}_{\alpha a;i}. \quad \square$$

We continue our study of $\nabla\tilde{\phi}$.

2.7.2 Lemma. *Let (M, g) be a connected pseudo-Riemannian manifold of signature (p, q), where $q \geq 5$. Assume that we have ${}^gR_P = C(P)R_{\phi(P)}$, where $\phi^2 = \pm\,\mathrm{Id}$, where $\phi = \phi^*$, and where $C \neq 0$.*

(1) *If $\nu^- \geq 2$, then $C_{;a} = C\tilde{\phi}_{\alpha a;\alpha}$.*

(2) *If $\nu^- \geq 3$, then $C_{;a} = 0$ and $C_{;\alpha} = -2C\tilde{\phi}_{a\alpha;a}$.*

(3) *If $\nu^+ \geq 2$, then $C_{;\alpha} = -C\tilde{\phi}_{a\alpha;a}$.*

(4) *If $\nu^+ \geq 3$, then $C_{;\alpha} = 0$ and $C_{;a} = +2C\tilde{\phi}_{a\alpha;\alpha}$.*

Proof. We apply Equation (2.7.1.a). If $\nu^- \geq 2$, then we may choose $a \neq b$ and use Lemma 2.7.1 and the second Bianchi identity to prove Assertion (1) by computing:

$$\begin{aligned} 0 =& {}^gR_{a\alpha\alpha a;b} + {}^gR_{a\alpha ab;\alpha} + {}^gR_{a\alpha b\alpha;a} \\ =& -\varepsilon C_{;b} + \varepsilon C\tilde{\phi}_{\alpha b;\alpha} + 0. \end{aligned}$$

If $\nu^- \geq 3$, then we may choose distinct indices a, b, and c. We apply Lemma 2.7.1 and the second Bianchi identity to see that:

$$\begin{aligned} 0 =& {}^gR_{cbbc;a} + {}^gR_{cbca;b} + {}^gR_{cbab;c} = \varepsilon C_{;a} + 0 + 0 \text{ and} \\ 0 =& {}^gR_{cbbc;\alpha} + {}^gR_{cbc\alpha;b} + {}^gR_{cb\alpha b;c} = \varepsilon C_{;\alpha} + \varepsilon C\tilde{\phi}_{b\alpha;b} + \varepsilon C\tilde{\phi}_{c\alpha;c}. \end{aligned}$$

Thus, $C_{;a} = 0$ and $C_{;\alpha} = -C(\tilde{\phi}_{b\alpha;b} + \tilde{\phi}_{c\alpha;c})$. Similarly, it can be shown that:

$$C_{;\alpha} = -C(\tilde{\phi}_{a\alpha;a} + \tilde{\phi}_{c\alpha;c}).$$

Thus, $\tilde{\phi}_{b\alpha;b} = \tilde{\phi}_{a\alpha;a} = \tilde{\phi}_{c\alpha;c}$ and Assertion (2) follows. We replace ϕ and $\tilde{\phi}$ by $-\phi$ and $-\tilde{\phi}$ to interchange the roles of $\mathcal{F}^{\pm}$ and thereby derive Assertions (3) and (4) from Assertions (1) and (2) with an appropriate change of signs. □

We can now show that $\phi^2 = \mathrm{Id}$ if $C \neq 0$.

Lemma 2.7.3. *Let (M, g) be a connected pseudo-Riemannian manifold of signature (p, q), where $q \geq 5$. Assume that we have ${}^gR_P = C(P)R_{\phi(P)}$, where $\phi^2 = \pm\,\mathrm{Id}$, where $\phi = \phi^*$, and where $C \neq 0$.*

(1) *We do not have $\nabla\tilde{\phi} = 0$. Furthermore, either $\nu^- \leq 2$ or $\nu^+ \leq 2$.*
(2) *We have $\phi^2 = \mathrm{Id}$. Furthermore either $\nu^- \leq 1$ or $\nu^+ \leq 1$.*

Proof. Assume that $\nabla\tilde{\phi} = 0$. We use Lemma 2.7.1 to see that

$$\Gamma_{ia\alpha} = -\tfrac{1}{2}\tilde{\phi}_{a\alpha;i} = 0.$$

Thus, $\nabla e_a \in \mathcal{F}^-$ so ${}^gR(e_a, e_\alpha)e_a \in \mathcal{F}^-$. This shows:

$$0 = {}^gR(e_a, e_\alpha, e_a, e_\alpha) = \varepsilon C.$$

Consequently, $C = 0$ which is false. Thus, $\nabla\tilde{\phi} \neq 0$.

Next, suppose that $\nu^- \geq 3$ and $\nu^+ \geq 3$. We may then apply Lemma 2.7.2 to see that $dC = 0$ and that $\nabla\tilde{\phi} = 0$ which is false. This completes the proof of Assertion (1).

Suppose that $\phi^2 = -\,\mathrm{Id}$. Then $(\phi v, \phi v) = (\phi^2 v, v) = -(v, v)$ so ϕ interchanges the roles of spacelike and timelike vectors. Thus, $p = q \geq 5$. Since $\tilde{\phi} = \sqrt{-1}\phi$, $\tilde{\phi}$ is purely imaginary. Thus, conjugation interchanges the distributions $\mathcal{F}^+$ and $\mathcal{F}^-$ so $\nu^+ = \nu^- = p \geq 5$. This contradicts Assertion (1). Consequently, $\phi^2 = \mathrm{Id}$.

By replacing ϕ by $-\phi$, we may assume $\nu^- \leq \nu^+$. To establish Assertion (2), we must rule out the case $\nu^- = 2$. Suppose, to the contrary, that $2 \leq \nu^- \leq \nu^+$. We use Lemma 2.7.2. Since $\nu^- \geq 2$ and $\nu^+ \geq 3$, we have $C_{;\alpha} = 0$, $C_{;a} = 2C\tilde{\phi}_{a\alpha;\alpha}$ and $C_{;a} = \tilde{\phi}_{a\alpha;\alpha}$. Thus, $dC = 0$. Thus, again $\nabla\tilde{\phi} = 0$ which contradicts Assertion (1). □

Assertion (1) of Theorem 2.1.7 follows from Lemma 2.7.3 and from:

Lemma 2.7.4. *Let (M, g) be a connected pseudo-Riemannian manifold of signature (p, q) for $q \geq 5$. Assume that (M, g) is rank 2 spacelike Jordan IP. Then either ${}^gR$ is nilpotent for all points of M or ${}^gR$ is nilpotent at no point of M.*

Proof. Let S_1 be the set of all points of M, where ${}^gR$ is nilpotent and let S_2 be the complementary subset of all points of M, where ${}^gR$ is not nilpotent. We assume that both S_1 and S_2 are non-empty and argue for a contradiction. Since S_1 is closed, since S_2 is open, and since M is connected, we may conclude that S_2 is not closed. Thus, we may choose points $P_i \in S_2$ so that $P_i \to P_\infty \in S_1$. Let R_n be the curvature tensor at P_n and let V_n be the tangent space of M at P_n.

Let $\mathcal{B} := \{f_1^-, ..., f_p^-, f_1^+, ..., f_q^+\}$ be a normalized local orthonormal frame for the real tangent bundle near P_∞ and let $V^\pm$ be the associated maximal spacelike and timelike distributions. We set $V_n^\pm := V^\pm(P_n)$. Since $P_n \in S_2$, we can express $R_n = C_n R_{\phi_n}$. Since $\mathrm{Tr}\{R_n(\pi)^2\} = -C_n^2$ for any spacelike 2 plane π, we have $C_n \to 0$ as R_∞ is nilpotent.

We apply Lemma 2.7.3 to see $\phi_n^2 = \mathrm{Id}$ and thus we do not need to complexify to define the sub-bundles $\mathcal{F}_n^\pm$ of V_n. By replacing ϕ_n by $-\phi_n$ if necessary, we may apply Lemma 2.7.3 to see $\dim \mathcal{F}_n^- \leq 1$. Thus,

$$\dim \mathcal{F}_n^+ \geq p + q - 1 \text{ so } \dim \mathcal{F}_n^+ \cap V_n^+ \geq q - 1 \geq 4.$$

Choose elements $v_n, w_n \in V_n^+ \cap \mathcal{F}_n^+$ so that $\{v_n, w_n\}$ forms an orthonormal spacelike set. We choose a compact neighborhood of P_∞ over which $S(V^+)$ is compact. By passing to a subsequence, we can suppose that $v_n \to v_\infty$ and $w_n \to w_\infty$. Let $\pi_n := \mathrm{span}\{v_n, w_n)$. Then

$$\pi_n \to \pi_\infty := \mathrm{span}\{v_\infty, w_\infty\}.$$

The planes π_n and π_∞ are spacelike. Let $z_\infty \in V_\infty$. Choose elements $z_n \in V_n$ so $z_n \to z_\infty$. Since $\phi_n v_n = v_n$ and $\phi_n w_n = w_n$, we may compute:

$$\begin{aligned} R(\pi_\infty) z_\infty &= \lim_{n\to\infty} R_n(v_n, w_n) z_n \\ &= \lim_{n\to\infty} C_n\{(\phi w_n, z_n)\phi v_n - (\phi v_n, z_n)\phi w_n\} \\ &= \lim_{n\to\infty} C_n\{(w_n, z_n) v_n - (v_n, z_n) w_n\} \\ &= \{\lim_{n\to\infty} C_n\} \cdot \{(w_\infty, z_\infty) v_\infty - (v_\infty, z_\infty) w_\infty\} = 0. \end{aligned}$$

This contradicts the assumption that R has spacelike rank 2. □

Since $\phi^2 = \mathrm{Id}$, the distributions $\mathcal{F}^\pm$ are real. By replacing ϕ by $-\phi$ if necessary, we may suppose that $\nu^- \leq \nu^+$. Thus, $\nu^- = 0$ or $\nu^- = 1$.

If $\nu^- = 0$, then $\phi = \mathrm{Id}$ and (M, g) has constant sectional curvature; this proves Assertion (2) of Theorem 2.1.7.

We therefore suppose $\nu^- = 1$. The distributions $\mathcal{F}^\pm$ are non-degenerate since $(\mathcal{F}^\pm)^\perp = \mathcal{F}^\mp$ and $\mathcal{F}^+ \cap \mathcal{F}^- = \{0\}$. Let $\{e_1\}$ be a local orthonormal section to $\mathcal{F}^-$ and let $\{e_2, ..., e_m\}$ be a normalized local orthonormal frame for $\mathcal{F}^+$; since we are working over $\mathbb{R}$, $(e_i, e_i) = \varepsilon_i = \pm 1$. By replacing g by $-g$, we may assume without loss of generality that e_1 is spacelike. If $\alpha \neq \beta$, then we may apply Lemma 2.7.1 to compute:

$$\begin{aligned}([e_\alpha, e_\beta], e_1) =& \Gamma_{\alpha\beta 1} - \Gamma_{\beta\alpha 1} = \Gamma_{\beta 1\alpha} - \Gamma_{\alpha 1\beta} \\ =& -\tfrac{1}{2}(\tilde{\phi}_{1\alpha;\beta} - \tilde{\phi}_{1\beta;\alpha}) = 0.\end{aligned}$$

Thus, the distribution $\mathcal{F}^+$ is integrable. Let $y = (y_1, ..., y_{m-1})$ be local coordinates on a leaf of this foliation. We define geodesic tubular coordinates on M by setting:

$$T(t, y) := \exp_y(te_1(y)). \tag{2.7.4.a}$$

Assertion (3) of Theorem 2.1.7 will follow from:

Lemma 2.7.5. *Let (M, g) be a connected pseudo-Riemannian manifold of signature (p, q) for $q \geq 5$. Assume that ${}^gR = CR_\phi$, where $\phi^2 = \mathrm{Id}$ and $C \neq 0$. Assume that $\nu^- = 1$ and that $\mathcal{F}^-$ is spacelike. Let T be defined by equation (2.7.4.a). Then:*

(1) *For fixed y_0, the curves $t \to T(t, y_0)$ are unit speed geodesics in (M, g) which are perpendicular to the leaves of the foliation $\mathcal{F}^+$.*
(2) *For fixed t_0, the surfaces $T(t_0, y)$ are leaves of the foliation $\mathcal{F}^+$ and inherit metrics of constant sectional curvature.*
(3) *Locally $ds^2 = dt^2 + f ds_K^2$, where $f(t)$ is non-zero smooth function and ds_K^2 is a metric of constant sectional curvature K.*
(4) *The warping function is given by $f(t) = Kt^2 + At + B$, where we have $A^2 - 4KB \neq 0$.*

Proof. We choose a local orthonormal frame $\{e_i\}$ for TM so e_1 spans $\mathcal{F}^-$ and $\{e_2, ..., e_m\}$ spans $\mathcal{F}^+$. We set $\varepsilon_i := (e_i, e_i) = \pm 1$; by assumption $\varepsilon_1 = +1$. We have $\dim(\mathcal{F}^+) = m - 1 \geq 3$. Taking into account the fact that $(e_\alpha, e_\alpha) = \pm 1$, we may apply Lemmas 2.7.1 and 2.7.2 to see that

$$\begin{aligned}&C_{;\alpha} = 0, \\ &\Gamma_{\alpha 1\beta} = -\tfrac{1}{2}\tilde{\phi}_{1\beta;\alpha} = -(e_\alpha, e_\beta)\tfrac{C_{;1}}{C}, \text{ and} \\ &\Gamma_{11\alpha} = -\tfrac{1}{2}\tilde{\phi}_{1\alpha;1} = \tfrac{C_{;\alpha}}{C} = 0.\end{aligned}$$

Let $\gamma(t, y_0)$ be an integral curve for e_1 starting at a point y_0 on the leaf of the foliation. Since e_1 is a unit vector, $\Gamma_{111} = 0$. As $\Gamma_{11\alpha} = 0$, γ is a geodesic. Thus, $\gamma(t, y_0) = T(t, y_0)$ so $\partial_t = e_1$. We compute

$$\partial_t(\partial_t, \partial_\alpha^y) = (\partial_t, \nabla_{\partial_t}\partial_\alpha^y) = (\partial_t, \nabla_{\partial_\alpha^y}\partial_t) = \tfrac{1}{2}\partial_\alpha^y(\partial_t, \partial_t) = 0.$$

This shows $(\partial_t, \partial_\alpha^y) = 0$ so the ∂_α^y span the perpendicular distribution $\mathcal{F}^+$ and the manifolds $T(t_0, y)$ are leaves of the foliation $\mathcal{F}^+$. Since $C_{;\alpha} = 0$, C is constant on the leaves of $\mathcal{F}^+$. Let $\tilde{R}$ be the curvature of the induced metric on some leaf of $\mathcal{F}^+$. We show that $\tilde{R}$ has constant sectional curvature by computing:

$$\begin{aligned}
&\tilde{R}(e_\alpha, e_\beta, e_\gamma, e_\sigma)\\
=&R(e_\alpha, e_\beta, e_\gamma, e_\sigma) - \Gamma_{\alpha 1\sigma}\Gamma_{\beta\gamma}{}^1 + \Gamma_{\beta 1\sigma}\Gamma_{\alpha\gamma}{}^1\\
=&\{C + (\tfrac{C_{;1}}{C})^2\}[t] \cdot \{(e_\beta, e_\gamma)(e_\alpha, e_\sigma) - (e_\alpha, e_\gamma)(e_\beta, e_\sigma)\}.
\end{aligned}$$

Let $\partial_\alpha^y = \sum_\gamma a_{\alpha\gamma}e_\gamma$. We demonstrate that the metric is a warped product:

$$\begin{aligned}
(\nabla_{\partial_t}\partial_\alpha^y, \partial_\beta^y) =&(\nabla_{\partial_\alpha^y}\partial_t, \partial_\beta^y) = \textstyle\sum_{\gamma\sigma} a_{\alpha\gamma}a_{\beta\sigma}(\nabla_{e_\gamma}\partial_t, e_\sigma)\\
=&-\textstyle\sum_{\gamma\sigma} a_{\alpha\gamma}a_{\beta\sigma}(e_\gamma, e_\sigma)\frac{C_{;1}}{C} = -\frac{C_{;1}}{C}g_{\alpha\beta} \text{ so}\\
\partial_t g_{\alpha\beta} =&(\nabla_{\partial_t}\partial_\alpha^y, \partial_\beta^y) + (\partial_\alpha^y, \nabla_{\partial_t}\partial_\beta^y) = -\frac{2C_{;1}}{C}g_{\alpha\beta}.
\end{aligned}$$

We use the argument given to prove Lemma 2.6.3 to show the warping function is quadratic. □

The rank 2 spacelike Jordan IP pseudo-Riemannian manifolds with curvature operators which are not nilpotent for at least one point of M are classified in Theorem 2.1.7. We focus our attention on the balanced setting $p = q$ and present some preliminary results:

Lemma 2.7.6. *Let (M, g) be a connected pseudo-Riemannian manifold of signature (p, p) for $p \geq 5$. Assume that (M, g) is spacelike rank 2 nilpotent Jordan IP.*

(1) *We have ${}^gR = \pm R_\phi$, where ϕ is a self-adjoint endomorphism of the tangent bundle which satisfies $\ker\phi = \operatorname{range}\phi$.*

(2) *The range of ϕ is an integrable distribution.*

Proof. We apply Theorem 2.1.6 to write $R = \pm R_\phi$, where we normalize ϕ by requiring that $C = \pm 1$. The map $P \to \phi$ can then be chosen to vary smoothly with P, at least locally, by Lemma 2.4.2. Since $\ker\phi$ contains no spacelike vectors, $\dim\{\ker(\phi)\} \leq p$. Since ϕ is self-adjoint and $\phi^2 = 0$, we show $\text{range}(\phi) = \ker(\phi)$ and complete the proof of Assertion (1) by computing:

$$\begin{aligned}
&\text{range}(\phi) \subset \ker(\phi),\\
&\dim\{\text{range}(\phi)\} \leq \dim\{\ker(\phi)\}, \text{ and}\\
&2p = \dim\{\text{range}(\phi)\} + \dim\{\ker(\phi)\} \leq 2\dim\{\ker(\phi)\} \leq 2p.
\end{aligned}$$

Let $\mathcal{K} := \text{range}\,\phi$ and let $\mathcal{S}$ be a maximal local spacelike distribution. Since $\mathcal{K}$ is totally isotropic, $\mathcal{K} \cap \mathcal{S} = \{0\}$. Thus, $\mathcal{K} = \text{range}\,\phi = \phi\mathcal{S}$ and $TM = \mathcal{S} \oplus \phi\mathcal{S}$. Let $L(\cdot,\cdot) := (\phi\cdot,\cdot)$ be the associated bilinear form. Then:

$$(\phi s, \phi\tilde{s}) = (s, \phi^2\tilde{s}) = 0.$$

Thus, if $0 \neq s \in \mathcal{S}$, then there must exist $\tilde{s} \in \mathcal{S}$ so that $(\phi s, \tilde{s}) \neq 0$. This shows that L is a non-degenerate bilinear form on $\mathcal{S}$. We complexify and choose a frame $\{s_a\}$ for $\mathcal{S}_{\mathbb{C}} := \mathcal{S} \otimes \mathbb{C}$ so:

$$(\phi e_a, e_b) = \delta_{ab}.$$

Let Roman indices a, b, etc. range from 1 to p and index this frame for $\mathcal{S}_{\mathbb{C}}$. Let Greek indices α, β, etc. range from $p+1$ to $2p$ and index the frame $e_\alpha := \phi(e_{\alpha-p})$ for the complexification $\mathcal{K}_{\mathbb{C}} := \mathcal{K} \otimes \mathbb{C}$. Let Roman indices i, j, etc. range from 1 to $2p$ and index the frame $\{e_1, ..., e_p, \phi e_1, ..., \phi e_p\}$. We have $\phi_{ab} = \delta_{ab}$, $\phi_{a\beta} = 0$, and $\phi_{\alpha\beta} = 0$. Let α, β, and i be indices which need not be distinct. We choose $a \neq i$. We use Equation (2.7.1.a) and the second Bianchi identity to compute:

$$\begin{aligned}
R_{ijkl;n} =&\phi_{il;n}\phi_{jk} + \phi_{il}\phi_{jk;n} - \phi_{ik;n}\phi_{jl} - \phi_{ik}\phi_{jl;n},\\
0 =&R_{a\alpha\beta a;i} + R_{a\alpha i\beta;a} + R_{a\alpha ai;\beta} = \phi_{\alpha\beta;i} + 0 - \phi_{\alpha i;\beta}, \text{ and}\\
\phi_{i\alpha;\beta} =&\phi_{\alpha i;\beta} = \phi_{\alpha\beta;i} = (\nabla_{e_i}\phi(\phi e_a), \phi e_b) - (\phi\nabla_{e_i}\phi e_a, \phi e_b) = 0.
\end{aligned}$$

We clear the previous notation. Let $\alpha = a + p$, $\beta = b + p$, and $\gamma = c + p$ be indices which are not necessarily distinct. We show that $\mathcal{K}_{\mathbb{C}}$, and hence also that $\mathcal{K}$, is an integrable distribution by computing:

$$\begin{aligned}
&([e_\alpha, e_\beta], e_\gamma) = (\nabla_{e_\alpha}\phi e_b, \phi e_c) - (\nabla_{e_\beta}\phi e_a, \phi e_c)\\
=&((\nabla_{e_\alpha}\phi - \phi\nabla_{e_\alpha})e_b, \phi e_c) - ((\nabla_{e_\beta}\phi - \phi\nabla_{e_\alpha})e_a, \phi e_c)\\
=&\phi_{b\gamma;\alpha} - \phi_{a\gamma;\beta} = 0 \text{ so } [e_\alpha, e_\beta] \in C^\infty({\mathcal{K}_{\mathbb{C}}}^\perp) = C^\infty(\mathcal{K}_{\mathbb{C}}). \quad \square
\end{aligned}$$

2.8 Four dimensional geometry

Ivanov and Petrova [97] exhibited an IP algebraic curvature tensor on $\mathbb{R}^4$ of rank 4 that is the only rank 4 IP algebraic curvature tensor on $\mathbb{R}^4$, up to orthogonal equivalence and rescaling. They also showed that this tensor was not geometrically realizable by an IP manifold. Subsequently, Zhang[163] constructed a similar tensor on $\mathbb{R}^{(2,2)}$. We used these tensors in the proof of Lemma 2.2.4.

Let $\{e_1^+, e_2^+, e_3^+, e_4^+\}$ be the standard normalized orthonormal basis for $\mathbb{R}^4$. Let R be the alternating tensor on $\mathbb{R}^4$ whose non-zero components are represented by:

$$\begin{array}{llll} R_{1212} = a_1, & R_{1234} = a_2, & R_{1313} = a_2, & R_{1324} = -a_1, \\ R_{1414} = a_2, & R_{1423} = a_1, & R_{2323} = a_2, & R_{2314} = a_1, \\ R_{2424} = a_2, & R_{2413} = -a_1, & R_{3434} = a_1, & R_{3412} = a_2. \end{array}$$

Similarly, let $\{e_1^-, e_2^-, e_3^+, e_4^+\}$ be the standard normalized orthonormal basis for $\mathbb{R}^{(2,2)}$. Let $\mathfrak{R}$ be the alternating 4 tensor on $\mathbb{R}^{(2,2)}$ whose non-zero components are represented by:

$$\begin{array}{llll} \mathfrak{R}_{1212} = a_1, & \mathfrak{R}_{1234} = -a_2, & \mathfrak{R}_{1313} = -a_2, & \mathfrak{R}_{1324} = a_1, \\ \mathfrak{R}_{1414} = -a_2, & \mathfrak{R}_{1423} = -a_1, & \mathfrak{R}_{2323} = -a_2, & \mathfrak{R}_{2314} = -a_1, \\ \mathfrak{R}_{2424} = -a_2, & \mathfrak{R}_{2413} = a_1, & \mathfrak{R}_{3434} = a_1, & \mathfrak{R}_{3412} = -a_2. \end{array}$$

2.8.1 Theorem. *Let $a_1, a_2 \in \mathbb{R}$ satisfy $a_2 + 2a_1 = 0$ and $a_1 \neq 0$. The tensors R and $\mathfrak{R}$ defined above are IP algebraic curvature tensors of rank 4 on $\mathbb{R}^4$ and $\mathbb{R}^{(2,2)}$, respectively.*

Proof. We follow the discussion in Gilkey and Semmelman [84] to study R. The curvature symmetry of Equation (1.5.1.e) is immediate since we assume R is alternating. We verify curvature symmetry (1.5.1.f):

$$\begin{aligned} R_{1234} &= a_2 = R_{3412}, \\ R_{1324} &= -a_1 = R_{2413}, \text{ and} \\ R_{1423} &= a_1 = R_{2314}. \end{aligned}$$

We use the condition $2a_1 + a_2 = 0$ to verify the Bianchi identity:

$$\begin{aligned} R_{1234} + R_{3124} + R_{2314} &= a_2 + a_1 + a_1, \\ R_{1243} + R_{4123} + R_{2413} &= -a_2 - a_1 - a_1, \\ R_{1342} + R_{3412} + R_{4132} &= a_1 + a_2 + a_1, \\ R_{2341} + R_{3421} + R_{4231} &= -a_1 - a_2 - a_1. \end{aligned}$$

It is more difficult to show that R is IP of rank 4. We shall give a different construction of this tensor using the algebra of the quaternions; the

IP property will be immediate from the construction and we shall complete the proof by checking the two definitions coincide.

Let $\{1, i, j, k\}$ be the usual basis for the quaternions, $\mathbb{H}$. We decompose an element x of $\mathbb{H}$ in the form:

$$x = x_0 + x_1 i + x_2 j + x_3 k.$$

The *real part* of x, the *complex conjugate* of x, and the inner product are then defined by:

$$\begin{aligned}
&\Re(x) := x_0 \text{ and } \bar{x} := x_0 - x_1 i - x_2 j - x_3 k \\
&(x, y) = \Re(x\bar{y}) = x_0 y_0 + x_1 y_1 + x_2 y_2 + x_3 y_3.
\end{aligned}$$

The *Clifford algebra*, discussed in Section 1.4, is the universal unital algebra generated by $\mathbb{R}^4$, subject to the *Clifford commutation relations*:

$$x * y + y * x = -2(x, y).$$

Let $(u, v) \in \mathbb{H}_+ \oplus \mathbb{H}_-$ and let $x \in \mathbb{R}^4 = \mathbb{H}$. We define a representation c of this algebra on the direct sum of two copies $\mathbb{H}_\pm$ of the quaternions by:

$$\begin{aligned}
&c(x)(u, v) := (xv, -\bar{x}u); \\
&c(x)^2(u, v) = (x(-\bar{x}u), -\bar{x}(xv)) = -|x|^2(u, v).
\end{aligned}$$

We polarize last identity to obtain the Clifford commutation relations and the norm relation:

$$\begin{aligned}
c(x)c(y) + c(y)c(x) &= -2(x, y)\,\mathrm{Id} \\
|c(x)(u, v)|^2 &= ((xv, -\bar{x}u), (xv, -\bar{x}u)) \\
&= \Re(\bar{x}u\bar{u}x + xv\bar{v}\bar{x}) = |x|^2(|u|^2 + |v|^2).
\end{aligned}$$

Thus, if $|x| = 1$, then $c(x)$ is an isometry of $\mathbb{H}_+ \oplus \mathbb{H}_-$. Since $c(x)^2 = -1$, $c(x)$ is skew-adjoint.

We define a *bilinear Clifford module structure* by setting:

$$c(x, y) := c(x)c(y) + (x, y)\,\mathrm{Id} = \tfrac{1}{2}\{c(x)c(y) - c(y)c(x)\}.$$

Since $c(x)$ and $c(y)$ interchange the chiral factors $\mathbb{H}_\pm$, their product preserves chirality. Thus, $c(x, y)$ restricts to define a map from $\mathbb{H}_+$ to itself. If $\{x, y\}$ is an orthonormal basis for an oriented 2 plane $\pi \subset \mathbb{R}^4$, then

$$c(x, y)^2 = c(x)c(y)c(x)c(y) = -c(x)c(x)c(y)c(y) = -\,\mathrm{Id}\,.$$

Since $c(x)$ and $c(y)$ are skew-adjoint isometries and since these endomorphisms anti-commute, $c(\pi) := c(x,y)$ is a skew-adjoint isometry of $\mathbb{H}_+$. Thus, $c : \mathbb{H} \otimes \mathbb{H} \to \mathfrak{so}(\mathbb{H}_+)$ is an alternating mixed tensor.

Let us fix the element $\xi = (0, i)$ of $0 \oplus \mathbb{H}_-$. We have:

$$\begin{aligned}(c(x)\xi, c(y)\xi) =&(xi, yi) = \Re(xi(-i)\bar{y}) = 0,\\ (c(x)\xi, c(x)\xi) =&(xi, xi) = \Re(xi(-i)\bar{x}) = 1, \text{ and}\\ (c(y)\xi, c(y)\xi) =&(yi, yi) = \Re(yi(-i)\bar{y}) = 1.\end{aligned}$$

Consequently, $\{c(x)\xi, c(y)\xi\}$ is an orthonormal set. We define a $c(\pi)$ invariant orthogonal splitting $\mathbb{H}_+ = E_1(\pi) \oplus E_2(\pi)$, corresponding operators, and a tensor $\mathcal{R}$ by setting:

$$\begin{aligned}&E_1(\pi) := \operatorname{span}\{c(x)\xi, c(y)\xi\}, \quad &&E_2(\pi) := E_1(\pi)^\perp,\\ &R_1(\pi) := c(\pi)|_{E_1(\pi)}, &&R_2(\pi) := c(\pi)|_{E_2(\pi)},\\ &\mathcal{R} := a_1R_1 + a_2R_2.\end{aligned}$$

If $\{x, y\}$ is an orthonormal set, then

$$\begin{aligned}R_1(x, y, z, w) =&(c(x)\xi, z)(c(x)c(y)c(x)\xi, w)\\ &+ (c(y)\xi, z)(c(x)c(y)c(y)\xi, w)\\ =&(c(x)\xi, z)(c(y)\xi, w) - (c(y)\xi, z)(c(x)\xi, w).\end{aligned}$$

Thus, R_1 can be regarded as an alternating 4 tensor. Furthermore, since

$$R_1(x, y, z, w) + R_2(x, y, z, w) = (c(x, y)z, w),$$

$R_1 + R_2$ can also be regarded as an alternating 4 tensor. This shows that $\mathcal{R}$ is an alternating 4 tensor. Furthermore, if $\pi := \operatorname{span}\{x, y\}$, then the planes $E_1(\pi)$ and $E_2(\pi)$ are invariant subspaces. We show that $\mathcal{R}$ is IP by noting that the eigenvalues of $R(\pi)$ on these planes are

$$\pm\sqrt{-1}|a_1| \text{ and } \pm\sqrt{-1}|a_2|.$$

We complete the proof by showing that $R = \mathcal{R}$. Let $e_1 = 1$, $e_2 = i$, $e_3 = j$, and $e_4 = k$ be the standard orthonormal basis for $\mathbb{H}$. Let

$$\begin{aligned}&E_{1,\mu\sigma} := \operatorname{span}\{e_\mu \cdot i, e_\sigma \cdot i\},\ E_{2,\mu\sigma} := E_{1,\mu\sigma}^\perp,\\ &R_{1,\mu\sigma} := R_1(e_\mu, e_\sigma), \qquad \text{and } R_{2,\mu\sigma} := R_2(e_\mu, e_\sigma).\end{aligned}$$

We then have:

$$\begin{array}{llll}
E_{1,12} = \text{span}\{i, 1\}, & R_{1;12}i = -1, & E_{2,12} = \text{span}\{j, k\}, & R_{2;12}j = k, \\
E_{1,13} = \text{span}\{i, k\}, & R_{1;13}i = -k, & E_{2,13} = \text{span}\{1, j\}, & R_{2;13}1 = j, \\
E_{1,14} = \text{span}\{i, j\}, & R_{1;14}i = \ \ j, & E_{2,14} = \text{span}\{1, k\}, & R_{2;14}1 = k, \\
E_{1,23} = \text{span}\{k, 1\}, & R_{1;23}1 = \ \ k, & E_{2,23} = \text{span}\{i, j\}, & R_{2;23}i = j, \\
E_{1,24} = \text{span}\{1, j\}, & R_{1;24}1 = -j, & E_{2,24} = \text{span}\{i, k\}, & R_{2;24}i = k, \\
E_{1,34} = \text{span}\{k, j\}, & R_{1;34}k = -j, & E_{2,34} = \text{span}\{1, i\}, & R_{2;34}1 = i.
\end{array}$$

The non-zero entries of $\mathcal{R}$, modulo the curvature symmetry of Equation (1.5.1.e), are given by:

$$\begin{array}{llll}
\mathcal{R}_{1212} = a_1, & \mathcal{R}_{1234} = \ \ a_2, & \mathcal{R}_{1313} = a_2, & \mathcal{R}_{1324} = -a_1, \\
\mathcal{R}_{1414} = a_2, & \mathcal{R}_{1423} = \ \ a_1, & \mathcal{R}_{2323} = a_2, & \mathcal{R}_{2314} = \ \ a_1, \\
\mathcal{R}_{2424} = a_2, & \mathcal{R}_{2413} = -a_1, & \mathcal{R}_{3434} = a_1, & \mathcal{R}_{3412} = \ \ a_2.
\end{array}$$

This shows that $\mathcal{R} = R$ and hence R is IP. It is also possible to express this tensor as a linear combination of the curvature tensor of the Fubini-Study metric on $\mathbb{CP}^2$ and the curvature tensor of constant sectional curvature; we refer to [84] for details.

We follow the discussion in Zhang [163] to express $\mathfrak{R}$ in terms of R. We complexify to extend R to $\otimes^4\mathbb{C}^4$. As in the proof of Lemma 1.10.5, we use *analytic continuation* to see that R is IP on the non-degenerate 2 planes. Let

$$\mathfrak{V} := \text{span}\{\sqrt{-1}e_1, \sqrt{-1}e_2, e_3, e_4\}.$$

The induced inner product on $\mathfrak{V}$ is real and has signature $(2, 2)$. Then $\mathfrak{R}$ is the restriction of R to $\mathfrak{V}$. It now follows that $\mathfrak{R}$ is an IP algebraic curvature tensor of rank 4. □

2.8.2 Remark. If we try to construct an IP algebraic curvature tensor of rank 4 on $\mathbb{R}^{(1,3)}$ by restricting R to $\text{span}\{\sqrt{-1}e_1, e_2, e_3, e_4\}$, then the resulting algebraic curvature tensor is not real. It is not known if there is an algebraic curvature tensor of rank 4 in signature $(1, 3)$.

2.9 Seven dimensional geometry

In this section, we shall, for the most part, ignore the first Bianchi identity, which is given in Equation (1.5.1.g), and focus instead on the first two identities given in Equations (1.5.1.e) and (1.5.1.f). We shall also restrict to the Riemannian setting.

Let R be an alternating 4 tensor; this means that R satisfies the symmetry of Equation (1.5.1.e). We may use Equation (1.6.1.a) to define a linear map $\Phi = \Phi_R : \Lambda^2(\mathbb{R}^m) \to \Lambda^2(\mathbb{R}^m)$ which is characterized by the identity:

$$(\Phi(x \wedge y), z \wedge w) = R(x, y, z, w).$$

Conversely, any linear map Φ from $\Lambda^2(\mathbb{R}^m)$ to $\Lambda^2(\mathbb{R}^m)$ determines an alternating 4 tensor $R = R_\Phi$; the tensor R satisfies Equation (1.5.1.f) if and only if the associated linear map Φ is self-adjoint.

Let R be an alternating 4 tensor. We use Lemma 1.9.1 to see that $R(\pi) := R(x, y)$ is independent of the particular oriented orthonormal basis $\{x, y\}$ chosen for a 2 plane π. We say that R is an *alternating IP* tensor if the eigenvalues of $R(\cdot)$ are constant on the Grassmannian $\mathrm{Gr}_2^+(\mathbb{R}^m)$ of oriented planes in $\mathbb{R}^m$.

We can use Theorem 4.1.4 to generalize Theorem 2.1.1 and see that an alternating IP tensor on $\mathbb{R}^m$ for $m \geq 5$ and $m \neq 7, 8$ has rank 2; Equations (1.5.1.f) and (1.5.1.g) play no role in the analysis. In this section, we will follow the discussion of Gilkey and Semmelman [84] and use the exceptional Lie group G_2 to construct alternating IP tensors of higher rank on $\mathbb{R}^7$. These tensors will satisfy Equations (1.5.1.e) and (1.5.1.f). However, they will not satisfy the first Bianchi identity which is given in Equation (1.5.1.g) and thus will not be algebraic curvature tensors.

The following construction of G_2 is due to Bryant and Salamon [36]. Let $\{e_1, ..., e_7\}$ be the standard basis for $\mathbb{R}^7$. Define $e_{ij} := e_i \wedge e_j$ and $e_{ijk} := e_i \wedge e_j \wedge e_k$. We set

$$\omega_1 := e_{45} - e_{67},\ \omega_2 := e_{46} + e_{57},\ \omega_3 := e_{47} - e_{56}, \text{ and}$$
$$\omega := e_{123} + e_1 \wedge \omega_1 + e_2 \wedge \omega_2 + e_3 \wedge \omega_3.$$

The exceptional Lie group G_2 is the subgroup of $SO(7)$ which preserves the 3 form ω:

$$G_2 := \{g \in \mathrm{SO}(7) : g^*\omega = \omega\}.$$

Let interior multiplication 'int' be the dual of exterior multiplication:

$$(\mathrm{int}(\xi)\theta_1, \theta_2) = (\theta_1, \xi \wedge \theta_2).$$

We define $\Lambda_7^+ := \mathrm{int}(\mathbb{R}^7)\omega = \mathrm{span}\{u_i\}$, where:

$$\begin{aligned}
u_1 &:= e_{23} + e_{45} - e_{67}, & u_2 &:= -e_{13} + e_{46} + e_{57},\\
u_3 &:= e_{12} + e_{47} - e_{56}, & u_4 &:= -e_{15} - e_{26} - e_{37},\\
u_5 &:= e_{14} - e_{27} + e_{36}, & u_6 &:= e_{17} + e_{24} - e_{35},\\
u_7 &:= -e_{16} + e_{25} + e_{34}.
\end{aligned}$$

We note that $(u_i, u_j) = 3\delta_{ij}$. Let Π be orthogonal projection on Λ_7^+;

$$\Pi\omega = \tfrac{1}{3} \textstyle\sum_{1 \leq i \leq 7} (\omega, u_i) u_i.$$

2.9.1 Theorem. *Let a and b be constants. Let $\Phi := 3a\Pi + b\,\mathrm{Id}$ and let $R := R_\Phi$; $R(x,y,z,w) = a\sum_{1\le i\le 7}(x\wedge y, u_i)(z\wedge w, u_i) + b(x\wedge y, z\wedge w)$.*

(1) *The tensor R satisfies equations* (1.5.1.e) *and* (1.5.1.f). *It is an alternating IP tensor of rank* 6 *for $a+b\neq 0$ and $a\neq 0$, of rank* 4 *for $a+b=0$ and $a\neq 0$, and of rank* 2 *for $b\neq 0$ and $a=0$.*

(2) *The tensor R does not satisfy equation* (1.5.1.g) *if* $\operatorname{rank} R > 2$.

Proof. It is immediate from the definition that R is alternating. Since Φ is self-adjoint, R satisfies Equation (1.5.1.f).

The difficult task is, as always, to show R is IP. We use Friedrich, Kath, Moroianu, and Semmelman [57] and Harvey and Lawson [94] to see that G_2 acts transitively on $Gr_2^+(\mathbb{R}^7)$. Since G_2 preserves ω, G_2 commutes with Φ so the rank and eigenvalues of R_Φ are constant on $\mathrm{Gr}_2^+(\mathbb{R}^7)$; thus R is IP. Since R is IP, to determine the rank, it suffices to compute $\operatorname{rank}\{R(\pi)\}$ for the 2 plane $\pi := \operatorname{span}\{e_1, e_2\}$. We have:

$$\begin{aligned}(R(\pi)e_i, e_j) =&(3a\Pi e_{12} + be_{12}, e_{ij})\\ =&a(e_{12}+e_{47}-e_{56}, e_{ij}) + b(e_{12}, e_{ij}).\end{aligned}$$

We complete the proof of Assertion (1) by using this identity to see:

$$\begin{array}{lll} R(\pi)e_1 = (a+b)e_2, & R(\pi)e_2 = -(a+b)e_1, & R(\pi)e_3 = 0,\\ R(\pi)e_4 = ae_7, & R(\pi)e_7 = -ae_4, & R(\pi)e_5 = -ae_6,\\ R(\pi)e_6 = ae_5. & & \end{array}$$

To show that R does not satisfy the Bianchi identity if $a\neq 0$, we compute:

$$\begin{aligned}0 &= R(e_1,e_2,e_4,e_7) + R(e_2,e_4,e_1,e_7) + R(e_4,e_1,e_2,e_7)\\ &= 3a\{(\Pi e_{12}, e_{47}) + (\Pi e_{24}, e_{17}) - (\Pi e_{14}, e_{27})\}\\ &= a(1+1+1). \quad \square\end{aligned}$$

2.9.2 Remark. It is tempting to try to construct an IP manifold of higher rank in dimension 7 using a homogeneous G_2 manifold. Such a manifold would necessarily be Osserman since G_2 acts transitively on the 1 planes. Hence, the manifold would have constant sectional curvature by Theorem 3.1.5 and would not be IP of rank greater than 2.

2.10 Eight dimensional geometry

In Section 2.8, we constructed a rank 4 Riemannian IP algebraic curvature tensor. In Section 2.9, we exhibited a rank 6 Riemannian mixed alternating IP tensor which was symmetric but which did not satisfy the first Bianchi identity. This section is devoted to the proof of Theorem 2.1.3 which shows that a non-trivial Riemannian IP algebraic curvature tensor on $\mathbb{R}^8$ has rank 2. We must give a geometric proof of this fact; it does not follow from the topological arguments used in Chapter 4 to study dimensions $m = 5, 6$ and $m \geq 9$.

We shall work in the Riemannian setting throughout this section. Let $S^7 := S(\mathbb{R}^8)$ be the unit sphere in $\mathbb{R}^8$ and let $\mathbb{RP}^7 := \mathbb{RP}(\mathbb{R}^8)$ be the associated projective space. Let R be an IP algebraic curvature tensor of rank r on $\mathbb{R}^8$. If π is an oriented 2 plane in $\mathbb{R}^8$, then $R(\pi)^2$ is a self-adjoint map of $\mathbb{R}^8$ and hence is diagonalizable. We let $\{\lambda_1, ..., \lambda_\ell\}$ be the eigenvalues of $R(\pi)^2$ and we let $\{m_1, ..., m_\ell\}$ be the associated multiplicities. We choose the notation so that $m_1 \leq ... \leq m_\ell$; this is not the notational convention employed in Theorem 4.1.4 but it is one that is more convenient for our present purposes. We suppose that $r > 2$ and apply Theorem 4.1.4 (4) to see that the possible multiplicities are

$$\vec{m} = (8) \text{ or } \vec{m} = (2, 6).$$

We follow the discussion in [74] to prove Theorem 2.1.3 by eliminating each of these possibilities. In Lemma 2.10.1, we show that the eigenvalue structure with multiplicity $\vec{m} = (8)$ is impossible; the proof of this Lemma is relatively straightforward and requires no advance preparation.

Before dealing with the remaining multiplicity structure $\vec{m} = (2, 6)$, we must establish a number of technical results. Here is a brief outline of our strategy. We suppose given an IP algebraic curvature tensor with multiplicity structure $\vec{m} = (2, 6)$. In 2.10.2, we establish our basic notational conventions. We begin our investigation in Lemma 2.10.3 by proving some orthogonality results. In Lemma 2.10.4, we show the map $\eta \rightarrow \ker(R(\xi, \eta)^2 - \lambda_1 \operatorname{Id})$ is an injective map from $\mathbb{P}(\xi^\perp)$ to $\operatorname{Gr}_2(\mathbb{R}^8)$ for any $\xi \in S^7$. Given an orthonormal set $\{\xi_1, \xi_2, \xi_3\}$ in $\mathbb{R}^8$, we shall exhibit in Lemma 2.10.5 a normalized orthonormal base $\{e_1, ..., e_8\}$ for $\mathbb{R}^8$ relative to which the actions of $R(\xi_1, \xi_2)$ and $R(\xi_1, \xi_3)$ are given in a particularly simple form. We will set

$$\Phi_1(\xi) := \cap_{\eta \in S(\xi^\perp)} \ker(R(\xi, \eta)^2 - \lambda_1 \operatorname{Id}) \text{ for } \xi \in S^7.$$

In Lemmas 2.10.6, 2.10.7, and 2.10.8, we shall show that $\dim \Phi_1(\xi) = 1$ so Φ_1 takes values in $\mathbb{RP}^7$. Since $\Phi_1(-\xi) = \Phi_1(\xi)$, Φ_1 induces a corresponding map $\Psi : \mathbb{RP}^7 \to \mathbb{RP}^7$ which we show is a homeomorphism in Lemma 2.10.9. In Lemma 2.10.10, we construct an isometry ψ of $\mathbb{R}^8$ which linearizes Ψ. In Lemma 2.10.11, we use ψ and R to construct a bilinear Clifford module structure $c(\cdot,\cdot)$ on $\mathbb{R}^8$; in Lemma 2.10.12, we use this structure to show that $\lambda_1 = 0$. We conclude the section by showing in Lemma 2.10.13 that multiplicity structure $\vec{m} = (2,6)$ is impossible.

We begin the proof of Theorem 2.1.3 by eliminating the multiplicity structure $\vec{m} = (8)$.

2.10.1 Lemma. *If R is an IP algebraic curvature tensor on $\mathbb{R}^8$, then R does not have the multiplicity structure $\vec{m} = (8)$.*

Proof. We suppose the contrary and argue for a contradiction. Let R be an IP algebraic curvature tensor on $\mathbb{R}^8$ so that $R(\pi)^2 = \lambda_1 \operatorname{Id}$, where $\lambda_1 \neq 0$. Let $\lambda_1 = -\mu_1^2$. By replacing R by $\mu_1^{-1}R$, we may suppose without loss of generality that $\lambda_1 = -1$ and hence that $R(\pi)^2 = -1$ for any 2 plane π.

Fix $e_1 \in S^7$. Since $R(e_2, e_1)^2 = -\operatorname{Id}$ if $|e_2| = 1$ and $e_2 \perp e_1$,

$$R(e_2,e_1)^2 = -|e_2|^2 \operatorname{Id} \text{ for any } e_2 \in e_1^\perp. \tag{2.10.1.a}$$

We polarize Equation (2.10.1.a) to derive the Clifford commutation relations:

$$\begin{aligned} &R(e_2,e_1)R(e_3,e_1) + R(e_3,e_1)R(e_2,e_1) = -2(e_2,e_3)\operatorname{Id} \\ &\text{if } e_1 \perp e_2 \text{ and } e_1 \perp e_3. \end{aligned} \tag{2.10.1.b}$$

The Jacobi operator $e_2 \to R(e_2,e_1)e_1$ is a self-adjoint operator which preserves $e_1^\perp$. We note that $R(e_1,e_1)e_1 = 0$. We diagonalize the Jacobi operator to find an orthonormal basis for $\mathbb{R}^8$ so that

$$R(e_i,e_1)e_1 = \sigma_i e_i \text{ for } 1 \le i \le 8 \text{ and } \sigma_i \in \mathbb{R}. \tag{2.10.1.c}$$

Let $i > 1$. Since $\{e_i, e_1\}$ is an orthonormal set, $R(e_i,e_1)$ is a skew-adjoint almost complex structure on $\mathbb{R}^8$ and hence is an isometry; it now follows that $\sigma_i = \pm 1$ for $i > 1$. Since $R(e_2,e_3)$ is an isometry,

$$|R(e_2,e_3)e_1| = 1. \tag{2.10.1.d}$$

We use Equations (2.10.1.b), (2.10.1.c), and (2.10.1.d) together with the first Bianchi identity to compute:

$$\begin{aligned} 1 =&|R(e_2,e_3)e_1| = |R(e_3,e_1)e_2 + R(e_1,e_2)e_3| \\ =&|\sigma_2 R(e_3,e_1)R(e_2,e_1)e_1 - \sigma_3 R(e_2,e_1)R(e_3,e_1)e_1| \\ =&|(\sigma_2+\sigma_3)R(e_3,e_1)R(e_2,e_1)e_1| \\ =&|\sigma_2+\sigma_3|. \end{aligned}$$

This is not possible as $\sigma_2 + \sigma_3 = (\pm 1) + (\pm 1)$ takes on the values $0, \pm 2$. $\square$

2.10.2 Notational conventions. For the remainder of this section, we assume that R is an IP algebraic curvature tensor on $\mathbb{R}^8$ with rank$(R) > 2$. We assume that $R(\pi)^2$ has two distinct negative eigenvalues λ_1 and λ_2 for $\pi \in \mathrm{Gr}_2^+(\mathbb{R}^8)$. Let W_u be the corresponding eigen bundles;

$$W_u(\pi) := \{e \in \mathbb{R}^8 : R(\pi)^2 e = \lambda_u e\} \text{ for } u = 1, 2;$$
$$\dim W_1 = 2 \text{ and } \dim W_2 = 6.$$

If $\lambda_2 = 0$, then rank $R \leq 2$ and thus $\lambda_2 \neq 0$. Let $\lambda_2 = -\mu_2^2$. By replacing R by $\mu_2^{-1} R$ if necessary, we shall assume henceforth without loss of generality that $\lambda_2 = -1$. We regard R as a map from $\otimes^2 \mathbb{R}^8$ to $\mathfrak{so}(8)$. It is convenient to decouple the domain from the range and to distinguish two copies of $\mathbb{R}^8$. We shall for the most part use Greek letters ξ, η, etc. for vectors which are arguments of the curvature operator $R(\cdot, \cdot)$. We shall use Roman letters e, f, etc. for vectors upon which the curvature operator R acts. If $\{\xi, \eta\}$ is an orthonormal set, then we will set

$$W_u(\xi, \eta) := W_u(\mathrm{span}\{\xi, \eta\}).$$

If $\{\xi_1, ..., \xi_\ell\}$ is an orthonormal subset of $\mathbb{R}^8$, then we shall let

$$r_{ij} := R(\xi_i, \xi_j) \text{ and } W_{u,ij} := W_u(\xi_i, \xi_j)$$
$$\text{for } 1 \leq i, j \leq \ell, i \neq j, u = 1, 2.$$

We begin by establishing some facts concerning the eigenbundles W_u. For the moment, we only assume R is IP and alternating.

2.10.3 Lemma. *Let R be an IP alternating tensor on $\mathbb{R}^8$. Let $\{\lambda_1, \lambda_2\}$ be the eigenvalues of $R(\cdot)^2$ on $\mathrm{Gr}_2(\mathbb{R}^8)$. We suppose that λ_1 has multiplicity 2, that λ_2 has multiplicity 6, and that $\lambda_2 = -1$. Let $\{\xi_1, \xi_2, \xi_3\}$ be an orthonormal subset of $\mathbb{R}^8$. Let $e \in S^7$. Then:*

(1) $(r_{12}^2 e, e) \in [\min(\lambda_1, \lambda_2), \max(\lambda_1, \lambda_2)] \subset \mathbb{R}$.
(2) *$e \in W_{u,12}$ if and only if $(r_{12}^2 e, e) = \lambda_u$.*
(3) *If $\lambda_1 = 0$, then $r_{12}^3 = -r_{12}$.*
(4) *If $e \in W_{u,12}$, then $r_{12}e \perp r_{13}e$.*
(5) *If $e \in W_{u,12} \cap W_{u,13}$, then $(r_{12}r_{13} + r_{13}r_{12})e = 0$.*

Proof. As r_{12}^2 is symmetric and has exactly two eigenvalues, Assertions (1) and (2) follow from Lemma 1.3.2. Suppose that $\lambda_1 = 0$. If $e \in W_{1,12}$, then

$$(r_{12}e, r_{12}e) = -(r_{12}^2 e, e) = 0 \text{ so } r_{12}e = 0.$$

Thus, $r_{12}|_{W_{1,12}} = 0$. Since $r_{12}^2 = -1$ and $r_{12}^3 = -r_{12}$ on $W_{2,12}$, Assertion (3) now follows.

To prove Assertions (4) and (5), we consider the 1 parameter family

$$r(\theta) := R(\xi_1, \cos\theta\xi_2 + \sin\theta\xi_3).$$

If $e \in W_{u,12}$, then we expand $(r(\theta)^2 e, e)$ in a Taylor series about $\theta = 0$:

$$\begin{aligned}&(r(\theta)^2 e, e)\\ =&\lambda_u \cos^2\theta - 4\cos\theta\sin\theta((r_{12}r_{13} + r_{13}r_{12})e, e) + \sin^2\theta(r_{13}^2 e, e)\\ =&\lambda_u - 4\theta((r_{12}e, r_{13}e) + O(\theta^2).\end{aligned}$$

By Assertion (1), λ_u is an extremal value of this inner product. This implies that $(r_{12}e, r_{13}e) = 0$ which proves Assertion (4).

If $e \in W_{u,12} \cap W_{u,13}$, then $r_{13}^2 e = \lambda_u e$. Thus, by Assertion (4),

$$(r(\theta)^2 e, e) = (\{\lambda_u + \cos\theta\sin\theta(r_{12}r_{13} + r_{13}r_{12})\}e, e) = \lambda_u.$$

By Assertion (2), $r(\theta)^2 e = \lambda_u e$. $\square$

We continue our study of the eigenbundles:

2.10.4 Lemma. *Let R be an IP alternating tensor on $\mathbb{R}^8$. Let $\{\lambda_1, \lambda_2\}$ be the eigenvalues of $R(\cdot)^2$ on $\mathrm{Gr}_2(\mathbb{R}^8)$. We suppose that λ_1 has multiplicity 2, that λ_2 has multiplicity 6, and that $\lambda_2 = -1$. Suppose that $\xi \in S^7$, and that $\eta_1, \eta_2 \in S(\xi^\perp)$. The following assertions are equivalent:*

(1) $\eta_1 = \pm\eta_2$.
(2) $W_1(\xi, \eta_1) = W_1(\xi, \eta_2)$.
(3) $W_2(\xi, \eta_1) = W_2(\xi, \eta_2)$.

Proof. Assertion (1) implies Assertion (2). Since $W_2^\perp = W_1$, Assertions (2) and (3) are equivalent. Suppose that Assertion (3) holds but that Assertion (1) fails; we argue for a contradiction. Let $\xi_1 := \xi$, let $\xi_2 := \eta_1$ and choose ξ_3 so that $\{\xi_1, \xi_2, \xi_3\}$ is an orthonormal basis for $\mathrm{span}\{\xi, \eta_1, \eta_2\}$. Express

$$\eta_2 = \cos\theta\xi_2 + \sin\theta\xi_3, \text{ where } \sin\theta \neq 0.$$

Let $e \in W_2(\xi_1, \xi_2) = W_2(\xi_1, \eta_2)$. By Lemma 2.10.3, $r_{12}e \perp r_{13}e$. Thus,

$$-1 = (R(\xi, \eta_2)^2 e, e) = -\cos\theta^2 + \sin^2\theta(r_{13}^2 e, e).$$

Since $\sin\theta \neq 0$, $(r_{13}^2 e, e) = -1$. We use Lemma 2.10.3 to see $e \in W_{2,13}$. Thus, $W_{2,12} \subset W_{2,13}$. Since $\dim W_{2,12} = \dim W_{2,13} = 6$, we may conclude that $W_{2,12} = W_{2,13}$. We have that $r_{12}^2 = -1$ and $r_{13}^2 = -1$ on $W_{2,12}$. We use Lemma 2.10.3 to see that

$$r_{12}r_{13} + r_{13}r_{12} = 0 \text{ on } W_{2,12} \cap W_{2,13} = W_{2,12}.$$

Thus, $i \to r_{12}$, $j \to r_{13}$, and $k \to r_{12}r_{13}$ defines a representation of the quaternions $\mathbb{H}$ on $W_{2,12}$. This is not possible since $\dim W_{2,12} = 6$. This contradiction establishes Lemma 2.10.4. $\square$

We regard $R : \otimes^2\mathbb{R}^8 \to \mathfrak{so}(8)$. Given a basis for the domain, we must define a normalized orthonormal basis for the range. Let $\lambda_1 = -\mu_1^2$.

2.10.5 Lemma. *Let R be an IP alternating tensor on $\mathbb{R}^8$. Let $\{\lambda_1, \lambda_2\}$ be the eigenvalues of $R(\cdot)^2$ on $\mathrm{Gr}_2(\mathbb{R}^8)$. We suppose that λ_1 has multiplicity 2, that λ_2 has multiplicity 6, and that $\lambda_2 = -1$. Let $\{\xi_1, \xi_2, \xi_3\}$ be an orthonormal subset of $\mathbb{R}^8$. Then there exists an orthonormal basis $\{e_1, ..., e_8\}$ for $\mathbb{R}^8$ with the following properties:*

(1) $\mathrm{span}\{e_1, e_2, e_3, e_4\} \subset W_{2,12} \cap W_{2,13}$,
$r_{12}e_1 = \ \ e_2$, $r_{12}e_2 = -e_1$, $r_{12}e_3 = -e_4$, $r_{12}e_4 = \ \ e_3$,
$r_{13}e_1 = \ \ e_3$, $r_{13}e_2 = \ \ e_4$, $r_{13}e_3 = -e_1$, $r_{13}e_4 = -e_2$,
$r_{12}r_{13}e_i = -r_{13}r_{12}e_i$ *for* $1 \leq i \leq 4$.
(2) $\dim\{W_{2,12} \cap W_{2,13}\} > 4$.
(3) $W_{2,12} = \mathrm{span}\{e_1, e_2, e_3, e_4, e_5, e_6\}$, $e_6 = r_{12}e_5$,
$W_{2,13} = \mathrm{span}\{e_1, e_2, e_3, e_4, e_5, e_7\}$, $e_7 = r_{13}e_5$.
(4) $W_{1,12} = \mathrm{span}\{e_7, e_8\}$, $W_{1,13} = \mathrm{span}\{e_6, e_8\}$, *and* $r_{12}r_{13}e_5 = \pm\mu_1 e_8$.

Proof. Let $V := W_{2,12} \cap W_{2,13}$. Since $\dim W_{2,12} = \dim W_{2,13} = 6$ and since $\dim(\mathbb{R}^8) = 8$, we may estimate:

$$\dim V \geq 6 + 6 - 8 = 4.$$

Since r_{12} preserves $W_{2,12}$ and since r_{12} is non-singular on $W_{2,12}$, we have $r_{12}V$ is a 4 dimensional subspace of $W_{2,12}$. Thus, we have the inequality

$$\dim(V \cap r_{12}V) \geq 4 + 4 - 6 = 2.$$

Choose a unit vector e_1 in $V \cap r_{12}V$. Let

$$e_2 := r_{12}e_1, \ e_3 := r_{13}e_1, \text{ and } e_4 := r_{13}r_{12}e_1.$$

We use Lemma 2.10.3 to prove Assertion (1). Since $e_1 \in V$ and $e_2 \in V$, we have:

$$(r_{12}r_{13} + r_{13}r_{12})e_1 = 0, \ r_{12}^2e_1 = -e_1, \ r_{12}^2e_2 = -e_2$$
$$(r_{12}r_{13} + r_{13}r_{12})e_2 = 0, \ r_{13}^2e_1 = -e_1, \ r_{13}^2e_2 = -e_2.$$

We show that $e_3 \in V$ by computing:

$$r_{13}^2 e_3 = r_{13}^3 e_1 = -r_{13}e_1 = -e_3, \text{ and}$$
$$r_{12}^2 e_3 = r_{12}^2 r_{13} e_1 = -r_{12}r_{13}r_{12}e_1 = -r_{12}r_{13}e_2 = r_{13}r_{12}e_2 = -e_3.$$

Since $e_3 \in V$, $(r_{12}r_{13} + r_{13}r_{12})e_3 = 0$. We show that $e_4 \in V$ by computing:

$$r_{13}^2 e_4 = r_{13}^2 r_{13} e_2 = -r_{13}e_2 = -e_4$$
$$r_{12}^2 e_4 = -r_{12}^2 r_{13} e_2 = -r_{12}r_{13}r_{12}e_2 = r_{12}r_{13}e_1 = -r_{13}r_{12}e_1 = -e_4.$$

We complete the proof of Assertion (1) by noting

$$(r_{12}r_{13} + r_{13}r_{12})e_4 = 0.$$

We now prove Assertion (2). We use Assertion (1) to define:

$$\begin{aligned} &\mathcal{S} := \operatorname{span}\{e_1, e_2, e_3, e_4\} \subset W_{2,12} \cap W_{2,13}, \\ &\tilde{W}_{2,12} := W_{2,12} \cap \mathcal{S}^\perp, \text{ and} \\ &\tilde{W}_{2,13} := W_{2,13} \cap \mathcal{S}^\perp. \end{aligned} \tag{2.10.5.a}$$

Note that $r_{1i}W_{2,1i} \subset W_{2,1i}$ and that r_{1i} is skew-adjoint. Thus, Display (2.10.5.a) implies that:

$$\begin{aligned} &r_{12}\mathcal{S} \subset \mathcal{S}, && r_{13}\mathcal{S} \subset \mathcal{S}, \\ &r_{12}\mathcal{S}^\perp \subset \mathcal{S}^\perp, && r_{13}\mathcal{S}^\perp \subset \mathcal{S}^\perp, \\ &r_{12}\tilde{W}_{2,12} \subset \tilde{W}_{2,12}, \text{ and} && r_{13}\tilde{W}_{2,13} \subset \tilde{W}_{2,13}. \end{aligned} \tag{2.10.5.b}$$

To prove Assertion (2), we must show that

$$\tilde{W}_{2,12} \cap \tilde{W}_{2,13} \neq \{0\}. \tag{2.10.5.c}$$

Let $f \in S(\tilde{W}_{2,12})$. We use Display (2.10.5.b) and the fact that $\lambda_2 = -1$ to see that $r_{12}|_{\tilde{W}_{2,12}} \in \mathfrak{so}(\tilde{W}_{2,12})$ has rank 2. Consequently:

$$\tilde{W}_{2,12} = \operatorname{span}\{f, r_{12}f\}. \tag{2.10.5.d}$$

Since $f \in W_{2,12}$, we use Lemma 2.10.3 to see that $r_{13}f \perp r_{12}f$. Since r_{13} is skew-adjoint, $r_{13}f \perp f$. Thus, $r_{13}\tilde{W}_{2,12} \perp \tilde{W}_{2,12}$. By Display (2.10.5.b), $r_{13}\mathcal{S} \subset \mathcal{S}$. As r_{13} is skew-adjoint, $\tilde{W}_{2,12} \perp r_{13}\mathcal{S}$ implies $r_{13}\tilde{W}_{2,12} \perp \mathcal{S}$. Thus,

$$r_{13}\tilde{W}_{2,12} \subset \mathcal{S}^\perp \cap \tilde{W}_{2,12}^\perp = (\mathcal{S} + \tilde{W}_{2,12})^\perp = W_{1,12}. \tag{2.10.5.e}$$

We divide the proof of Assertion (2) into two cases. We first suppose that $\lambda_1 \neq 0$ and hence R has constant rank 8. Let $f \in S(W_{1,12})$. Since $\lambda_1 \neq 0$, $r_{12}|_{W_{1,12}} \in \mathfrak{so}(W_{1,12})$ has rank 2. Thus,

$$W_{1,12} = \operatorname{span}\{f, r_{12}f\}.$$

By Lemma 2.10.3, $r_{13}f \perp r_{12}f$. Since r_{13} is skew-adjoint, $r_{13}f \perp f$. Thus, $r_{13}W_{1,12} \perp W_{1,12}$. Since $r_{13}\mathcal{S} \subset \mathcal{S}$, $r_{13}W_{1,12} \perp \mathcal{S}$. Thus, we also have that

$$\text{(2.10.5.g)} \qquad r_{13}W_{1,12} \subset \mathcal{S}^\perp \cap W_{1,12}^\perp = (\mathcal{S} + W_{1,12})^\perp = \tilde{W}_{2,12}.$$

We use Equations (2.10.5.e) and (2.10.5.g) to see:

$$r_{13}^2 \tilde{W}_{2,12} \subset \tilde{W}_{2,12}.$$

Since r_{13}^2 is self-adjoint and preserves $\tilde{W}_{2,12}$, $r_{13}^2|_{\tilde{W}_{2,12}}$ is diagonalizable. Let σ_1 and σ_2 be the eigenvalues of $r_{13}^2|_{\tilde{W}_{2,12}}$; $\sigma_i \in \{\lambda_1, \lambda_2\}$ for $i = 1, 2$. If $\sigma_i = \lambda_2$ for some i, then

$$\tilde{W}_{2,12} \cap \tilde{W}_{2,13} \neq \{0\}$$

as desired. We therefore suppose $\sigma_1 = \sigma_2 = \lambda_1$ i.e. that $\tilde{W}_{2,12} = W_{1,13}$. As

$$r_{12}|_{W_{1,13}} \in \mathfrak{so}(W_{1,13}) \text{ and } r_{13}|_{W_{1,13}} \in \mathfrak{so}(W_{1,3})$$

are operators of rank 2, by Lemma 1.2.15 there is a constant c so that

$$r_{12}|_{W_{1,13}} = c r_{13}|_{W_{1,13}}.$$

It then follows that $R(\xi_1, \xi_2 - c\xi_3) = 0$ on $W_{1,13}$; this is false as R has constant rank 8 since $\lambda_1 \neq 0$; Assertion (2) now follows.

We now deal with the second case in the proof of Assertion (2) by assuming $\lambda_1 = 0$. We may then apply Lemma 2.10.3 to see that $r_{ij}^3 = -r_{ij}$. We use Equation (2.10.5.e) to see $r_{13}\tilde{W}_{2,12} \subset W_{1,12}$. By Equation (2.10.5.d),

$$\tilde{W}_{2,12} = \operatorname{span}\{f, r_{12}f\} \text{ if } f \in S(\tilde{W}_{2,12}).$$

We have $\ker(r_{13}) = W_{1,13}$. If $\tilde{W}_{2,12} \cap W_{1,13} = \{0\}$, then $r_{13}f$ and $r_{13}r_{12}f$ are linearly independent so

$$W_{1,12} = \operatorname{span}\{r_{13}f, r_{13}r_{12}f\}.$$

Since $f \in W_{2,12}$, we use Lemma 2.10.3 to see $r_{12}f \perp r_{13}f$. We compute

$$(r_{13}^2 f, r_{13}f) = -(r_{13}^3 f, f) = (r_{13}f, f) = 0, \text{ and}$$
$$(r_{13}^2 f, r_{13}r_{12}f) = -(r_{13}^3 f, r_{12}f) = (r_{13}f, r_{12}f) = 0.$$

This shows that $r_{13}^2 f \perp \operatorname{span}\{r_{13}f, r_{13}r_{12}f\} = W_{1,12}$. Since $r_{13}^2 S \subset S$ and $f \perp S$, we also have $r_{13}^2 f \perp S$. Thus,

$$r_{13}^2 f \in \mathcal{S}^\perp \cap W_{1,12}^\perp = \tilde{W}_{2,12} \quad \text{so} \quad r_{13}^2 \tilde{W}_{2,12} \subset \tilde{W}_{2,12}.$$

Since $r_{13}f \neq 0$, $(f, r_{13}^2 f) = (r_{13}f, r_{13}f) \neq 0$ so $r_{13}^2 f \neq 0$. Since we choose an arbitrary element $f \in S(\tilde{W}_{2,12})$, r_{13}^2 is an isomorphism of $\tilde{W}_{2,12}$. Since r_{13}^2 has only two eigenvalues, we must therefore have

$$\tilde{W}_{2,12} = \tilde{W}_{2,13} \text{ so } W_{2,12} = W_{2,13}$$

which contradicts Lemma 2.10.4. Consequently, we have

$$\tilde{W}_{2,12} \cap W_{1,13} \neq \{0\}.$$

We define an orthonormal basis $\{f_1, f_2\}$ for $\tilde{W}_{2,12}$ by taking:

$$f_1 \in S(\tilde{W}_{2,12} \cap W_{1,13}) \text{ and } f_2 := -r_{12}f_1 \in \tilde{W}_{2,12}.$$

We then have $r_{12}f_2 = f_1$. Thus,

$$r_{13}r_{12}f_2 = r_{13}f_1 = 0.$$

Since $r_{13}f_2 \perp f_2$, $r_{13}f_2 \perp r_{12}f_2 = f_1$, and $r_{13}f_2 \perp S$,

$$r_{13}f_2 \in W_{1,12} \text{ so } r_{12}r_{13}f_2 = 0.$$

Define a smooth 1 parameter family of operators $r(\theta) \in \mathfrak{so}(\mathbb{R}^8)$ by setting:

$$r(\theta) = R(\xi_1, \cos\theta\xi_2 + \sin\theta\xi_3).$$

By Lemma 2.10.3, $r(\theta)^3 = -r(\theta)$. Since $0 = r_{12}r_{13}f_2 = -r_{13}r_{12}f_2$ and since $r(\theta)^3 = -r(\theta)$, we may compute:

$$\begin{aligned}
&-r(\theta)f_2 = r(\theta)^3 f_2 \\
&=\{\cos^3\theta r_{12}^3 + \cos^2\theta\sin\theta(r_{12}^2 r_{13} + r_{12}r_{13}r_{12} + r_{13}r_{12}^2) \\
&+\cos\theta\sin^2\theta(r_{12}r_{13}^2 + r_{13}r_{12}r_{13} + r_{13}^2 r_{12}) + \sin^3\theta r_{13}^3\}f_2 \\
&=\{-\cos^3\theta r_{12} - \cos^2\theta\sin\theta r_{13} + \cos\theta\sin^2\theta r_{12}r_{13}^2 - \sin^3\theta r_{13}\}f_2 \\
&=\{-r(\theta) + \cos\theta\sin^2\theta(r_{12}r_{13}^2 + r_{12})\}f_2.
\end{aligned}$$

This shows that $r_{12}r_{13}^2 f_2 = -r_{12}f_2$. Since $|r_{12}f_2| = 1$ and since r_{12} is a linear operator of norm 1, we conclude $|r_{13}^2 f_2| \geq 1$. Since the eigenvalue 1 is an extremal value, we use Lemma 1.3.2 to see that $f_2 \in W_{2,13}$ and hence $W_{2,13} \cap W_{2,13} \neq \{0\}$. This completes the proof of Assertion (2).

We use Assertion (2) to prove Assertion (3) by setting:

$$e_5 \in S(\tilde{W}_{2,12} \cap \tilde{W}_{2,13}), \; e_6 := r_{12}e_5, \; e_7 := r_{13}e_5.$$

We then have $\tilde{W}_{2,12} = \operatorname{span}\{e_5, e_6\}$, and $\tilde{W}_{2,13} = \operatorname{span}\{e_5, e_7\}$. By Lemma 2.10.3, $r_{12}e_5 \perp r_{13}e_5$. Thus, $\{e_5, e_6, e_7\}$ is an orthonormal set and Assertion (3) now follows.

To prove Assertion (4), we complete the orthonormal basis by choosing

$$e_8 \in S(\operatorname{span}\{e_1, ..., e_7\}^\perp).$$

We use Assertion (3) to see $e_7, e_8 \perp W_{2,12}$ and $e_6, e_8 \perp W_{2,13}$. Thus,

$$W_{2,12}^\perp = W_{1,12} = \operatorname{span}\{e_7, e_8\}, \text{ and}$$
$$W_{2,13}^\perp = W_{1,13} = \operatorname{span}\{e_6, e_8\}.$$

Since $r_{12}r_{13}e_5 = r_{12}e_7$ and since $e_7 \in W_{1,12}$, we have $r_{12}e_7 \in W_{1,12}$. Since $r_{12}e_7 \perp e_7$, there exists a constant c so that $r_{12}r_{13}e_5 = r_{12}e_7 = ce_8$. Since $r_{12}^2 e_7 = \lambda_1 e_7$, we may complete the proof of Assertion (4) by computing:

$$c^2 = (ce_8, ce_8) = (r_{12}e_7, r_{12}e_7) = -(r_{12}^2 e_7, e_7) = -\lambda_1 = \mu_1^2. \quad \square$$

In Section 2.3, we classified the alternating mixed tensors of constant rank 2. In the Riemannian setting, the definition given in Equation (2.3.2.a) is equivalent to the definition:

$$\Phi(a) := \cap_{b \in S(a^\perp)} \operatorname{range} T(a, b).$$

A similar map played a crucial role in the proof of Lemma 1.8.6. In our present investigation, it is more useful to study the eigenspaces rather than just the range. We have two distinct eigenvalues λ_1 and λ_2. The following spaces will be important to our continued investigations. If $\xi \in S^7$, then let:

$$\text{(2.10.5.g)} \qquad \Phi_1(\xi) := \cap_{\eta \in S(\xi^\perp)} W_1(\xi, \eta) \text{ and } \Phi_2(\xi) := \cap_{\eta \in S(\xi^\perp)} W_2(\xi, \eta).$$

So far, we have only used the curvature symmetry of Equation (1.5.1.e). We shall use the remaining two curvature symmetries of Equations (1.5.1.f) and (1.5.1.g) in the proof of the following:

2.10.6 Lemma. *Let R be an IP algebraic curvature on $\mathbb{R}^8$. Let $\{\lambda_1, \lambda_2\}$ be the eigenvalues of $R(\cdot)^2$ on $\mathrm{Gr}_2(\mathbb{R}^8)$. We suppose that λ_1 has multiplicity 2, that λ_2 has multiplicity 6, and that $\lambda_2 = -1$. Let $u = 1, 2$.*

(1) *If $\xi \in S^7$ and if $\xi \in \Phi_u(\xi)$, then $\lambda_u = 0$.*

(2) *We have $\cap_{\xi \in S^7} \Phi_u(\xi) = \{0\}$.*

Proof. We suppose that Assertion (1) is false and argue for a contradiction. Choose $\xi \in S^7$ so that $\xi \in \Phi_u(\xi)$. Let $\eta \in S(\xi^\perp)$. Let $\lambda_u = -\mu_u^2$. Let

$$\tilde{\mathcal{J}} := \mu_u^{-1} \mathcal{J}_R(\xi) : \eta \to \mu^{-1} R(\eta, \xi)\xi$$

be the normalized *Jacobi operator.* This is a self-adjoint operator which preserves $\xi^\perp$. Let $\eta \in S(\xi^\perp)$. Since $\xi \in W_u(R(\xi, \eta))$, $R(\eta, \xi)^2 \xi = -\mu_u^2 \xi$. We show that $\tilde{\mathcal{J}}$ is an isometry of $\xi^\perp$ by computing:

$$\begin{aligned}(\tilde{\mathcal{J}}\eta, \tilde{\mathcal{J}}\eta) &= \mu_u^{-2}(R(\eta, \xi)\xi, R(\eta, \xi)\xi) \\ &= -\mu_u^{-2}(R(\xi, \eta)^2 \xi, \xi) = -\mu_u^{-2}\lambda_u = 1.\end{aligned}$$

We can therefore choose an orthonormal basis $\{\eta_1, ..., \eta_7\}$ for $\xi^\perp$ and we may choose signs $\varepsilon_i = \pm 1$ so that

$$\tilde{\mathcal{J}}\eta_i = \varepsilon_i \eta_i \text{ for } 1 \le i \le 7.$$

Let $r_i := R(\xi, \eta_i)$. As $\xi \in W_u(\xi, \eta_i)$, we may use Lemma 2.10.3 to see that

$$(r_i r_j + r_j r_i)\xi = 0 \text{ for } 1 \le i < j \le 7. \qquad (2.10.6.a)$$

Let $\{i, j, k\}$ be distinct indices ranging between 1 and 7. We use the first Bianchi identity, the fact that r_i is skew-adjoint, and the Clifford commutation relations of Equation (2.10.6.a) to compute:

$$\begin{aligned}0 =&(r_i \eta_j, \eta_k) + (r_j \eta_k, \eta_i) + (r_k \eta_i, \eta_j) \\ =&\mu_u^{-2}\{\varepsilon_j \varepsilon_k (r_i r_j \xi, r_k \xi) + \varepsilon_k \varepsilon_i (r_j r_k \xi, r_i \xi) + \varepsilon_i \varepsilon_j (r_k r_i \xi, r_j \xi)\} \\ =&\mu_u^{-2}\{(\varepsilon_j \varepsilon_k + \varepsilon_k \varepsilon_i + \varepsilon_i \varepsilon_j)(r_i r_j \xi, r_k \xi).\end{aligned}$$

Since $\mu_u^{-2}(\pm 1 \pm 1 \pm 1) \neq 0$, we see that $r_i r_j \xi \perp r_k \xi$. We also have $r_i r_j \xi \perp r_j \xi$ and $r_i r_j \xi = -r_j r_i \xi \perp r_i \xi$. Since $\xi \in W_u(\xi, \eta_i)$, $r_i r_j \xi \perp \xi$ by Lemma 2.10.3. Since $r_i r_j \xi$ is perpendicular to $\{\xi, \eta_1, ..., \eta_7\}$, we have

$$r_i r_j \xi = 0 \text{ for } 1 \le i < j \le 7.$$

Since η_j is a non-zero multiple of $r_j \xi$, $\eta_j \in \ker r_i$ for $i \neq j$. Thus,

$$\mathrm{span}\{\eta_2, ..., \eta_7\} \subset \ker R(\xi, \eta_1) \text{ so } \dim \ker R(\xi, \eta_1) \ge 6.$$

Thus, $\mathrm{rank}\, R(\xi, \eta_1) \le 2$ which is false. This contradiction establishes Assertion (1).

Suppose that Assertion (2) fails. We argue for a contradiction. Choose $\xi \in S(\cap_\eta \Phi_u(\eta))$. Then in particular, $\xi \in \Phi_u(\xi)$ and hence $\lambda_u = 0$ by Assertion (1). This implies $u = 1$.

Choose an orthonormal basis $\{\xi_1, ..., \xi_8\}$ for $\mathbb{R}^8$ so that $\xi = \xi_1$. Let i and j be distinct indices with $1 \le i, j \le 8$. Since $\xi_1 \in \Phi_1(\xi_i) \subset W_1(\xi_i, \xi_j)$, we have $R(\xi_i, \xi_j)\xi_1 = 0$. Trivially, of course, $R(\xi_i, \xi_i)\xi_1 = 0$. We use the curvature symmetry (1.5.1.f) to see that for any j, k:

$$(R(\xi_1, \xi_2)\xi_j, \xi_k) = (R(\xi_j, \xi_k)\xi_1, \xi_2) = 0$$

and thus $R(\xi_1, \xi_2)\xi_j = 0$ so $R(\xi_1, \xi_2)$ vanishes identically which is false; this contradiction establishes Assertion (2). □

The following is a technical fact that we will use presently to prove that Φ_1 takes values in $\mathbb{RP}^7$.

2.10.7 Lemma. *Let R be an IP algebraic curvature on $\mathbb{R}^8$. Let $\{\lambda_1, \lambda_2\}$ be the eigenvalues of $R(\cdot)^2$ on $\mathrm{Gr}_2(\mathbb{R}^8)$. We suppose that λ_1 has multiplicity 2, that λ_2 has multiplicity 6, and that $\lambda_2 = -1$. Let $\{\xi_1, ..., \xi_8\}$ be an orthonormal basis for $\mathbb{R}^8$. If $W_{1,12} \cap W_{1,13} \cap W_{1,14} \ne \{0\}$, then $\dim \Phi_1(\xi_1) = 1$.*

Proof. Assume that $W_{1,12} \cap W_{1,13} \cap W_{1,14} \ne \{0\}$. Choose

$$f_1 \in S(W_{1,12} \cap W_{1,13} \cap W_{1,14}).$$

Choose elements f_2, f_3, f_4 so that $\{f_1, f_i\}$ is an orthonormal basis for $W_{1,1i}$ for $i = 2, 3, 4$. We apply Lemma 2.10.5 to the orthonormal set $\{\xi_1, \xi_2, \xi_3\}$. Since $f_1 \in W_{1,12} \cap W_{1,13}$, we have $f_1 = \pm e_8$ and $f_2 = \pm e_7 \in W_{2,13}$. Similarly, by applying Lemma 2.10.5 to the orthonormal set $\{\xi_1, \xi_2, \xi_4\}$, we also have $f_2 \in W_{2,14}$. Thus, $f_2 \in W_{1,12} \cap W_{2,13} \cap W_{2,14}$. Arguing similarly with f_3 and f_4, we see that we have the relations:

$$f_1 \in W_{1,12} \cap W_{1,13} \cap W_{1,14},$$
$$f_2 \in W_{1,12} \cap W_{2,13} \cap W_{2,14},$$
$$f_3 \in W_{2,12} \cap W_{1,13} \cap W_{2,14},$$
$$f_4 \in W_{2,12} \cap W_{2,13} \cap W_{1,14}.$$

Thus, $\{f_1, f_2, f_3, f_4\}$ is an orthonormal set which we can extend to an orthonormal basis $\{f_1, ..., f_8\}$ for $\mathbb{R}^8$.

By Lemma 2.10.5, $\dim(W_{1,1i} \cap W_{1,1j}) = 1$ for $i \ne j$. Thus, we may choose elements

$$g_{ik} = a_{ik} f_1 + b_{ik} f_i \in S(W_{1,1i} \cap W_{1,1k}) \text{ for } 2 \le i \le 4, 5 \le k \le 8.$$

If the coefficients b_{ik} are non-zero for $i = 2, 3, 4$, then $\{g_{2k}, g_{3k}, g_{4k}\}$ is a linearly independent subset of $W_{1,1k}$; this is not possible as $\dim W_{1,1k} = 2$. Thus, $b_{ik} = 0$ for at least one value of i and hence $f_1 \in W_{1,1k}$. Therefore:

$$f_1 \in W_{1,1i} \text{ for } 2 \le i \le 8.$$

Since $f_1 \in W_{1,1i} \cap W_{1,1j}$, we use Lemma 2.10.3 to see that

$$(r_{1i}r_{1j} + r_{1j}r_{1i})f_1 = 0 \text{ for } 2 \le i < j \le 8.$$

We expand $\eta \in S(\xi_1^\perp)$ in the form $\eta = b_2\xi_2 + ... + b_8\xi_8$ to demonstrate:

$$R(\xi_1, \eta)^2 f_1 = \textstyle\sum_i b_i^2 r_{1i}^2 f_1 + \sum_{i<j} b_i b_j (r_{1i}r_{1j} + r_{1j}r_{1i}) f_1 = \lambda_1 f_1.$$

This shows that $f_1 \in W_1(\xi_1, \eta)$ for all $\eta \perp \xi_1$. We complete the proof by using Lemma 2.10.5 to compute:

$$\mathrm{span}\{f_1\} \subset \Phi_1(\xi_1) \subset W_{1,12} \cap W_{1,13} = \mathrm{span}\{f_1\}. \quad \square$$

We can now show that Φ_1 takes values in $\mathbb{RP}^7$.

2.10.8 Lemma. *Let R be an IP algebraic curvature on $\mathbb{R}^8$. Let $\{\lambda_1, \lambda_2\}$ be the eigenvalues of $R(\cdot)^2$ on $\mathrm{Gr}_2(\mathbb{R}^8)$. We suppose that λ_1 has multiplicity 2, that λ_2 has multiplicity 6, and that $\lambda_2 = -1$.*

(1) *If $\xi \in S^7$, then $\dim \Phi_1(\xi) = 1$.*
(2) *If $\{\xi_1, \xi_2, \xi_3\}$ is a linearly independent set with $\xi_1 \in S^7$ and with $\xi_2, \xi_3 \in S(\xi_1^\perp)$, then $\Phi_1(\xi_1) = W_1(\mathrm{span}\{\xi_1, \xi_2\}) \cap W_1(\mathrm{span}\{\xi_1, \xi_3\})$.*

Proof. We suppose that Assertion (1) fails and argue for a contradiction. Choose $\eta_1 \in S^7$ so that $\dim \Phi_1(\eta_1) \ne 1$. Choose η_2, η_3, and η_4 so that $\{\eta_1, \eta_2, \eta_3, \eta_4\}$ is an orthonormal set. Since $\dim W_1(\eta_1, \eta_2) = 2$, we have that $\dim \Phi_1(\eta_1) \le 2$. If $\dim \Phi_1(\eta_1) = 2$, then

$$\Phi_1(\eta_1) = W_1(\eta_1, \eta_i) \text{ for } i = 2, 3$$

which contradicts Lemma 2.10.4. Thus, we must have that $\dim \Phi_1(\eta_1) = 0$. We may now use Lemma 2.10.7 to see that

$$W_{1,12} \cap W_{1,13} \cap W_{1,14} = \{0\}. \tag{2.10.8.a}$$

We apply Lemma 2.10.5 to the orthonormal set $\{\eta_1, \eta_2, \eta_3\}$. Let $f_1 := e_8$, $f_2 := e_6$, and $f_3 := e_7$. Then $\{f_1, f_2, f_3\}$ is an orthonormal set with

$$W_{1,12} = \mathrm{span}\{f_1, f_2\} \text{ and } W_{1,13} = \mathrm{span}\{f_1, f_3\}.$$

By Lemma 2.10.5, $\dim\{W_{1,1i}\cap W_{1,1j}\}=1$ for $2\le i<j\le 4$. Let

$$g_i=a_if_1+b_if_i\in S(W_{1,1i}\cap W_{1,14})\text{ for }i=2,3.$$

If $\{g_1,g_2\}$ is a linearly dependent set, then $g_i=\pm_i f_1$ since $\{f_1,f_2,f_3\}$ is an orthogonal set. This would imply

$$f_1\in W_{1,12}\cap W_{1,13}\cap W_{1,14}$$

which would contradict Equation (2.10.8.a). Thus, g_2 and g_3 are linearly independent vectors so

$$W_{1,14}=\operatorname{span}\{g_2,g_3\}\subset W_{1,12}+W_{1,13}.$$

Let $E:=W_{1,12}+W_{1,13}$. Since $\dim\{W_{1,12}\cap W_{1,13}\}=1$, $\dim E=3$. The only condition imposed on η_4 was $\eta_4\in S(\operatorname{span}\{\eta_1,\eta_2,\eta_3\}^\perp)$. We apply Lemma 4.2.7 to see the map $\alpha:\eta\to W_1(R(\eta_1,\eta))$ is a smooth map:

$$\alpha:\mathbb{P}(\operatorname{span}\{\eta_1,\eta_2,\eta_3\}^\perp)\to\operatorname{Gr}_2(E).$$

By Lemma 2.10.4, α is injective. We use Theorem 4.3.2 to derive the desired contradiction which proves Assertion (1) by computing:

$$4=\dim\mathbb{P}(\operatorname{span}\{\eta_1,\eta_2,\eta_3\}^\perp)\le\dim\operatorname{Gr}_2(E)=2.$$

We use Assertion (1) to prove Assertion (2). Let $\xi_1\in S^7$ and let $\{\xi_1,\xi_2,\xi_3\}$ be a linearly independent set with $\xi_2,\xi_3\in S(\xi_1^\perp)$. We use Assertion (1) to see that $\dim\Phi(\xi_1)=1$. By Lemma 2.10.5,

$$W_1(\operatorname{span}\{\xi_1,\xi_2\})\neq W_1(\operatorname{span}\{\xi_1,\xi_3\}).$$

This shows $\dim\big\{W_1(\operatorname{span}\{\xi_1,\xi_2\})\cap W_1(\operatorname{span}\{\xi_1,\xi_3\})\big\}=1$. Assertion (2) now follows since

$$\Phi(\xi_1)\subset W_1(\operatorname{span}\{\xi_1,\xi_2\})\cap W_1(\operatorname{span}\{\xi_1,\xi_3\}).\quad\square$$

Since $\Phi_1(-\xi)=\Phi_1(\xi)$, we may define a map $\Psi:\mathbb{RP}^7\to\mathbb{RP}^7$ by setting:

$$\Psi([\xi])=\Phi_1(\xi).$$

In our analysis of alternating mixed tensors, which was performed in Section 2.3, we showed that a similar map was either injective or constant. The case in which Ψ is constant is ruled out by Lemma 2.10.6. In the following Lemma, we show Ψ is a homeomorphism.

2.10.9 Lemma. *Let R be an IP algebraic curvature on $\mathbb{R}^8$. Let $\{\lambda_1, \lambda_2\}$ be the eigenvalues of $R(\cdot)^2$ on $\mathrm{Gr}_2(\mathbb{R}^8)$. We suppose that λ_1 has multiplicity 2, that λ_2 has multiplicity 6, and that $\lambda_2 = -1$. Then:*

(1) *The map $\Psi : [\xi] \to \Phi_1(\xi)$ is a homeomorphism from $\mathbb{RP}^7$ onto $\mathbb{RP}^7$.*

(2) *If $\{\xi_1, \xi_2, \xi_3\}$ is an orthonormal set, then $W_{1,12} \cap W_{1,13} \cap W_{1,23} = \{0\}$.*

Proof. Let $\{\xi_1, \xi_2, \xi_3\}$ be an orthonormal set. Let

$$\mathcal{O} := \{\xi \in S^7 : |\xi - \xi_1| < 1\}.$$

If $\xi \in \mathcal{O}$, then $\{\xi, \xi_2, \xi_3\}$ is a linearly independent set. By Lemma 2.10.8,

$$\Phi_1(\xi) = W_1(\xi, \xi_2) \cap W_1(\xi, \xi_3).$$

It now follows from Lemma 4.2.7 that the map $\xi \to \Phi_1(\xi)$ is continuous on $\mathcal{O}$. As $\mathcal{O}$ is an open neighborhood of ξ_1 and as ξ_1 was arbitrary, the map Φ_1 is continuous. As $\Phi_1(-\xi) = \Phi_1(\xi)$, Ψ is well defined and continous.

Suppose that Ψ is not $1-1$. We can then find vectors $\eta_1, \eta_2 \in S^7$ so that $\Phi_1(\eta_1) = \Phi_1(\eta_2)$ but so that $\eta_1 \neq \pm\eta_2$. Let $\eta_3 \in S^7$ be arbitrary. We choose an orthonormal set $\{\xi_1, \xi_2\}$ so that $\xi_i \perp \eta_j$ for $i = 1, 2$ and $j = 1, 2, 3$. Since $\Phi_1(\eta_1) = \Phi_1(\eta_2)$, we use Lemma 2.10.8 to see:

$$\begin{aligned}
&\Phi_1(\eta_1) = \Phi_1(\eta_i) \subset W_1(\xi_j, \eta_i) \text{ for } i = 1, 2 \text{ and } j = 1, 2,\\
&\Phi_1(\eta_1) \subset W_1(\xi_j, \eta_1) \cap W_1(\xi_j, \eta_2) = \Phi_1(\xi_j) \text{ for } j = 1, 2,\\
&\Phi_1(\eta_1) = \Phi_1(\xi_j) \subset W_1(\xi_j, \eta_3) \text{ for } j = 1, 2,\\
&\Phi_1(\eta_1) \subset W_1(\xi_1, \eta_3) \cap W_1(\xi_2, \eta_3) = \Phi_1(\eta_3).
\end{aligned}$$

This implies that Φ_1 is constant which contradicts Lemma 2.10.6.

We have shown that the map $\Psi : \mathbb{RP}^7 \to \mathbb{RP}^7$ is continuous and 1-1. By Theorem 4.3.2, the range of Ψ is open. Since $\mathbb{RP}^7$ is compact, the range of Ψ is closed. Since $\mathbb{RP}^7$ is connected, Ψ is surjective. Assertion (1) now follows from Theorem 4.3.1.

Suppose that Assertion (2) fails. We can then choose

$$f \in S(W_{1,12} \cap W_{1,13} \cap W_{1,23}).$$

We then use Lemma 2.10.8 to see that $f \in S(\Phi_1(\xi_1) \cap \Phi_1(\xi_2) \cap \Phi_1(\xi_3))$ so $\Phi_1(\xi_1) = \Phi_1(\xi_2) = \Phi_1(\xi_3)$. This contradicts Assertion (1) and thereby completes the proof of Assertion (2). □

Lift Ψ from $\mathbb{RP}^7$ to the universal cover S^7 to define a map $\psi : S^7 \to S^7$ so that $[\psi\xi] = \Phi_1(\xi)$. Since Ψ is a homeomorphism, ψ is a homeomorphism. Note that the distinguished vector e_8 of Lemma 2.10.5 is $\pm\psi\xi_1$.

2.10.10 Lemma. *Let R be an IP algebraic curvature on $\mathbb{R}^8$. Let $\{\lambda_1, \lambda_2\}$ be the eigenvalues of $R(\cdot)^2$ on $\mathrm{Gr}_2(\mathbb{R}^8)$. We suppose that λ_1 has multiplicity 2, that λ_2 has multiplicity 6, and that $\lambda_2 = -1$. There exists a universal choice of sign $\varepsilon = \pm 1$ and there exists $\mu_1 \in \mathbb{R}$ with $\lambda_1 = -\mu_1^2$ so that if $\{\xi_1, \xi_2, \xi_3\}$ is any orthonormal set, then:*

(1) *$\{\psi\xi_1, \psi\xi_2\}$ is an orthonormal basis for $W_{1,12}$.*

(2) *$r_{12}\psi\xi_1 = \mu_1\psi\xi_2$ and $\psi\xi_3 = \varepsilon r_{12} r_{13}\psi\xi_2$.*

(3) *The map ψ extends to a linear isometry of $\mathbb{R}^8$.*

Proof. We apply Lemma 2.10.5 to the orthonormal set $\{\xi_1, \xi_2, \xi_3\}$ to choose an orthonormal basis $\{e_1, ..., e_8\}$ for $\mathbb{R}^8$ so that:

$$\begin{aligned} & e_6 = r_{12}e_5, && e_7 = r_{13}e_5, \\ \text{(2.10.10.a)}\quad & W_{2,12} = \mathrm{span}\{e_1, e_2, e_3, e_4, e_5, e_6\}, && W_{1,12} = \mathrm{span}\{e_7, e_8\}, \\ & W_{2,13} = \mathrm{span}\{e_1, e_2, e_3, e_4, e_5, e_7\}, && W_{1,13} = \mathrm{span}\{e_6, e_8\}. \end{aligned}$$

Similarly, we apply Lemma 2.10.5 to the orthonormal set $\{\xi_2, \xi_1, \xi_3\}$ to choose an orthonormal basis $\{\tilde{e}_1, ..., \tilde{e}_8\}$ for $\mathbb{R}^8$ so that:

$$\begin{aligned} & \tilde{e}_6 = -r_{12}\tilde{e}_5, && \tilde{e}_7 = r_{23}\tilde{e}_5, \\ \text{(2.10.10.b)}\quad & W_{2,12} = \mathrm{span}\{\tilde{e}_1, \tilde{e}_2, \tilde{e}_3, \tilde{e}_4, \tilde{e}_5, \tilde{e}_6\}, && W_{1,12} = \mathrm{span}\{\tilde{e}_7, \tilde{e}_8\}, \\ & W_{2,23} = \mathrm{span}\{\tilde{e}_1, \tilde{e}_2, \tilde{e}_3, \tilde{e}_4, \tilde{e}_5, \tilde{e}_7\}, && W_{1,23} = \mathrm{span}\{\tilde{e}_6, \tilde{e}_8\} \end{aligned}$$

Since $\psi\xi_1 \in W_{1,12} \cap W_{1,13}$, we have $\psi\xi_1 = \pm e_8$. Since $\psi\xi_i \in W_{1,1i}$ for $i = 2, 3$, we may expand

$$\psi\xi_2 = a_2\psi\xi_1 + b_2e_7 \text{ and } \psi\xi_3 = a_3\psi\xi_1 + b_3e_6.$$

By Lemma 2.10.9, Ψ is injective. Thus, $\psi\xi_i \neq \pm\psi\xi_j$ for $i \neq j$. This shows that $b_2 \neq 0$ and $b_3 \neq 0$. Thus, the vectors $\{\psi\xi_1, \psi\xi_2, \psi\xi_3\}$ are linearly independent. Since $\psi\xi_i \in W_{1,23}$ for $i = 2, 3$, we have

$$\text{(2.10.10.c)}\qquad W_{1,23} = \mathrm{span}\{\psi\xi_2, \psi\xi_3\} \subset W_{1,12} + W_{1,13}.$$

Let $E := (W_{1,12} + W_{1,13}) \cap W_{1,23}^{\perp}$. We use Equation (2.10.10.c) to see $\dim E \leq 1$. Since $\psi\xi_2 \in W_{1,12} \cap W_{1,23}$, we have $\psi\xi_2 = \pm\tilde{e}_8$. Since

$$\{\psi\xi_2 = a_2\psi\xi_1 + b_2e_7, -b_2e_7 + a_2\psi\xi_1\}$$

is an orthonormal basis for $W_{1,12}$, we have $-b_2e_7 + a_2\psi\xi_1 = \pm\tilde{e}_7 \perp W_{1,23}$. Similarly, we argue to see that $-b_3e_6 + a_3\psi\xi_1 \perp W_{1,23}$. Thus, we have

$$-b_2\psi\xi_1 + a_2f_2 \in E \text{ and } -b_3\psi\xi_1 + a_3f_3 \in E.$$

Since $\dim E = 1$, these two vectors are multiples of each other. Since $f_2 \perp f_3$, we have $a_2 = a_3 = 0$. Therefore,

$$\psi\xi_1 = \pm e_8, \psi\xi_2 = \pm e_7, \text{ and } \psi\xi_3 = \pm e_6.$$

Since $e_7 \perp e_8$, we have that $\psi\xi_1 \perp \psi\xi_2$, which proves Assertion (1). Since $\{\psi\xi_1, \psi\xi_2\}$ is an orthonormal basis for $W_{1,12}$, we have $r_{12}\psi\xi_1 = \pm\mu_1\psi\xi_2$. If $\mu_1 = 0$, then we do not need to worry about the sign. If $\mu_1 \neq 0$, then the sign δ is uniquely determined and is a continuous function on the set of all orthonormal pairs $\{\xi_1, \xi_2\}$. Since this domain is connected and δ is continuous, δ is constant; we replace μ_1 by $-\mu_1$ if necessary.

Similarly, we have

$$r_{12}r_{13}\psi\xi_2 = \pm r_{12}r_{13}e_7 = \pm r_{12}r_{13}r_{13}e_5 = \pm r_{12}e_5 = \pm e_6 = \pm\psi\xi_3.$$

Again, by continuity, the sign is universal. This proves Assertion (2).

Let $\{\eta_1, \eta_2\}$ be an orthonormal set. Let $a_1^2 + a_2^2 = 1$. Choose ξ_1, ξ_2 so that $\{\xi_1, \xi_2, \eta_1, \eta_2\}$ is an orthonormal set. We apply Assertion (2) to see

$$\begin{aligned}(2.10.10.d)\qquad &\psi(a_1\eta_1 + a_2\eta_2)\\ =&\varepsilon R(\xi_1, \xi_2)R(\xi_1, a_1\eta_1 + a_2\eta_2)\psi(\xi_2)\\ =&a_1\varepsilon R(\xi_1, \xi_2)R(\xi_1, \eta_1)\psi(\xi_2) + a_2\varepsilon R(\xi_1, \xi_2)R(\xi_1, \eta_2)\psi(\xi_2)\\ =&a_1\psi\eta_1 + a_2\psi\eta_2.\end{aligned}$$

Since ψ is the lift of the homeomorphism Ψ, ψ is a homeomorphism which is $\mathbb{Z}_2$ equivariant, i.e. $\psi(-x) = -\psi(x)$. We may extend ψ from S^7 to $\mathbb{R}^8$ radially by defining $\psi(r\eta) = r\psi\eta$ for $r \in \mathbb{R}$. We use Equation (2.10.10.d) to see this extension is linear. Since $|\eta| = 1$ implies $|\psi\eta| = 1$, ψ is an isometry. This completes the proof of Assertion (3). □

At this point, we shall introduce a *bilinear Clifford module structure* by renormalizing the action of R on W_1. This will define an alternating bilinear map $c(\xi, \eta)$ so that $c(\xi, \eta)^2 = -1$ if $\{\xi, \eta\}$ forms an orthonormal set. This is quite similar to the map introduced in Section 2.8 to discuss the four dimensional setting. Although we shall not be able to express $c(\xi, \eta) = \frac{1}{2}\{c(\xi)c(\eta) - c(\eta)c(\xi)\}$, where $c(\cdot)$ is a Clifford module structure, many of the arguments we shall give are closely related to those given in Section 2.8. Thus, once again, Clifford modules will play an important role.

2.10.11 Lemma. *Let R be an IP algebraic curvature on $\mathbb{R}^8$. Let $\{\lambda_1, \lambda_2\}$ be the eigenvalues of $R(\cdot)^2$ on $\mathrm{Gr}_2(\mathbb{R}^8)$. We suppose that λ_1 has multiplicity 2, that λ_2 has multiplicity 6, and that $\lambda_2 = -1$. Let ψ be the linear isometry* *of* $\mathbb{R}^8$ *defined in* Lemma 2.10.10. *Set* $p(\xi,\eta)e := (\psi\eta, e)\psi\xi - (\psi\xi, e)\psi\eta$ *to define an alternating map* $p : \otimes^2\mathbb{R}^8 \to \mathfrak{so}(8)$.

(1) *There exists* $\mu_1 \in \mathbb{R}$ *with* $\mu_1^2 = -\lambda_1$ *so that* $R(\xi_1,\xi_2)e = \mu_1 p(\xi_1,\xi_2)e$ *for any orthonormal set* $\{\xi_1,\xi_2\}$ *and any* $e \in W_{1,12}$.

(2) *Set* $c(\xi,\eta) := R(\xi,\eta) + (1-\mu_1)p(\xi,\eta)$. *The operators* $R(\xi,\eta)$, $c(\xi,\eta)$, *and* $p(\xi,\eta)$ *commute.*

(3) *Let* $\xi_1 \in S^7$, *let* $\xi_2, \xi_3 \in S(\xi_1^\perp)$, *and set* $c_{ij} := c(\xi_i,\xi_j)$. *Then we have that* $c_{12} \in \mathfrak{so}(8)$, *that* $c_{12}^2 = -\operatorname{Id}$, *that* $\psi\xi_1 = c_{12}\psi\xi_2$, *and that* $c_{12}c_{13} + c_{13}c_{12} = -2(\xi_2,\xi_3)\operatorname{Id}$.

Proof. Let $\{\xi_1,\xi_2\}$ be an orthonormal set. We apply Lemma 1.2.15 to see that there exists a constant $\mu_1 = \mu_1(\xi_1,\xi_2)$ so that $R|_{W_{1,12}} = \mu_1 p|_{W_{1,12}}$. Since $-\mu_1^2 = \lambda_1$, μ_1 is independent of the particular orthonormal set chosen; this proves Assertion (1).

To prove Assertion (2), we note that:

$$c(\xi_1,\xi_2)e = \begin{cases} p(\xi_1,\xi_2)e & \text{if } e \in W_{1,12}, \\ R(\xi_1,\xi_2)e & \text{if } e \in W_{2,12}. \end{cases}$$

To prove Assertion (3), we observe that $c(\xi_1,\xi_2) \in \mathfrak{so}(8)$ and we observe that $c(\xi_1,\xi_2)^2 = -\operatorname{Id}$. Since $\psi\xi_1 \in W_{1,12}$,

$$c(\xi_1,\xi_2)\psi\xi_2 = p_{12}\psi\xi_2 = (\psi\xi_2,\psi\xi_2)\psi\xi_1 - (\psi\xi_2,\psi\xi_1)\psi\xi_2 = \psi\xi_1.$$

We polarize the identity $c(\xi_1,\xi_2)^2 = -\operatorname{Id}$ to see

$$c_{12}c_{13} + c_{13}c_{12} = -2(\xi_2,\xi_3)\operatorname{Id}. \quad \square$$

The following is a somewhat technical Lemma we shall need to study the operators defined in Lemma 2.10.11.

2.10.12 Lemma. *Let* R *be an IP algebraic curvature on* $\mathbb{R}^8$. *Let* $\{\lambda_1,\lambda_2\}$ *be the eigenvalues of* $R(\cdot)^2$ *on* $\operatorname{Gr}_2(\mathbb{R}^8)$. *We suppose that* λ_1 *has multiplicity* 2, *that* λ_2 *has multiplicity* 6, *and that* $\lambda_2 = -1$. *Adopt the notation of Lemma* 2.10.11. *If* $\xi \in S^7$, *then define*

$$\sigma(\xi) := \xi + (\mu_1 - 1)(\xi,\psi\xi)\psi\xi \text{ and}$$
$$Z(\xi) := \{\eta : (\xi,\eta) = 0,\ (\psi\xi,\eta) = 0,\ (\xi,\psi\eta) = 0\}.$$

(1) *If* $\psi\xi \neq \pm\xi$, *there exists* $\Xi \in S(\xi^\perp)$ *and* $\theta \in \mathbb{R}$ *so* $\sin\theta \neq 0$ *with*
 1a) $\psi\xi = \{\cos\theta + \sin\theta c(\xi,\Xi)\}\xi$
 1b) $Z(\xi) = \{\eta : (\xi,\eta) = 0,\ (\psi\xi,\eta) = 0,\ (\Xi,\eta) = 0\}$.

(2) *If* $\psi\xi = \pm\xi$, *then* $Z(\xi) = \{\eta : (\xi, \eta) = 0\}$.
(3) *If* $\eta \in Z(\xi)$, *then* $R(\xi, \eta)\xi = c(\xi, \eta)\sigma(\xi)$.
(4) *The Jacobi operator* $\mathcal{J}_R(\xi) : \eta \mapsto R(\eta, \xi)\xi$ *preserves* $Z(\xi)$.
(5) *We have* $\sigma(\xi) = 0$, $\lambda_1 = 0$, *and* $\psi = \pm\,\mathrm{Id}$.

Proof. Let $\xi \in S^7$. To prove Assertion (1), we suppose that $\xi \neq \pm\psi\xi$. Let $\{\xi, \zeta\}$ be an orthonormal basis for $\mathrm{span}\{\xi, \psi\xi\}$. We may then find $\theta \in \mathbb{R}$ with $\sin\theta \neq 0$ so that $\psi\xi = \cos\theta\xi + \sin\theta\zeta$.

Let $\Xi \in S(\xi^\perp)$. Since $c(\xi, \Xi) \in \mathfrak{so}(8)$ and since $c(\xi, \Xi)^2 = -\,\mathrm{Id}$, the Jacobi operator $\mathcal{J}_c(\xi) : \Xi \to c(\Xi, \xi)\xi$ is an isometry of $\xi^\perp$. Thus, we can choose $\Xi \in S(\xi^\perp)$ so that $\mathcal{J}_c(\xi)\Xi = -\zeta$. We then have

$$\psi\xi = \cos\theta\xi + \sin\theta c(\xi, \Xi)\xi.$$

Let $\eta \in S(\xi^\perp)$. We use Lemma 2.10.11 (3) to compute:

$$\begin{aligned}(\xi, \psi\eta) =& -(\xi, c(\xi, \eta)\psi\xi) = -(\xi, c(\xi, \eta)\{\cos\theta + \sin\theta c(\xi, \Xi)\}\xi)\\ =& -\sin\theta(\xi, c(\xi, \eta)c(\xi, \Xi)\xi)\\ =& -\tfrac{\sin\theta}{2}(\{c(\xi, \Xi)c(\xi, \eta) + c(\xi, \eta)c(\xi, \Xi)\}\xi, \xi)\\ =& \sin\theta(\Xi, \eta)(\xi, \xi) = \sin\theta(\Xi, \eta).\end{aligned}$$

Since $\sin\theta \neq 0$, $(\xi, \psi\eta) = 0$ if and only if $(\Xi, \eta) = 0$. This completes the proof of Assertion (1).

Let $\xi \in S^7$. To prove Assertion (2), we suppose that $\psi\xi = \pm\xi$. Clearly $Z(\xi) \subset S(\xi^\perp)$. Conversely, let $\eta \in S(\xi^\perp)$. Then trivially $\eta \perp \psi\xi$. By Lemma 2.10.11,

$$\psi\eta = -c(\xi, \eta)\psi\xi = \pm c(\xi, \eta)\xi.$$

We complete the proof of Assertion (2) by checking:

$$\psi\eta \perp \xi \text{ so } \eta \in Z(\xi).$$

Let $\xi \in S^7$ and let $\eta \in S(Z(\xi))$. We prove Assertion (3) by using the definitions given in Lemmas 2.10.11 and 2.10.12 to compute:

$$\begin{aligned}R(\xi, \eta)\xi =& (\mu_1 - 1)\{(\xi, \psi\eta)\psi\xi - (\xi, \psi\xi)\psi\eta\} + c(\xi, \eta)\xi\\ =& -(\mu_1 - 1)(\xi, \psi\xi)\psi\eta + c(\xi, \eta)\xi\\ =& (\mu_1 - 1)(\xi, \psi\xi)c(\xi, \eta)\psi\xi + c(\xi, \eta)\xi\\ =& c(\xi, \eta)\{\xi + (\mu_1 - 1)(\xi, \psi\xi)\psi\xi\}\\ =& c(\xi, \eta)\sigma(\xi).\end{aligned}$$

Let $\xi \in S^7$ and let $\eta \in S(Z(\xi))$. To prove Assertion (4), we must show $\mathcal{J}_R(\xi)\eta \in Z(\xi)$. If $\xi = \pm\psi\xi$, then $Z(\xi) = \xi^\perp$. Since $R(\eta,\xi)\xi \perp \xi$, we have $\mathcal{J}_R(\xi)$ preserves $Z(\xi)$. This completes the proof of Assertion (4) in this special case.

Next, we suppose that $\xi \neq \pm\psi\xi$. Let $\eta \in Z(\xi)$. We use Assertion (1) to see that $\mathcal{J}_R(\xi)\eta \in Z(\xi)$ if we have the relations:

$$\mathcal{J}_R(\xi)\eta \perp \xi, \mathcal{J}_R(\xi)\eta \perp \psi\xi, \text{ and } \mathcal{J}_R(\xi)\eta \perp \Xi.$$

We always have $\mathcal{J}_R(\xi)\eta \perp \xi$. By assumption $\xi \perp \psi\eta$. We show that $\mathcal{J}_R(\xi)\eta \perp \psi\xi$ by using Assertion (3) and Lemma 2.10.11 to compute:

$$\begin{aligned}
0 =&(\xi, \psi\eta) = -(\xi, c(\xi,\eta)\psi\xi) = (c(\xi,\eta)\xi, \psi\xi)\\
=&(c(\xi,\eta)\xi, \psi\xi) + (\xi,\psi\xi)(\mu_1 - 1)(c(\xi,\eta)\psi\xi,\psi\xi)\\
=&(c(\xi,\eta)\sigma(\xi),\psi\xi) = (R(\xi,\eta)\xi,\psi\xi) = -(\mathcal{J}_R(\xi)\eta,\psi\xi).
\end{aligned}$$

We now check that $\mathcal{J}_R(\xi)\eta \perp \Xi$. Since $\sin\theta \neq 0$,

$$\pi := \operatorname{span}\{\xi,\psi\xi\} = \operatorname{span}\{\xi, c(\xi,\Xi)\xi\}.$$

As $c(\xi,\Xi)^2 = -\operatorname{Id}$, $c(\xi,\Xi)\pi \subset \pi$. Since $\psi\xi \in \pi$, $\pi = \operatorname{span}\{\psi\xi, c(\xi,\Xi)\psi\xi\}$. Consequently, we have:

$$\begin{aligned}
&\eta \perp \xi,\ \eta \perp \psi\xi,\\
&\eta \perp \pi,\ \eta \perp c(\xi,\Xi)\xi, \text{ and}\\
&\eta \perp c(\xi,\Xi)\psi\xi = -\psi\Xi.
\end{aligned}$$

We complete the proof of Assertion (4) by using Lemma 2.10.11 to compute:

$$\begin{aligned}
0 =&(\mu_1 - 1)\{(\psi\Xi,\xi)(\psi\xi,\eta) - (\psi\Xi,\eta)(\psi\xi,\xi)\} + (c(\xi,\Xi)\xi,\eta)\\
=&(R(\xi,\Xi)\xi,\eta) = (R(\xi,\eta)\xi,\Xi) = -(\mathcal{J}_R(\xi)\eta,\Xi).
\end{aligned}$$

Suppose that Assertion (5) fails, i.e. $\sigma(\xi_1) \neq 0$ for some $\xi_1 \in S^7$. Let $\eta \in S(Z(\xi_1))$. We then have

$$|\mathcal{J}_R(\xi_1)\eta| = |R(\xi_1,\eta)\xi_1| = |c(\xi_1,\eta)\sigma| = |\sigma|.$$

We use this equation and Assertion (4) to see that the map $\frac{1}{|\sigma|}\mathcal{J}_R(\xi_1)|_{Z(\xi_1)}$ is a self-adjoint isometry of $Z(\xi_1)$. Consequently, we may diagonalize the restriction of $\mathcal{J}_R(\xi_1)$ to $Z(\xi_1)$ to decompose

$$Z(\xi_1) = Z_+(\xi_1) \oplus Z_-(\xi_1), \text{ where } \mathcal{J}_R(\xi_1) = \pm|\sigma| \text{ on } Z_\pm(\xi_1).$$

We consider the following two cases:

(1) Suppose $\dim Z_\varepsilon(\xi_1) \geq 4$ for $\varepsilon = \pm 1$. Choose $\xi_2 \in S(Z_\varepsilon(\xi_1))$ and impose 3 additional conditions to choose $\xi_3 \in S(Z_\varepsilon(\xi_1))$ so that $\xi_2 \perp \xi_3$, $\xi_2 \perp \psi\xi_3$, and $\psi\xi_2 \perp \xi_3$.

(2) Suppose $\dim Z_\varepsilon(\xi_1) = 3$ and $\dim Z_{-\varepsilon}(\xi_1) = 2$ for $\varepsilon = \pm 1$. We choose $\xi_2 \in S(Z_{-\varepsilon}(\xi_1))$. We can impose 2 conditions to choose $\xi_3 \in S(Z_\varepsilon(\xi_1))$ so $\xi_2 \perp \psi\xi_3$ and $\psi\xi_2 \perp \xi_3$. Since ξ_2 and ξ_3 belong to different eigenspaces of a self-adjoint operator, we automatically have $\xi_2 \perp \xi_3$.

This shows that we can choose ξ_2 and ξ_3 so that:

$$\begin{aligned}
&\xi_1 \perp \xi_2,\ \xi_1 \perp \psi\xi_2,\ \psi\xi_1 \perp \xi_2,\\
&\xi_1 \perp \xi_3,\ \xi_1 \perp \psi\xi_3,\ \psi\xi_1 \perp \xi_3,\\
&\xi_2 \perp \xi_3,\ \xi_2 \perp \psi\xi_3,\ \psi\xi_2 \perp \xi_3.
\end{aligned}$$

We use the first Bianchi identity to see:

$$R(\xi_2, \xi_3)\xi_1 = -R(\xi_3, \xi_1)\xi_2 - R(\xi_1, \xi_2)\xi_3. \tag{2.10.12.a}$$

Since $\xi_1 \perp \psi\xi_2$ and since $\xi_1 \perp \psi\xi_3$, we have that $\xi_1 \perp W_1(\xi_2, \xi_3)$; consequently, $\xi_1 \in W_2(\xi_2, \xi_3)$ so:

$$|R(\xi_2, \xi_3)\xi_1| = 1. \tag{2.10.12.b}$$

Similarly, we have that $\xi_3 \in W_2(\xi_1, \xi_2)$ and that $\xi_2 \in W_2(\xi_1, \xi_3)$. Thus,

$$R(\xi_1, \xi_2)\xi_3 = c(\xi_1, \xi_2)\xi_3 \text{ and } R(\xi_1, \xi_3)\xi_2 = c(\xi_1, \xi_3)\xi_2. \tag{2.10.12.c}$$

Since ξ_2 and ξ_3 belong to $Z(\xi_1)$, we may use Assertion (3) to see that

$$\xi_2 = c(\xi_1, \xi_2)\sigma \text{ and } \xi_3 = c(\xi_1, \xi_3)\sigma \text{ where } \sigma = \sigma(\xi_1). \tag{2.10.12.d}$$

Since $\xi_2, \xi_3 \in Z_\pm(\xi)$, we can find $\varepsilon_i = \pm 1$ so

$$\xi_i = \varepsilon_i |\sigma|^{-1} R(\xi_i, \xi)\xi \text{ for } i = 1, 2. \tag{2.10.12.e}$$

By Equations (2.10.12.a), (2.10.12.c), (2.10.12.d), and (2.10.12.e):

$$\begin{aligned}
R(\xi_2, \xi_3)\xi_1 =& -R(\xi_3, \xi_1)\xi_2 - R(\xi_1, \xi_2)\xi_3 = c(\xi_1, \xi_3)\xi_2 - c(\xi_1, \xi_2)\xi_3\\
=&|\sigma|^{-1}\{\varepsilon_2 c(\xi_1, \xi_3)R(\xi_2, \xi_1) + \varepsilon_3 c(\xi_1, \xi_2)R(\xi_3, \xi_1)\}\xi_1\\
=&|\sigma|^{-1}\{\varepsilon_2 c(\xi_1, \xi_3)c(\xi, \xi_2) + \varepsilon_3 c(\xi_1, \xi_2)c(\xi_3, \xi_1)\}\sigma.
\end{aligned}$$

We use the Clifford commutation relations to demonstrate:

$$R(\xi_2,\xi_3)\xi_1=\begin{cases} 2|\sigma|^{-1}c(\xi_1,\xi_2)c(\xi_1,\xi_3)\sigma & \text{if } \varepsilon_2=1,\ \varepsilon_3=-1\\ -2|\sigma|^{-1}c(\xi_1,\xi_2)c(\xi_1,\xi_3)\sigma & \text{if } \varepsilon_2=-1,\ \varepsilon_3=1\\ 0 & \text{if } \varepsilon_2=\varepsilon_3\end{cases}$$

Since $c(\xi_1,\xi_i)$ is an isometry for $i=2,3$, we have either that $|R(\xi_2,\xi_3)\xi_1|=2$ or that $|R(\xi_2,\xi_3)\xi_1|=0$. This contradicts Equation (2.10.12.b). Thus, $\sigma=0$.

Since $\sigma=0$, $\xi+(\mu_1-1)(\xi,\psi\xi)\psi\xi=0$. Consequently, $\psi\xi=\pm\xi$ and $\mu_1=0$; thus $\lambda_1=0$. We use continuity to see the sign is universal so $\psi=\pm\text{Id}$. □

2.10.13. *The eigenvalue structure* $\vec{m}=(2,6)$ *is impossible.*

Proof. We use Lemma 2.10.12. Let $\{\xi_1,\xi_2,\xi_3\}$ be an orthonormal set. We adopt the notation of Lemma 2.10.5. Since $\{\xi_1=\pm\psi\xi_1,\xi_2=\pm\psi\xi_2\}$ is an orthonormal basis for $W_{1,12}$ and since $\{\xi_1=\pm\psi\xi_1,\xi_3=\pm\psi\xi_3\}$ is an orthonormal basis for $W_{1,13}$, we have $\xi_1=\pm e_8$, $\xi_2=\pm e_7$, and $\xi_3=\pm e_6$. Consequently, $r_{13}\xi_2=\pm r_{12}\xi_3=\pm e_5$. Similarly, $r_{23}\xi_1=\pm r_{12}\xi_3=\pm e_5$. We use the first Bianchi identity to establish the final contradiction by computing:

$$0=r_{12}\xi_3+r_{23}\xi_1+r_{31}\xi_2=(\pm1\pm1\pm1)e_5. \quad \square$$

2.11 Almost complex IP tensors

We begin by reviewing, briefly, some definitions that we will be using in this section. Let V be a vector space of signature $(2p,2q)$ which is equipped with a Hermitian almost complex structure J. An orthonormal basis $\mathcal{B}$ for V is said to be a *normalized complex basis* if we have

(2.11.1.a) $$\mathcal{B}:=\{e_1^-,Je_1^-,...,e_p^-,Je_p^-,e_1^+,Je_1^+,...,e_q^+,Je_q^+\},$$

where the vectors $\{e_1^-,Je_1^-,...,e_p^-,Je_p^-\}$ are timelike and where the vectors $\{e_1^+,Je_1^+,...,e_q^+,Je_q^+\}$ are spacelike. We say that a 2 plane $\pi\subset V$ is a *complex line* if $J\pi=\pi$; let $\mathbb{CP}(V)\subset\text{Gr}_2(V)$ be the projective space of complex lines and let $\mathbb{CP}^\pm(V)$ be the projective spaces of complex spacelike $(+)$ and timelike $(-)$ lines; there are no mixed complex lines. We say that a linear transformation T of V is *complex* if $JT=TJ$.

We say that an algebraic curvature tensor R on V is *almost complex* if $R(x,Jx)$ is complex for every $x\in V$; we refer to Lemma 1.6.2 for other equivalent characterizations. We say that an algebraic curvature tensor R on V is *almost complex spacelike/timelike IP* if R is almost complex and if the eigenvalues of $R(\pi)$, viewed as a complex linear transformation of V, are constant on $\mathbb{CP}^{\pm}(V)$; by Lemma 1.10.7, these are equivalent notions so we shall simply say that R is *almost complex IP*.

The following Lemma will be used in Section 4.1 to complete the proof of Theorem 2.1.8. In the Riemannian setting, if R is almost complex, then $JR(x, Jx)$ is self-adjoint, when regarded as a complex map, and hence is diagonalizable with real eigenvalues.

2.11.2 Lemma. *Let R be an almost complex IP algebraic curvature tensor on a Riemannian vector space V. Let $\lambda \in \mathbb{R}$ be an extremal eigenvalue of $JR(\cdot)$ on $\mathbb{CP}(V)$. Let $x, y \in S(V)$. Then $R(x, Jx)y = \lambda Jy$ if and only if $R(y, Jy)x = \lambda Jx$.*

Proof. Let $A(x) := -JR(x, Jx)$; this is a self-adjoint operator on V. Suppose $\lambda \in \{\lambda_{\min}, \lambda_{\max}\} \subset \mathbb{R}$ is an extremal eigenvalue of A. Let $x, y \in S(V)$. By Lemma 1.3.2, $y \in E_\lambda(x)$ if and only if $(A(x)y, y) = \lambda$, i.e.:

$$\lambda = (-JR(x, Jx)y, y) = R(x, Jx, y, Jy).$$

The curvature symmetry $R(x, Jx, y, Jy) = R(y, Jy, x, Jx)$ then shows that $y \in E_\lambda(x)$ if and only if $x \in E_\lambda(y)$ and completes the proof. □

The eigenvalue structure does not control the Jordan form in the higher signature setting. Let R be an almost complex algebraic curvature tensor on R. We say that R is *almost complex spacelike/timelike Jordan IP* if the Jordan form of $R(\pi)$, regarded as a complex linear mapping, is constant on $\mathbb{CP}^{\pm}(V)$. We use Lemma 1.10.7 to see that any almost complex spacelike (or timelike) Jordan IP algebraic curvature tensor is an almost complex IP algebraic curvature tensor. The converse need not hold. Furthermore, there exist algebraic curvature tensors which are almost complex spacelike Jordan IP but which are not almost complex timelike Jordan IP.

2.11.3 Lemma. *Let J be a pseudo-Hermitian almost complex structure on a vector space V of signature $(2p, 2q)$.*

(1) *If $q \geq 2$ and if $p \geq 2$, then there exists an algebraic curvature tensor R on V which is almost complex IP, but which is neither almost complex spacelike Jordan IP nor almost complex timelike Jordan IP.*

(2) *If $p > q \geq 1$, then there exists an algebraic curvature tensor R on V which is almost complex spacelike Jordan IP, but which is not almost complex timelike Jordan IP.*

Proof. Let ϕ be a self-adjoint complex linear map of V with $\phi^2 = 0$. Set

$$R_\phi(x, y)z := (\phi y, z)\phi x - (\phi x, z)\phi y.$$

By Lemma 1.8.1, R_ϕ is an almost complex algebraic curvature tensor on V and $R_\phi(x, y)^2 = 0$ for any (x, y). Thus, R_ϕ is IP as 0 is the only eigenvalue of $R_\phi(x, y)$. This shows that R_ϕ is almost complex IP. By Lemma 1.2.10, the Jordan form of $R_\phi(\pi)$ is determined by rank $R_\phi(\pi)$. By Lemma 1.2.2, $R_\phi(\pi)$ has rank 2 if and only if $\ker \phi \cap \pi = \{0\}$. Thus, R_ϕ will be an almost complex spacelike (resp. timelike) Jordan IP if and only if $\ker \phi$ contains no spacelike (resp. timelike) vectors. Let $\mathcal{B}$ be a normalized complex orthonormal basis for V.

To prove Assertion (1), we construct a self-adjoint complex map ϕ with $\phi^2 = 0$ so that $\ker \phi$ contains both spacelike (resp. timelike) vectors and so that there exists at least one spacelike (resp. timelike) complex line π so $\pi \cap \ker \phi = \{0\}$ by setting:

$$\begin{aligned} &\phi(e_1^\pm) = \pm(e_1^+ + e_1^-), \quad \phi(Je_1^\pm) = \pm(Je_1^+ + Je_1^-), \\ &\phi(e_i^\pm) = 0 \text{ for } i > 1, \quad \phi(Je_i^\pm) = 0 \text{ for } i > 1. \end{aligned}$$

To prove Assertion (2), we construct a self-adjoint complex map ϕ with $\phi^2 = 0$ so that $\ker \phi$ contains no spacelike vectors, and so that $\ker \phi$ contains timelike vectors by setting

$$\begin{aligned} &\phi(e_j^\pm) = \pm(e_j^+ + e_j^-), \quad \phi(Je_j^\pm) = \pm(Je_j^+ + Je_j^-) \quad \text{for } 1 \le j \le q, \\ &\phi(e_i^-) = 0, \qquad\qquad\qquad \phi(Je_i^-) = 0, \qquad\qquad\qquad \text{for } i > q. \qquad \square \end{aligned}$$

In dimension 4, Kath [110] has classified the algebraic curvature tensors which are both almost complex spacelike Jordan IP and almost complex timelike Jordan IP. Apart from her results, there are no general classification results for almost complex IP algebraic curvature tensors and metrics. It is useful, therefore, to present some examples.

Recall that we have defined two families of algebraic curvature tensors:

$$R_\phi(x, y)z := \begin{cases} (\phi y, z)\phi x - (\phi x, z)\phi y & \text{if } \phi = \phi^*, \\ (\phi y, z)\phi x - (\phi x, z)\phi y - 2(\phi x, y)\phi z & \text{if } \phi = -\phi^*. \end{cases}$$

Let J be a pseudo-Hermitian almost complex structure on V. Assume $\phi J = \pm J\phi$. We may then use Lemma 1.8.1 to conclude that R_ϕ is an almost complex algebraic curvature tensor. We shall say that ϕ is *admissible* if

(1) Either $\phi = \phi^*$ and $\phi J = \pm J\phi$ or $\phi = -\phi^*$ and $J\phi = -\phi J$.

(2) Either $\phi^2 = \mathrm{Id}$, or $\phi^2 = -\mathrm{Id}$, or $\phi^2 = 0$ and $\ker \phi = \operatorname{range} \phi$.

A pair $\{\phi_1, \phi_2\}$ is said to be *admissible* if

(1) Both ϕ_1 and ϕ_2 are admissible.
(2) We have $\phi_1 J = J\phi_1$, $\phi_2 J = -J\phi_2$, and $\phi_1^*\phi_2 + \phi_2^*\phi_1 = 0$.
(3) If $\phi_1^2 = \phi_2^2 = 0$ and if π is any non-degenerate complex line, then we have that $\phi_1\pi \cap \phi_2\pi = \{0\}$.

2.11.4 Theorem. *Let J be a pseudo-Hermitian almost complex structure on a vector space V. Let ϕ, ϕ_1, and ϕ_2 be linear transformations of V, and let c, c_1, and c_2 be arbitrary real constants. Then:*

(1) *If ϕ is admissible, then cR_ϕ is an algebraic curvature tensor which is both almost complex spacelike Jordan IP and almost complex timelike Jordan IP.*
(2) *If $\{\phi_1, \phi_2\}$ is admissible, then $c_1 R_{\phi_1} + c_2 R_{\phi_2}$ is an algebraic curvature tensor which is both almost complex spacelike Jordan IP and almost complex timelike Jordan IP.*

We remark that Assertion (1) is not a special case of Assertion (2) since the structures described in Assertion (1) can not always be extended to the structure described in Assertion (2).

Proof. We follow the discussion in Gilkey and Ivanova [78] to prove this result. Let $x \in S^\pm(V)$ and let $\pi := \operatorname{span}\{x, Jx\} \in \mathbb{CP}^\pm(V)$.

To prove Assertion (1), we suppose that ϕ is admissible. Thus, as noted above, R_ϕ is an almost complex algebraic curvature tensor. If ϕ is skew-adjoint, then $J\phi = -\phi J$. Thus,

$$(\phi x, Jx) = (-J\phi x, x) = (\phi Jx, x) = -(Jx, \phi x) = 0.$$

This means that

$$R_\phi(x, Jx)z = \begin{cases} (\phi Jx, z)\phi x - (\phi x, z)\phi Jx & \text{if } \phi = \phi^*, \\ (\phi Jx, z)\phi x - (\phi x, z)\phi Jx - 0 & \text{if } \phi = -\phi^*. \end{cases} \tag{2.11.4.a}$$

Thus, it does not matter whether ϕ is self-adjoint or skew-adjoint. We use the identity $\phi J = \pm J\phi$ to see that

$$\phi\pi = \operatorname{span}\{\phi x, \phi Jx\} = \operatorname{span}\{\phi x, J\phi x\} \in \mathbb{CP}(V).$$

To prove Assertion (1), we must show that the Jordan normal form of $R_\phi(\pi)$ does not depend on the particular non-degenerate complex line π which has been chosen. We distinguish two cases:

(1) Suppose that $\phi^2 = \varepsilon\,\mathrm{Id}$ and that $\phi^* = \varrho\phi$, where $\varepsilon = \pm 1$ and where $\varrho = \pm 1$. Then $(\phi x, \phi y) = (\phi^*\phi x, y) = \varepsilon\varrho(x, y) = \pm(x, y)$ so either ϕ

is an isometry or ϕ is a para-isometry. Thus, as π is either spacelike or timelike, $\phi\pi$ is either spacelike or timelike. We use Equation (2.11.4.a) to see that $R_\phi(\pi)$ is a 90 degree rotation in the plane $\phi\pi$ and that $R_\phi(\pi)$ vanishes on $(\phi\pi)^\perp$. Consequently, the Jordan normal form of $R_\phi(\pi)$ is independent of π so R is Jordan IP.

(2) Suppose that $\phi^2 = 0$. We then have

$$(\phi x, \phi y) = (\phi^* \phi x, y) = \pm(\phi^2 x, y) = 0 \text{ for any } x, y$$

so the range of ϕ is totally isotropic. Since $\ker\phi = \operatorname{range}\phi$, the kernel of ϕ contains neither spacelike nor timelike vectors. Because π is either spacelike or timelike this means that $\ker\phi \cap \pi = \{0\}$. Consequently, ϕ is an isomorphism from π to $\phi\pi$. Thus, $\dim\phi\pi = 2$ and $\phi\pi = \operatorname{span}\{\phi x, J\phi x\} \subset \operatorname{range}\phi$ is a totally isotropic complex line. We use Equation (2.11.4.a) to prove that in this case $R_\phi(\pi)^2 = 0$ and that $\operatorname{rank}(R_\phi(\pi)) = 2$. We may now use Lemma 1.2.10 to see that the Jordan normal form of $R_\phi(\pi)$ is independent of π and hence R_ϕ is Jordan IP.

This completes the proof of Assertion (1) of Theorem 2.11.4.

To prove Assertion (2), we suppose that $\{\phi_1, \phi_2\}$ is an admissible pair and set $R := c_1 R_{\phi_1} + c_2 R_{\phi_2}$. Again, we use Lemma 1.8.1 to see that R is an almost complex algebraic curvature tensor. We use the given commutation relations to show that $\{\phi_1 x, \phi_1 Jx, \phi_2 x, \phi_2 Jx\}$ is an orthogonal set by computing:

$$\begin{aligned}
(\phi_1 x, \phi_1 Jx) &= (\phi_1 x, J\phi_1 x) = -(J\phi_1 x, \phi_1 x) = 0,\\
(\phi_1 x, \phi_2 x) &= (\phi_2^* \phi_1 x, x) = -(\phi_1^* \phi_2 x, x) = -(\phi_2 x, \phi_1 x) = 0,\\
(\phi_1 x, \phi_2 Jx) &= -(J\phi_2^* \phi_1 x, x) = 0,\\
(\phi_1 Jx, \phi_2 x) &= (\phi_2^* \phi_1 Jx, x) = -(J\phi_2^* \phi_1 x, x) = (\phi_2^* \phi_1 x, Jx)\\
&= -(\phi_1^* \phi_2 x, Jx) = -(\phi_2 x, \phi_1 Jx) = 0,\\
(\phi_1 Jx, \phi_2 Jx) &= (\phi_2^* \phi_1 Jx, Jx) = -(\phi_1^* \phi_2 Jx, Jx)\\
&= -(\phi_2 Jx, \phi_1 Jx) = 0,\\
(\phi_2 x, \phi_2 Jx) &= -(\phi_2 x, J\phi_2 x) = (J\phi_2 x, \phi_2 x) = 0.
\end{aligned}$$

Thus, the two complex lines $\phi_1\pi$ and $\phi_2\pi$ are orthogonal complex lines. To complete the proof, we must show that the Jordan normal form of $R(\pi)$ is independent of π. We distinguish three cases:

(1) Suppose that $\phi_1^2 = \pm\operatorname{Id}$ and that $\phi_2^2 = \pm\operatorname{Id}$. In proving Assertion (1), we showed that $R_{\phi_s}(\pi)$ is a 90 degree rotation in the plane $\phi_s\pi$ and $R_{\phi_s}(\pi)$ vanishes on $\phi_s\pi^\perp$ $(s = 1, 2)$. Since $\phi_1\pi$ and $\phi_2\pi$ are

orthogonal non-degenerate complex lines, the Jordan normal form of $R(\pi)$ is independent of π.

(2) Suppose that $\phi_1^2 = \pm\,\mathrm{Id}$ and that $\phi_2^2 = 0$; the argument if we have that $\phi_1^2 = 0$ and that $\phi_2^2 = \pm\,\mathrm{Id}$ is similar and is therefore omitted. In proving Assertion (1), we showed that $R_{\phi_1}(\pi)$ is a 90 degree rotation in the plane $\phi_1\pi$, that $\phi_1\pi$ is a timelike or spacelike complex line, and that $R_{\phi_1}(\pi)$ vanishes on $\phi_1\pi^\perp$. We also showed that range $R_{\phi_2}(\pi)$ is $\phi_2\pi$, that $\phi_2\pi$ is a totally isotropic complex line, and that $R_{\phi_2}^2 = 0$. Consequently, $\phi_1\pi \cap \phi_2\pi = \{0\}$ and the Jordan normal form of $R(\pi)$ is independent of π.

(3) Suppose that $\phi_1^2 = \phi_2^2 = 0$. In proving Assertion (1), we showed that range $R_{\phi_s}(\pi) = \phi_s\pi$, that $\phi_s\pi$ is a totally isotropic complex line, and that $R_{\phi_s}^2 = 0$ $(s = 1, 2)$. In this case, we have assumed that $\phi_1\pi \cap \phi_2\pi = \{0\}$. It now follows that $\mathrm{rank}(R_\phi(\pi)) = 4$ and that $R_\phi(\pi)^2 = 0$. The desired result now follows from Lemma 1.2.10. □

There is another family of examples that we can construct.

2.11.5 Theorem. *Let J be a pseudo-Hermitian almost complex structure on a vector space V. Let c, c_1, c_2, and c_3 be arbitrary real constants. We then have:*

(1) *The tensor $R := c_0 R_{\mathrm{Id}} + c_1 R_J$ is an algebraic curvature tensor which is both almost complex spacelike Jordan IP and almost complex timelike Jordan IP.*

(2) *Let ϕ be a skew-adjoint transformation of V with $\phi^2 = \pm\,\mathrm{Id}$ and $\phi J = -J\phi$. Then $R := c_0 R_{\mathrm{Id}} + c_1 R_J + c_2 R_\phi + c_3 R_{J\phi}$ is an algebraic curvature tensor which is both almost complex spacelike Jordan IP and almost complex timelike Jordan IP.*

Again, we remark that Assertion (1) is not a special case of Assertion (2) since the structure described in Assertion (1) can not always be extended to the structure described in Assertion (2). The structure in (2) is a quaternion structure if $\phi^2 = -\,\mathrm{Id}$ and a para-quaternion structure if $\phi^2 = \mathrm{Id}$.

Proof. We prove Assertions (1) and (2) at the same time by considering the tensors

$$R := c_0 R_{\mathrm{Id}} + c_1 R_J, \text{ or}$$
$$R := c_0 R_{\mathrm{Id}} + c_1 R_J + c_2 R_\phi + c_3 R_{J\phi}.$$

By Lemma 1.8.1, R is an almost complex algebraic curvature tensor. We show $\{x, Jx, \phi x, J\phi x\}$ is an orthogonal set by computing:

$$\begin{aligned}
&(x, Jx) = (-Jx, x) = 0, \\
&(x, \phi x) = (-\phi x, x) = 0, \\
&(x, J\phi x) = (\phi Jx, x) = -(J\phi x, x) = 0, \\
&(Jx, \phi x) = -(\phi Jx, x) = 0, \\
&(Jx, J\phi x) = (x, \phi x) = 0, \\
&(\phi x, J\phi x) = -(J\phi x, \phi x) = 0.
\end{aligned}$$

Let $\pi := \operatorname{span}\{x, Jx\}$. Let $(x, x) = \delta = \pm 1$ and let $\phi^2 = \varepsilon\,\mathrm{Id}$. We use the definition to see that

$$\begin{aligned}
R(\pi)y =& c_0\{(y, Jx)x - (y, x)Jx\} \\
&+ c_1\{(y, JJx)Jx - (y, Jx)JJx - 2(Jx, Jx)Jy\} \\
&+ c_2\{(y, \phi Jx)\phi x - (y, \phi x)\phi Jx - 2(\phi x, Jx)\phi y\} \\
&+ c_3\{(y, J\phi Jx)J\phi x - (y, J\phi x)J\phi Jx - 2(J\phi x, Jx)J\phi y\} \\
=& c_0\{(y, Jx)x - (y, x)Jx\} + c_1\{(y, Jx)x - (y, x)Jx - 2\delta Jy\} \\
&+ (c_2 + c_3)\{(y, \phi Jx)\phi x - (\phi x, y)\phi Jx\}
\end{aligned}$$

We complete the proof by computing:

$$\begin{aligned}
&R(\pi)x = -\delta(c_0 + 3c_1)Jx, \\
&R(\pi)Jx = \delta(c_0 + 3c_1)x, \\
&R(\pi)\phi x = -\delta(2c_1 + c_2\varepsilon + c_3\varepsilon)J\phi x, \qquad (2.11.5.a) \\
&R(\pi)J\phi x = \delta(2c_1 + c_2\varepsilon + c_3\varepsilon)\phi x, \\
&R(\pi)y = -\delta 2c_1 Jy \text{ if } y \perp \operatorname{span}\{x, Jx, \phi x, J\phi x\}. \quad \square
\end{aligned}$$

We now restrict to the Riemannian setting. Let J be the usual Hermitian almost complex structure on $\mathbb{R}^{(0,2s)} = \mathbb{C}^s$. Let $\{i, j, k\}$ be the usual skew-symmetric quaternion structure on $\mathbb{R}^{(0,4s)} = \mathbb{C}^{2s}$, where we take $J = i$ to define a Hermitian almost complex structure. If R is an almost complex IP algebraic curvature tensor, we let $\{\lambda_0, ..., \lambda_\ell\}$ and $\{\mu_0, ..., \mu_\ell\}$ be the corresponding eigenvalues and multiplicities of $R(\pi)$ viewed as a complex linear map. We order the multiplicities so $\mu_0 \geq ... \geq \mu_\ell$. We can use Theorem 2.11.5 to construct examples realizing all the possibilities of Theorem 2.1.8.

2.11.6 Theorem. *Let λ_0, λ_1, and λ_2 be distinct real numbers. Let $V = \mathbb{C}^s$ with the standard almost complex structure i.*

(1) *Let $R = c_0 R_{\mathrm{Id}} + c_1 R_i$. Then R is an almost complex IP algebraic curvature tensor on V. We can choose the constants c_0 and c_1 so that $R(\cdot)$ has eigenvalues $\sqrt{-1}\lambda_0$ and $\sqrt{-1}\lambda_1$ with multiplicities $\mu_0 = s-1$ and $\mu_1 = 1$ on $\mathbb{CP}(V)$, respectively.*

(2) *Suppose that $s = 2t$ is even. Let $\{i, j, k\}$ be the usual quaternion action on $\mathbb{C}^s = \mathbb{H}^t$ and let $R = c_0 R_{\mathrm{Id}} + c_1 R_i + c_2 R_j$. Then R is an almost complex IP algebraic curvature tensor on V. We can choose the constants c_0, c_1, and c_2 so that the eigenvalues of $R(\cdot)$ on $\mathbb{CP}(V)$ are given by either (2a) or by (2b) below:*
   2a) *$\sqrt{-1}\lambda_0$ and $\sqrt{-1}\lambda_1$ with multiplicities $\mu_0 = 2s-2$ and $\mu_1 = 2$*
   2b) *$\sqrt{-1}\lambda_0$, $\sqrt{-1}\lambda_1$, and $\sqrt{-1}\lambda_2$ with multiplicities $\mu_0 = 2s-2$, $\mu_1 = 1$, and $\mu_2 = 1$.*

Proof. We use Equation (2.11.5.a) to see $(c_0 + 3c_1)$ is an eigenvalue of multiplicity 1, that $(2c_1 - c_2 - c_3)$ is an eigenvalue of multiplicity 1, and that $2c_1$ is an eigenvalue of multiplicity $2s-2$ for the self-adjoint operator $JR(\pi)$. We choose the constants c_s for $s = 0, 1, 2, 3$ so that $\lambda_1 = c_0 + 3c_1$, $\lambda_2 = 2c_1 - c_2 - c_3$, and $\lambda_0 = 2c_1$ to complete the proof of Assertion (2); Assertion (1) follows similarly. □

2.12 Higher order IP tensors

Let R be an algebraic curvature tensor on a vector space V of signature (p, q). Let $\{e_1, ..., e_\nu\}$ be a basis for $\sigma \in \widetilde{\mathrm{Gr}}_\nu(V)$. We set $h_{ij} := (e_i, e_j)$ and let h^{ij} be the components of the inverse matrix. We use Theorem 1.9.4 to define the *higher order skew-symmetric curvature operator* by setting:

$$\mathfrak{S}_R(\sigma) := \textstyle\sum_{1 \le i,j,k,l \le \nu} h^{ik} h^{jl} R(e_i, e_j) R(e_k, e_l).$$

We shall say that an algebraic curvature tensor R is ν *IP* if $J_R(\cdot)$ has constant eigenvalues on $\widetilde{\mathrm{Gr}}_\nu(V)$ or, equivalently by Lemma 1.10.8, on $\mathrm{Gr}_{r,s}(V)$ for any admissible pair (r, s) with $r + s = \nu$. A pseudo-Riemannian manifold (M, g) is said to be ν IP if ${}^g R_P$ is ν IP at every point of the manifold. One has the following examples:

2.12.1 Theorem. *Let $1 \le \nu \le m-1$. Let $p + q = m$.*

(1) *Let V be a vector space of signature (p, q). Let ϕ be a self-adjoint map of V with $\phi^2 = \delta\,\mathrm{Id}$ for $\delta = \pm 1, 0$. If $1 \le \nu \le m-1$, then R_ϕ is a ν IP algebraic curvature tensor.*

(2) *The manifolds of dimension m described in Lemmas 2.6.1, 2.6.3, and 2.6.5 are ν IP.*

Proof. Assertion (2) will follow directly from Assertion (1). To prove Assertion (1), we suppose first $\delta = 0$. By Lemma 1.8.1, $R_\phi(x,y)R_\phi(\bar{x},\bar{y}) = 0$ for any vectors $x, y, \bar{x}, \bar{y} \in V$; thus $\mathfrak{S}_{R_\phi}(\sigma) = 0$ for any σ and Assertion (1) holds.

We therefore suppose $\phi^2 = \delta\,\mathrm{Id}$ for $\delta = \pm 1$. Let $\{e_1, ..., e_\nu\}$ be an orthonormal basis for $\sigma \in \widetilde{\mathrm{Gr}}_\nu(V)$. We extend this set to a full orthonormal basis $\{e_1, ..., e_m\}$ for V. Let $\epsilon_i = (e_i, e_i)$. Let $i \neq j$. We have:

$$\begin{aligned} R(e_i,e_j)\phi e_k =& (\phi e_k, \phi e_j)\phi e_i - (\phi e_k, \phi e_i)\phi e_j \\ =& \begin{cases} \delta\varepsilon_j \phi e_i & \text{if } k = j, \\ -\delta\varepsilon_i \phi e_j & \text{if } k = i, \\ 0 & \text{if } k \neq i, j, \end{cases} \\ R(e_i,e_j)^2\phi e_k =& \begin{cases} -\varepsilon_i\varepsilon_j \phi e_k & \text{if } k = i, j, \\ 0 & \text{if } k \neq i, j. \end{cases} \end{aligned}$$

We show that the eigenvalues of $\mathfrak{S}_{R_\phi}(\sigma)$ only depend on the dimension of σ and thereby complete the proof by computing:

$$\mathfrak{S}_{R_\phi}(\sigma)(\phi e_k) = \textstyle\sum_{1\le i,j\le\nu} \varepsilon_i\varepsilon_j R^2_{e_i,e_j} e_k = \begin{cases} -2(\nu-1)\phi e_k & \text{if } k \le \nu, \\ 0 & \text{if } k > \nu. \end{cases} \quad \square$$

2.12.2 Remark. Suppose that $\phi^2 = \pm\mathrm{Id}$. The argument given above shows that R_ϕ has constant Jordan normal form on $\mathrm{Gr}_{r,s}(V)$ and hence is higher order Jordan IP of type (r,s) for any admissible pair.

Chapter 3
The Jacobi Operator

3.1 Introduction

Let R be an algebraic curvature tensor on a vector space V of signature (p,q) and dimension $m = p+q$. In Chapter 3, we will study the geometry of three self-adjoint operators which are associated to R: the Jacobi operator, the higher order Jacobi operator, and the Szabó operator.

The *Jacobi operator* is defined by:

$$\mathcal{J}_R(x) : y \to R(y,x)x.$$

We say that R is *spacelike Osserman* or *timelike Osserman* if the eigenvalues of $\mathcal{J}_R$ are constant on the pseudo-spheres

$$S^+(V) := \{v \in V : (v,v) = +1\} \text{ or}$$
$$S^-(V) := \{v \in V : (v,v) = -1\},$$

of unit spacelike or unit timelike vectors, respectively. We say that R is *Osserman* if R is either spacelike Osserman or timelike Osserman; we showed in Lemma 1.10.1 that spacelike Osserman and timelike Osserman were equivalent notions.

We say that an algebraic curvature tensor R is *spacelike Jordan Osserman* or *timelike Jordan Osserman* if the Jordan normal form of the Jacobi operator $\mathcal{J}_R$ is constant on $S^+(V)$ or on $S^-(V)$, respectively. We say that R is *Jordan Osserman* if R is both spacelike Jordan Osserman and timelike Jordan Osserman.

The Jordan normal form of a linear map determines the eigenvalue structure. Thus, if R is spacelike or timelike Jordan Osserman, then R is Osserman; the reverse implication need not hold - there are Osserman algebraic curvature tensors which are neither spacelike Jordan Osserman nor timelike Jordan Osserman. Furthermore, there are algebraic curvature tensors which are spacelike Jordan Osserman but not timelike Jordan Osserman. In Section 3.2, we will explore the relations between these various concepts and exhibit examples showing that the Jordan normal form of a pseudo-Riemannian Jordan Osserman algebraic curvature tensor can be arbitrarily complicated.

More generally, let $\mathcal{B} := \{e_1, ..., e_k\}$ be a basis for a non-degenerate k dimensional subspace $\sigma \in \widetilde{\operatorname{Gr}}_k(V)$. Let $h_{ij} := (e_i, e_j)$ describe the induced metric on σ; the inverse matrix h^{ij} describes the induced metric on the dual space σ^*. We showed in Theorem 1.9.4 that the *higher order Jacobi operator*

$$\mathcal{J}_R(\sigma)y := \textstyle\sum_{i,j} h^{ij} R(y, e_i)e_j \tag{3.1.1.a}$$

is a self-adjoint linear transformation of V which is independent of the basis $\mathcal{B}$ which was chosen for σ. This operator has also been called the *Stanilov operator* by some authors. If $x \in S^\pm(V)$, then $\mathcal{J}_R(x) = \pm\mathcal{J}_R(\operatorname{span}\{x\})$ so these operators agree apart from a normalizing sign convention.

Let σ be spacelike. We apply Theorem 1.9.4 to see that $\mathcal{J}_R(\sigma)$ can be computed as a suitable average:

$$\mathcal{J}_R(\sigma) = \tfrac{k}{\operatorname{vol}(S^{k-1})} \textstyle\int_{x\in\sigma,\ |x|=1} \mathcal{J}_R(x)dx.$$

We say that (r, s) is an *admissible pair* if $\operatorname{Gr}_{r,s}(V)$ is non-trivial, i.e.

$$0 \le r \le p,\ 0 \le s \le q, \text{ and } 1 \le r + s \le m - 1.$$

For such a pair, we shall say that an algebraic curvature tensor R is *Osserman of type* (r, s) if the eigenvalues of the higher order Jacobi operator are constant on $\operatorname{Gr}_{r,s}(V)$. We showed in Lemma 1.10.2 that only the value $r + s$ is relevant i.e. if $r + s = \tilde{r} + \tilde{s} = k$, then R is Osserman of type (r, s) if and only if R is Osserman of type $(\tilde{r}, \tilde{s})$; such an algebraic curvature tensor is said to be k *Osserman*.

We say that R is *Jordan Osserman of type* (r, s) if the Jordan normal form of the higher order Jacobi operator is constant on $\operatorname{Gr}_{r,s}(V)$. In Lemma 1.10.3, we established the following two duality results:

(1) If R is k Osserman, then R is $m - k$ Osserman.
(2) If R is Jordan Osserman of type (r, s), then R is Jordan Osserman of type $(p - r, q - s)$.

In Sections 3.2 and 3.3, we will construct algebraic curvature tensors which are higher order Jordan Osserman for certain (but not all) values of (r,s). We summarize those results graphically below. Entries with '$\star$' will mark points where the tensor is Jordan Osserman, entries with '$\circ$' will mark points where the tensor is Osserman but not Jordan Osserman, entries with $\times$ will mark points where the tensor is not Osserman, and the two corners $(0,0)$ and (p,q) are excluded as inadmissible and marked with '$-$'. The r-axis $(0 \leq r \leq p)$ is horizontal and the s-axis $(0 \leq s \leq q)$ is vertical.

Let $\{e_1^-, ..., e_p^-, e_1^+, ..., e_q^+\}$ be a normalized orthonormal basis for a vector space V of signature (p,q). Let $2a \leq \min(p,q)$. Define $0 \neq \phi_a \in \mathfrak{so}(V)$ by

$$\phi_a e_k^\pm = \begin{cases} \pm(e_{2i}^- + e_{2i}^+) & \text{if } k = 2i-1 \leq 2a, \\ \mp(e_{2i-1}^- + e_{2i-1}^+) & \text{if } k = 2i \leq 2a, \\ 0 & \text{if } k > 2a. \end{cases}$$

Theorem 3.2.2, Gilkey and Ivanova [81], shows the picture for R_{ϕ_a} is:

$2a < p \leq q$

$\star$	$\circ$	...	$\circ$	$-$
$\circ$	$\circ$	...	$\circ$	$\circ$
...	...	...	...	...
$\circ$	$\circ$	...	$\circ$	$\circ$
$-$	$\circ$	...	$\circ$	$\star$

$2a = p < q$

$\star$	$\star$	...	$\star$	$-$
$\circ$	$\circ$	...	$\circ$	$\circ$
...	...	...	...	...
$\circ$	$\circ$	...	$\circ$	$\circ$
$-$	$\star$	...	$\star$	$\star$

$2a = p = q$

$\star$	$\star$	...	$\star$	$-$
$\star$	$\circ$	...	$\circ$	$\star$
...	...	...	...	...
$\star$	$\circ$	...	$\circ$	$\star$
$-$	$\star$	...	$\star$	$\star$

Let $\phi \in \mathfrak{so}(V)$ and let $\phi^2 = -\operatorname{Id}$. Theorem 3.3.1 Gilkey, Stanilov, and Videv [85] shows that the picture for $R := R_{\text{Id}} + R_\phi$, which is the curvature tensor for complex projective space, takes the form:

$\times$	$\times$	...	$\star$	$-$
$\times$	$\times$	...	$\times$	$\star$
...	...	...	...	...
$\star$	$\times$	...	$\times$	$\times$
$-$	$\star$	...	$\times$	$\times$

Finally, define a self-adjoint map Φ_a of V by setting:

$$\Phi_a e_i^\pm = \begin{cases} \pm(e_i^+ + e_i^-) & \text{if } i \leq a, \\ 0 & \text{if } i > a. \end{cases}$$

Theorem 3.3.2 of Stavrov [146] shows that the picture for R_{Φ_a} is given by:

$a=2<p=q$				
⋆	○	...	○	−
○	○	...	○	○
...	...	...	...	...
○	○	...	○	○
−	○	...	○	⋆

$a=3<p=q$				
⋆	⋆	...	○	−
⋆	○	...	○	○
...	...	...	...	...
○	○	...	○	⋆
−	○	...	⋆	⋆

$a=p=q$				
⋆	⋆	...	⋆	−
⋆	○	...	○	⋆
...	...	...	...	...
⋆	○	...	○	⋆
−	⋆	...	⋆	⋆

We restrict to the Riemannian setting ($p = 0$) for a bit. In Section 3.4, we prove the following duality result which is closely related to the notion of 'symmetry' used in the proof of Theorem 2.1.8.

3.1.2 Theorem. *Let R be an algebraic curvature tensor on a Riemannian vector space V of dimension m.*

(1) *Suppose that R is 1 Osserman. Let v_1 and v_2 be unit vectors in V. If $\mathcal{J}_R(v_1)v_2 = \lambda v_2$, then $\mathcal{J}_R(v_2)v_1 = \lambda v_1$.*

(2) *Let $1 \leq k \leq m-1$. Suppose that R is k Osserman. Let $\tau \in \operatorname{Gr}_k(V)$.*
   2a) *Let $v \in S(V)$. If $\mathcal{J}_R(\tau)v = \lambda v$, then $\mathcal{J}_R(v)\tau \subset \tau$.*
   2b) *Let σ be any subspace of V. If $\mathcal{J}_R(\tau)\sigma \subset \sigma$, then $\mathcal{J}_R(\sigma)\tau \subset \tau$.*

We shall use Lemma 1.3.3 to prove Assertion (2a). We shall then derive Assertions (1) and (2b) from Assertion (2a). Assertion (1) was first proved by Rakić [131] using a different approach so Theorem 3.1.2 is known in the literature as the Rakić duality principle.

Let V be a Riemannian vector space of dimension m and let R be an Osserman algebraic curvature tensor on V. We identify the associated projective space $\mathbb{RP}(V)$ with $\operatorname{Gr}_1(V)$. The cohomology algebra is given by

$$H^*(\mathbb{RP}(V); \mathbb{Z}_2) = \mathbb{Z}_2[x]/(x^m = 0).$$

If E is a real vector bundle over $\mathbb{RP}(V)$, then $w(E) \in H^*(\mathbb{RP}(V); \mathbb{Z}_2)$ is the total Stiefel-Whitney class; we refer to Section 4.3 for further details. In Section 4.1, we will derive following result from Theorems 3.1.2 and 4.1.3:

3.1.3 Theorem. *Let R be an Osserman algebraic curvature tensor on a Riemannian vector space V of dimension m. Let λ be an eigenvalue of $\mathcal{J}_R$ of multiplicity μ and let $E_\lambda := \{(\sigma, w) \in \mathbb{RP}(V) \times V : \mathcal{J}_R(\sigma)w = \lambda w\}$ be the associated eigenbundle over $\mathbb{RP}(V)$. If $2\mu \leq m$, then $w(E_i) = (1+x)^\mu$.*

Let R be a Riemannian Osserman algebraic curvature tensor. Since the Jacobi operator $\mathcal{J}_R(x)x = 0$ and since $\mathcal{J}_R(x)$ is self-adjoint, $\mathcal{J}_R(x)x^\perp \subset x^\perp$. Let the *reduced Jacobi operator* $\tilde{\mathcal{J}}_R(x)$ be the restriction of $\mathcal{J}_R(x)$ to $x^\perp$. This operator encodes the same information as the operator $\mathcal{J}_R(x)$; the eigenvalues and thus the Jordan normal form of $\mathcal{J}_R$ are determined by the eigenvalues of $\tilde{\mathcal{J}}_R$. There is no higher order analogue of this operator.

The operators $\mathcal{J}_R(x)$ and $\tilde{\mathcal{J}}_R(x)$ are diagonalizable. Let $\{\lambda_s, \mu_s\}_{0\le s\le \ell}$ be the eigenvalues and associated multiplicities of $\tilde{\mathcal{J}}_R(\cdot)$ on $S(V)$; we order the eigenvalues so $\mu_0 \ge ... \ge \mu_\ell$. Note that $1 + \mu_0 + ... + \mu_\ell = \dim V$. Set

$$E_s(x) := \{(x, y) \in S(V) \times V : x \perp y \text{ and } \tilde{\mathcal{J}}_R(x)y = \lambda_s y\}.$$

Since $\tilde{J}_R$ is a smooth function of x and since $\dim(E_s) = \mu_s$ is constant, we use Lemma 4.2.5 to see that the E_s are smooth vector bundles of dimension μ_s over $S(V)$. Because $T_xS(V) = x^\perp$, we obtain a decomposition of the tangent bundle of the sphere:

$$TS(V) = \oplus_s E_s.$$

Chi [42] noted that the result of Adams [1], given in Theorem 1.4.2, could be used to control the eigenvalue structure of a Riemannian Osserman algebraic curvature tensor. The following theorem is the analogue of Theorem 2.1.8 in this context. Recall that the *Adams number* $\nu(m)$ is defined by:

$$\nu(a2^s) = \nu(2^s) \text{ for } a \equiv 1 \bmod 2,$$
$$\nu(1) = 0,\ \nu(2) = 1,\ \nu(4) = 3,\ \nu(8) = 7, \text{ and } \nu(16\tilde{b}) = \nu(\tilde{b}) + 8.$$

3.1.4 Theorem. *Let R be an Osserman algebraic curvature tensor on a Riemannian vector space V of dimension m. Let $\{\lambda_s, \mu_s\}_{0\le s\le \ell}$ be the eigenvalues and multiplicities of the reduced Jacobi operator $\tilde{\mathcal{J}}_R(\cdot)$ on $S(V)$, where we order the multiplicities so that $\mu_0 \ge \mu_1 \ge ... \ge \mu_\ell$.*

(1) *We have that $\mu_1 + ... + \mu_\ell \le \nu(m)$.*
(2) *If $m \equiv 1$ mod 2, then $\tilde{\mathcal{J}}_R$ has only one eigenvalue.*
(3) *If $m \equiv 2$ mod 4, then $\tilde{\mathcal{J}}_R$ has at most two eigenvalues; if $\tilde{\mathcal{J}}_R$ has two distinct eigenvalues, then one eigenvalue has multiplicity* 1.

Proof. Assertion (1) follows from Theorem 1.4.2. Assertions (2) and (3) follow as $\nu(m) = 0$ if m is odd and as $\nu(m) = 1$ if $m \equiv 2 \bmod 4$. □

In Section 3.2, we will show that Theorem 3.1.4 is sharp by constructing Riemannian Osserman algebraic curvature tensors which exhibit the maximal eigenvalue structure permitted by Theorem 3.1.4.

We say that a pseudo-Riemannian manifold (M, g) is k Osserman (resp. Jordan Osserman of type (r, s)) if ${}^gR$ is k Osserman (resp. Jordan Osserman of type (r,s)) at every point of M and if the eigenvalues (resp. Jordan normal form) do not vary from point to point.

Suppose that (M,g) is a Riemannian manifold of dimension m. If (M,g) is a local rank 1 symmetric space or is flat, then the local isometries of (M,g) act transitively on the sphere bundle $S(M,g)$ of unit tangent vectors. Thus, the eigenvalues of $\mathcal{J}_R$ are constant on $S(M,g)$. Osserman [126] wondered if the converse held: if the eigenvalues of $\mathcal{J}_R$ are constant on $S(M,g)$, then does this imply that (M,g) is either a local rank 1 symmetric space or is a flat space? This question has been called the *Osserman conjecture* by later authors and, consequently, algebraic curvature tensors with this property are called Osserman. Chi [42] has established the Osserman conjecture if $m \equiv 1 \bmod 2$, if $m \equiv 2 \bmod 4$, or if $m = 4$ in the Riemannian setting:

3.1.5 Theorem. *Let (M,g) be an Osserman Riemannian manifold. Suppose that $m \equiv 1$ mod 2, that $m \equiv 2$ mod 4, or that $m = 4$. Then (M,g) is either flat or is a local rank 1 symmetric space.*

There are other partial results concerning Riemannian Osserman algebraic curvature tensors and Riemannian Osserman manifolds in dimensions $m \equiv 0 \bmod 4$ and $m > 4$, but the Osserman conjecture remains open; we refer to Dotti and Druetta [50] and Gilkey, Swann, and Vanhecke [87] for details.

Let ϕ be a Hermitian almost complex structure on V, i.e. $\phi \in \mathfrak{so}(V)$ and $\phi^2 = -\operatorname{Id}$. We use Lemma 1.8.1 to see that

$$R_{\mathrm{Id}}(x,y)z := (y,z)x - (x,z)y \text{ and}$$
$$R_\phi(x,y)z := (\phi y,z)\phi x - (\phi x,z)\phi y - 2(\phi x,y)\phi z$$

are algebraic curvature tensors. Suppose that R is a Riemannian Osserman algebraic curvature tensor such that the reduced Jacobi operator $\tilde{\mathcal{J}}_R$ either has only one eigenvalue or has two eigenvalues with one eigenvalue having multiplicity 1; this is always the case by Theorem 3.1.4 if $m \equiv 1 \bmod 2$ or if $m \equiv 2 \bmod 4$. Chi began the proof of Theorem 3.1.5 by classifying these tensors:

3.1.6 Theorem. *Let R be an Osserman algebraic curvature tensor on a Riemannian vector space V.*

(1) *If $\tilde{\mathcal{J}}_R$ has only one eigenvalue, then there exists a constant c_0 so that $R = c_0 R_{\mathrm{Id}}$.*

(2) *If $\tilde{\mathcal{J}}_R$ has two distinct eigenvalues and if one eigenvalue has multiplicity 1, then there exists a Hermitian almost complex structure ϕ on V and constants c_0 and c_1 so that $R = c_0 R_0 + c_1 R_\phi$.*

Suppose that $\tilde{\mathcal{J}}_R$ has only one eigenvalue c. This implies that if $\{x, y\}$ is any orthonormal set, then $R(y, x)x = cy$. Consequently, R has constant sectional curvature. Thus, by Lemma 1.6.4, $R = cR_{\mathrm{Id}}$ and Assertion (1) follows. We postpone the proof of Assertion (2) until Section 3.5. We note that Theorem 3.1.6 fails in the indefinite setting; in Section 3.2, we will construct algebraic curvature tensors so that 0 is the only eigenvalue of $\mathcal{J}_R$ and so that R does not have constant sectional curvature.

In Section 3.6, we shall complete the classification results which were begun in Sections 1.14 and 1.15. Suppose that (M, g) is a connected pseudo-Riemannian manifold of arbitrary signature. Recall that (M, g) is said to be a *space form* if for every $P \in M$, there is a real number $\kappa(P)$ so that

$$^gR = \kappa(P)R_{\mathrm{Id}}.$$

(M, g) is said to be a *complex space form* (resp. *para-complex space form*) if (M, g) is not a space form and if for every $P \in M$, there is a skew-adjoint map $\phi(P)$ of T_PM so that $\phi^2 = -\mathrm{Id}$ (resp. $\phi^2 = \mathrm{Id}$), and there are real numbers $c_0(P)$ and $c_1(P)$ so that

$$^gR = c_0(P)R_{\mathrm{Id}} + c_1(P)R_{\phi(P)}.$$

In Theorem 3.6.1, we classify the space forms of dimension $m \neq 2$. In Theorem 3.6.5, we classify the complex space forms of dimension $m \neq 2, 4$. In Theorem 3.6.7, we classify the para-complex space forms of dimension $m \neq 2, 4$. If (M, g) is a Riemannian Osserman manifold of dimension $m \equiv 1$ mod 2 or $m \equiv 2$ mod 4, then Theorem 3.1.6 implies (M, g) is a (complex) space form. Thus, Theorem 3.1.5 will follow from the classification results of Section 3.6 cited above. The case $m = 4$ in Theorem 3.1.5 is somewhat special and we refer to Chi [42] for details.

We will complete the classification of Riemannian k Osserman algebraic curvature tensors if $2 \leq k \leq m - 2$ in Section 3.7 by proving:

3.1.7 Theorem. *Let $m \neq 2, 4$ and let $2 \leq k \leq m - 2$.*

(1) *Let R be a k Osserman algebraic curvature tensor on a Riemannian vector space of dimension m. If m is odd, then there exists a constant c_0 so that $R = c_0R_0$. If m is even, then either there exists a constant c_0 so that $R = c_0R_0$ or there exists a constant c_1 and a Hermitian almost complex structure ϕ on V so that $R = c_1R_\phi$.*

(2) *If (M, g) is a k Osserman Riemannian manifold of dimension m, then (M, g) has constant sectional curvature.*

Let $m \neq 2, 4$. We can use the results of Section 1.15 to derive Assertion (2) from Assertion (1). Let (M, g) be a complex space form. We then have that ${}^gR = c_0 R_{\text{Id}} + c_1 R_\phi$, where $\phi^2 = -\text{Id}$. By Lemma 1.15.1, $c_0 = c_1$. Thus, in particular, ${}^gR$ can not have the form $c_1 R_\phi$ for $c_1 \neq 0$ and thus if (M, g) is k Osserman, then (M, g) has constant sectional curvature.

Every Lorentzian k Osserman algebraic curvature tensor has constant sectional curvature. The following result was first established when $k = 1$ Blažić, Bokan and Gilkey[19] and García-Rio, Kupeli, and Vázquez-Abal [64] and subsequently extended to the higher order setting by various authors [85, 86, 145].

3.1.8 Theorem. *Let $m \geq 3$ and let $1 \leq k \leq m - 1$.*

(1) *If R is a k Osserman algebraic curvature tensor on a Lorentzian vector space of dimension m, then R has constant sectional curvature.*
(2) *If (M, g) is a Lorentzian k Osserman manifold of dimension m, then (M, g) has constant sectional curvature.*

Proof. We follow the discussion of Gilkey and Stavrov [86]. We suppose R is a k Osserman algebraic curvature tensor on a Lorentzian vector space for $1 \leq k \leq m - 1$. By Lemma 1.10.3, if v is any null vector of V, then $\text{Tr}\{\mathcal{J}_R(v)^2\} = 0$. By Lemma 1.7.5, this implies R has constant sectional curvature. □

We remark that the situation is not as simple in the higher signature context. It is known that there are non-homogeneous Osserman manifolds of signature (p, q) if $p \geq 2$ and $q \geq 2$; see, for example, work of Blažić, Bokan, Gilkey, and Rakić [20] and also work of García-Rio, Vázquez-Abal and Vázquez-Lorenzo [67]. Alekseevsky, Blažić, Bokan, and Rakić have shown [4] that a manifold of signature (2,2) is Jordan Osserman if and only if it is (locally) self-dual and Einstein. See also related work in [23, 22, 21, 27, 28, 29, 65, 68]

We say that $\nabla R \in \otimes^5 V^*$ is an *algebraic covariant derivative curvature tensor* if ∇R satisfies the symmetries given in Assertion (2) of Lemma 1.12.1:

$$\begin{aligned}
&\nabla R(a, b, c, d; e) = -\nabla R(b, a, c, d; e) = \nabla R(c, d, a, b; e) \\
&\nabla R(a, b, c, d; e) + \nabla R(a, c, d, b; e) + \nabla R(a, d, b, c; e) = 0 \\
&\nabla R(a, b, c, d; e) + \nabla R(a, b, d, e; c) + \nabla R(a, b, e, c; d) = 0.
\end{aligned}$$

Here ∇R is to be understood in a purely formal sense. The *Szabó operator* $\mathcal{S}_{\nabla R}(x)$ is the self-adjoint operator which is defined by:

$$\mathcal{S}_{\nabla R}(x) : y \to (\nabla_x R)(y, x)x \text{ i.e. } (\mathcal{S}_{\nabla R}(x)y, z) = \nabla R(y, x, x, z; x).$$

We say that ∇R is *spacelike Szabó* (resp. *timelike Szabó*) if the eigenvalues of $\mathcal{S}_{\nabla R}$ are constant on $S^+(V)$ (resp. $S^-(V)$). In Lemma 1.10.9, we showed these two notions were equivalent and hence such tensors are simply denoted as Szabó. In Section 3.8, we will prove the following result that is due to Szabó [150] in the Riemannian setting and to Gilkey and Stavrov [86] in the Lorentzian setting:

3.1.9 Theorem. *Let ∇R be a Szabó algebraic covariant derivative curvature tensor on a Riemannian or Lorentzian vector space V. Then $\nabla R = 0$.*

We will show in Section 3.8 that Theorem 3.1.9 fails in the higher signature setting:

3.1.10 Theorem. *Let V be a vector space of signature (p, q). If $p \geq 2$ and if $q \geq 2$, then there exists a non-trivial Szabó algebraic covariant derivative curvature tensor on V so that $\mathcal{S}_{\nabla R}(\cdot)$ is nilpotent and so that $\mathcal{S}_{\nabla R}$ does not vanish identically.*

We note that the result of Stavrov [146], stated in Theorem 4.5.3, can be used to bound the rank of $\mathcal{S}_{\nabla R}$.

We say that a pseudo-Riemannian manifold (M, g) is a local 2 point homogeneous space if the local isometries of (M, g) act transitively on the pseudo-sphere bundles $S^\pm(M, g)$; this implies that $\mathcal{J}_R(\cdot)$ and $\mathcal{S}_{\nabla R}(\cdot)$ have constant eigenvalues on $S^\pm(M, g)$. Theorems 3.1.8 and 3.1.9 then yield the following well known geometrical results:

3.1.11 Theorem.

(1) *If (M, g) is a local 2 point homogeneous Riemannian manifold, then (M, g) is symmetric.*
(2) *If (M, g) is a local 2 point homogeneous Lorentzian manifold, then (M, g) has constant sectional curvature.*

3.2 Examples of Osserman tensors

In this section, we shall construct algebraic curvature tensors which illustrate the relationship between the various notions we shall be considering. In Chapter 2, the curvature tensors associated to a symmetric map ϕ played a central role in our analysis. In Chapter 3, the corresponding role is played, for the most part, by the skew-symmetric maps. We follow the discussion in Gilkey-Ivanova [81].

Let V be a vector space of signature (p,q) and let $\phi \in \mathfrak{so}(V)$. Let $x \in S^{\pm}(V)$ and let $y \perp x$. From the definitions given previously, we have:

$$\begin{aligned}
&R_\phi(x,y)z := (\phi y, z)\phi x - (\phi x, z)\phi y - 2(\phi x, y)\phi z,\\
&\mathcal{J}_{R_\phi}(x)y \quad = 3(y, \phi x)\phi x,\\
&R_{\mathrm{Id}}(x,y)z := (y,z)x - (x,z)y,\\
&\mathcal{J}_{R_{\mathrm{Id}}}(x)y \quad = (x,x)y.
\end{aligned} \tag{3.2.1.a}$$

Let $\{e_1^-, ..., e_p^-, e_1^+, ..., e_q^+\}$ be a normalized orthonormal basis for a vector space V of signature (p,q), where $p \geq 2$ and $q \geq 2$. If $2 \leq 2a \leq \min(p,q)$, then we define $0 \neq \phi_a \in \mathfrak{so}(V)$ with $\phi_a^2 = 0$ by setting:

$$\phi_a e_k^{\pm} = \begin{cases} \pm(e_{2i}^- + e_{2i}^+) & \text{if } k = 2i-1 \leq 2a,\\ \mp(e_{2i-1}^- + e_{2i-1}^+) & \text{if } k = 2i \leq 2a,\\ 0 & \text{if } k > 2a. \end{cases} \tag{3.2.1.b}$$

We can use Assertions (4) and (5) of the following Theorem to show that Osserman, spacelike Jordan Osserman, and timelike Jordan Osserman are inequivalent properties:

3.2.2 Theorem. *Let V be a vector space of signature (p,q). Let $\phi_a \in \mathfrak{so}(V)$ be defined by display (3.2.1.b). Let $R_a := R_{\phi_a}$, where $2 \leq 2a \leq \min(p,q)$. We have:*

(1) *R_a is m-stein. If $1 \leq k \leq p+q-1$, then R_a is k Osserman.*

(2) *If $2a < \min(p,q)$, then R_a is Jordan Osserman of types $(p,0)$ and $(0,q)$; R_a is not Jordan Osserman of type (r,s) otherwise.*

(3) *If $2a = p < q$, then R_a is Jordan Osserman of types $(p,0)$, $(0,q)$, $(r,0)$ and (r,q) for any $1 \leq r \leq p-1$; R_a is not Jordan Osserman otherwise.*

(4) *If $2a = q < p$, then R_a is Jordan Osserman of types $(p,0)$, $(0,q)$, $(0,s)$ and (p,s) for any $1 \leq s \leq q-1$; R_a is not Jordan Osserman otherwise.*

(5) *If $2a = p = q$, then R_a is Jordan Osserman of types $(p,0)$, $(0,q)$, $(r,0)$, (r,q), $(0,s)$, and (p,s) for $1 \leq r \leq p-1$ and for $1 \leq s \leq q-1$; R_a is not Jordan Osserman otherwise.*

(6) *If $p \geq 3$ and if $q \geq 3$, then R_1 is Osserman, but is neither spacelike Jordan Osserman nor timelike Jordan Osserman.*

(7) *If $p > q = 2\bar{q} > 0$, then $R_{\bar{q}}$ is spacelike Jordan Osserman but not timelike Jordan Osserman.*

The curvature tensors R_a described above have nilpotent Jacobi operators. Let $\lambda \in \mathbb{R}$ be given. Set $R_{\lambda,a} := \lambda R_{\mathrm{Id}} + R_a$. Then

$$\mathcal{J}_{R_{\lambda,a}}(x) - \lambda(x,x)\,\mathrm{Id} = \mathcal{J}_{R_a}(x) \text{ on } x^\perp.$$

Thus, these curvature tensors can also be used to provide examples of algebraic curvature tensors which are Osserman but which are not timelike and/or spacelike Jordan Osserman. This shows that the eigenvalue 0 does not play a distinguished role in this analysis. We further remark that the Jacobi operator of these curvature tensors is not diagonalizable.

We first prove Assertion (1) of Theorem 3.2.2. Let $\mathcal{J} := \mathcal{J}_{R_a}$. Since

$$(\phi_a x, \phi_a y) = -(\phi_a^2 x, y) = 0,$$

the range of ϕ_a is totally isotropic and, as $\phi_a \neq 0$, non-trivial. This implies that $p > 0$ and $q > 0$. We use Display (3.2.1.a) to see that

$$\mathcal{J}(x)\mathcal{J}(y)z = 9(z, \phi_a y)(\phi_a x, \phi_a y)\phi_a x = 0.$$

Thus, $\mathcal{J}(\sigma)^2 = 0$ for any non-degenerate k plane σ so, by Lemma 1.2.10, 0 is the only eigenvalue of $\mathcal{J}(\sigma)$. Consequently, R is k Osserman. We complete the proof of Assertion (1) by using Lemma 1.10.1 to see R_a is m-stein.

Before continuing with the proof of Theorem 3.2.2, we must establish the following technical Lemma.

3.2.3 Lemma. *Let V be a vector space of signature (p,q), where $p \geq 2$ and $q \geq 2$. Let $\phi_a \in \mathfrak{so}(V)$ be defined by display* (3.2.1.b) *and let $R_a := R_{\phi_a}$, where $2 \leq 2a \leq \min(p,q)$.*

(1) *R_a is not Jordan Osserman of type (r,s) if:*
   1a) *$2a < p$, $1 \leq r \leq p-1$, and $0 \leq s \leq q$.*
   1b) *$2a = p$, $1 \leq r \leq p-1$, and $1 \leq s \leq q-1$.*
   1c) *$2a < q$, $1 \leq s \leq q-1$, and $0 \leq r \leq p$.*
   1d) *$2a = q$, $1 \leq s \leq q-1$, and $1 \leq r \leq p-1$.*

(2) *R_a is Jordan Osserman of type (r,s) if:*
   2a) *$(r,s) = (p,0)$ or $(r,s) = (0,q)$.*
   2b) *$2a = p$, $s = 0$ or $s = q$, and $1 \leq r \leq p-1$.*
   2c) *$2a = q$, $r = 0$ or $r = p$, and $1 \leq s \leq q-1$.*

Proof. Let $\mathcal{J}_a := \mathcal{J}_{R_a}$. Let $\sigma \in \mathrm{Gr}_{r,s}(V)$. As $\mathcal{J}_a(\sigma)^2 = 0$, we use Lemma 1.2.10 to see that the Jordan normal form of $\mathcal{J}_R(\cdot)$ is constant on $\mathrm{Gr}_{r,s}(V)$ if and only $\mathrm{rank}\{\mathcal{J}_R(\cdot)\}$ is constant $\mathrm{Gr}_{r,s}(V)$.

If $\{e_1, ..., e_k\}$ is an orthonormal basis for a subspace $\sigma \in \mathrm{Gr}_{r,s}(V)$, then

$$\mathcal{J}_a(\sigma) = \textstyle\sum_{1\leq i\leq k}(e_i, e_i)\mathcal{J}_a(e_i).$$

To prove Assertions (1a) and (1b), we must find subspaces $\sigma_1, \sigma_2 \in \mathrm{Gr}_{r,s}(V)$ so that $\mathrm{rank}\,\mathcal{J}_a(\sigma_1) \neq \mathrm{rank}\,\mathcal{J}_a(\sigma_2)$. Set:

$$\tau := \begin{cases} \mathrm{span}\{e_2^-, ..., e_r^-, e_2^+, ..., e_s^+\} & \text{if } r \geq 2,\ s \geq 2, \\ \mathrm{span}\{e_2^-, ..., e_r^-\} & \text{if } r \geq 2,\ s \leq 1, \\ \mathrm{span}\{e_2^+, ..., e_s^+\} & \text{if } r \leq 1,\ s \geq 2, \\ \{0\} & \text{if } r \leq 1,\ s \leq 1. \end{cases}$$

To prove Assertion (1a), we suppose that $2a < p$, that $1 \leq r \leq p-1$, and that $0 \leq s \leq q$. We then set:

$$\sigma_1 := \begin{cases} \tau \oplus \mathrm{span}\{e_1^-, e_1^+\} & \text{if } s \geq 1, \\ \tau \oplus \mathrm{span}\{e_1^-\} & \text{if } s = 0, \end{cases}$$

$$\sigma_2 := \begin{cases} \tau \oplus \mathrm{span}\{e_p^-, e_1^+\} & \text{if } s \geq 1, \\ \tau \oplus \mathrm{span}\{e_p^-\} & \text{if } s = 0. \end{cases}$$

The index '1' does not appear among the indices comprising the basis for τ so there is no 'interaction'. Furthermore, since $2a < p$, $\mathcal{J}_a(e_p^-) = 0$. We have the cancellation $(e_1^-, e_1^-)\mathcal{J}_a(e_1^-) + (e_1^+, e_1^+)\mathcal{J}_a(e_1^+) = 0$. We may now use Lemma 1.2.2 to complete the proof of Assertion (1a) by showing that $\mathrm{rank}\{\mathcal{J}_a(\sigma_1)\} \neq \mathrm{rank}\{\mathcal{J}_a(\sigma_2)\}$:

$$\mathrm{rank}\{\mathcal{J}_a(\sigma_1)\} = \begin{cases} \mathrm{rank}\{\mathcal{J}_a(\tau)\} & \text{if } s \geq 1, \\ \mathrm{rank}\{\mathcal{J}_a(\tau)\} + 1 & \text{if } s = 0, \end{cases}$$

$$\mathrm{rank}\{\mathcal{J}_a(\sigma_2)\} = \begin{cases} \mathrm{rank}\{\mathcal{J}_a(\tau)\} + 1 & \text{if } s \geq 1, \\ \mathrm{rank}\{\mathcal{J}_a(\tau)\} & \text{if } s = 0. \end{cases}$$

To prove Assertion (1b), we suppose that $2a = p$, that $1 \leq r \leq p-1$, and that $1 \leq s \leq q-1$. We then set:

$$\sigma_1 := \quad \tau \oplus \mathrm{span}\{e_1^-, e_1^+\}, \quad \text{and}$$

$$\sigma_2 := \begin{cases} \tau \oplus \mathrm{span}\{e_{r+1}^-, e_1^+\} & \text{if } r \geq s, \\ \tau \oplus \mathrm{span}\{e_1^-, e_{s+1}^+\} & \text{if } r < s. \end{cases}$$

Again, since the index '1' does not belong to the indices $\{2, ..., \max(r,s)\}$ which appear in the basis for τ, there is no cancellation. If $r \geq s$ (resp.

$r < s$), then the index $r+1$ (resp. $s+1$) does not appear among these indices either. If $s+1 \leq p$, then $\mathcal{J}_a(e_{s+1}^+) \neq 0$; if $s+1 > p$, then $\mathcal{J}_a(e_{s+1}^+) = 0$. Since $r+1 \leq p = 2a$, $\mathcal{J}_a(e_{r+1}^-) \neq 0$. We prove Assertion (2) by computing:

$$\begin{aligned}\operatorname{rank}\{\mathcal{J}_a(\sigma_1)\} = &\quad \operatorname{rank}\{\mathcal{J}_a(\tau)\}\\ \operatorname{rank}\{\mathcal{J}_a(\sigma_2)\} = &\begin{cases}\operatorname{rank}\{\mathcal{J}_a(\tau)\}+2 & \text{if } r \geq s\\ \operatorname{rank}\{\mathcal{J}_a(\tau)\}+2 & \text{if } r < s < p,\\ \operatorname{rank}\{\mathcal{J}_a(\tau)\}+1 & \text{if } r < p \leq s.\end{cases}\end{aligned}$$

We interchange the roles of spacelike and timelike to prove Assertions (1c) and (1d) in exactly the same fashion that Assertions (1a) and (1b) were proved; we omit details in the interests of brevity.

We shall prove that R_a is Osserman of type $(p,0)$ by showing that $\operatorname{rank}\{\mathcal{J}_a(\sigma)\} = 2a$ for every maximal timelike subspace σ of V. Suppose that $\sigma \in \operatorname{Gr}_{p,0}(V)$. As

$$\operatorname{range}\{\mathcal{J}_a(\sigma)\} \subset \{\phi_a e_1^-, ..., \phi_a e_{2a}^-\},$$

we have $\operatorname{rank}\{\mathcal{J}_a(\sigma)\} \leq 2a$. We suppose that $\operatorname{rank}\{\mathcal{J}_a(\sigma)\} < 2a$ and argue for a contradiction. Let $V_a^- := \operatorname{span}\{e_1^-, ..., e_{2a}^-\}$. As $\dim V_a^- = 2a$ and as $\operatorname{rank}\{\mathcal{J}_a(\sigma)\} < 2a$,

$$\ker\{\mathcal{J}_a(\sigma)\} \cap V_a^- \neq \{0\}.$$

Thus, we may choose $0 \neq w \in V_a^-$ with $\mathcal{J}_a(\sigma)w = 0$. Let $\{v_1, ..., v_p\}$ be an orthonormal basis for σ. Then $\mathcal{J}_a(\sigma) = -\sum_i \mathcal{J}_a(v_i)$ so

$$\begin{aligned}0 =&(\mathcal{J}_a(\sigma)w, w) = -3\textstyle\sum_{1\leq i\leq p}(\phi_a v_i, w)^2 \text{ so}\\ 0 =&(\phi_a v_i, w) = -(v_i, \phi_a w) \quad \text{for } 1 \leq i \leq p.\end{aligned}$$

Consequently, $\phi_a w \perp \sigma$. Since σ is a maximal timelike subspace, either $\phi_a w = 0$ or $\phi_a w$ is spacelike. Because ϕ_a is skew-adjoint and $\phi_a^2 = 0$, the range of ϕ_a is totally isotropic. Consequently, $\phi_a w$ is a null vector, not a spacelike vector. Thus, we must have that $\phi_a w = 0$. This is false as ϕ_a is injective on V_a^-. This contradiction shows that R_a is Jordan Osserman of type $(p,0)$; it then follows dually by Lemma 1.10.3 that R_a is Jordan Osserman of type $(0,q)$. This establishes Assertion (2a).

We now prove Assertion (2b). Let $2a = p$ and let $1 \leq r \leq p-1$. If we can show that $\operatorname{rank} \mathcal{J}_a(\sigma) = r$ for every $\sigma \in \operatorname{Gr}_{r,0}$, then Lemma 1.2.10 will imply that R_a is Jordan Osserman of type $(r,0)$. The case $s = q$ will then follow dually from Lemma 1.10.3 since $1 \leq (p-r) \leq p-1$ as well.

Let $\{v_1, ..., v_r\}$ be an orthonormal basis for σ. Then:

$$\mathcal{J}_a(\sigma)y = -3\sum_{1\le i\le r}(y, \phi_a(v_i))\phi_a(v_i).$$

Thus, by Lemma 1.2.2, $\operatorname{rank}\mathcal{J}_a(\sigma) = r$ if $\ker\phi_a \cap \sigma = \{0\}$. We suppose the contrary and choose $0 \neq v \in \sigma$ so that $\phi_a(v) = 0$. We expand:

$$\begin{aligned} v &= c_1^- e_1^- + ... + c_p^- e_p^- + c_1^+ e_1^+ + ... + c_q^+ e_q^+, \\ \phi_a v &= (c_1^+ - c_1^-)\phi_a(e_1^+) + ... + (c_p^+ - c_p^-)\phi_a(e_p^+). \end{aligned}$$

Since $\phi_a v = 0$, we have $c_i^+ = c_i^-$ for $1 \le i \le p$. Thus,

$$\begin{aligned} (v, v) &= (c_1^+)^2 - (c_1^-)^2 + ... + (c_p^+)^2 - (c_p^-)^2 + (c_{p+1}^+)^2 + ... + (c_q^+)^2 \\ &= (c_{p+1}^+)^2 + ... + (c_q^+)^2 \ge 0. \end{aligned}$$

Since $v \neq 0$ and since σ is timelike, $(v, v) < 0$. This contradiction completes the proof of Assertion (2b). The proof of Assertion (2c) is similar and is therefore omitted. □

Proof of Theorem 3.2.2. We have already proved Assertion (1). By Lemma 3.2.3 (2a), R_a is Jordan Osserman of type $(p, 0)$ and of type $(0, q)$.

Suppose that $2a < \min(p, q)$. By Lemma 3.2.3 (1a), since $2a < p$, R_a is not Jordan Osserman of type (r, s) for $1 \le r \le p - 1$. Similarly, by Lemma 3.2.3 (1c), since $2a < q$, R_a is not Jordan Osserman of type (r, s) for $1 \le s \le q - 1$ and Assertion (2) follows.

Suppose that $2a = p < q$. By Lemma 3.2.3 (2b), R_a is Jordan Osserman of type $(r, 0)$ if $1 \le r \le p - 1$ and of type $(p - r, q)$. By Lemma 3.2.3 (1c), since $2a < q$, R_a is not Jordan Osserman of type (r, s) if $1 \le s \le q - 1$ and Assertion (3) follows; assertion (4) follows similarly.

Suppose that $2a = p = q$. Since $2a = p$, we use Lemma 3.2.3 (2b) to see that R_a is Jordan Osserman of type $(r, 0)$ if $1 \le r \le p - 1$. Since $2a = q$, we have by Lemma 3.2.3 (2c) that R_a is Jordan Osserman of type $(0, s)$ and (p, q) if $1 \le s \le q - 1$. If $1 \le r \le p - 1$ and $1 \le s \le q - 1$, then R_a is not Jordan Osserman by Lemma 3.2.3 (1b). This proves Assertion (5); Assertions (6) and (7) are immediate consequences of the previous assertions. □

The *Adams number* $\nu(m)$ is given by:

$$\begin{aligned} &\nu(a2^s) = \nu(2^s) \text{ for } a \equiv 1 \bmod 2, \\ &\nu(1) = 0,\ \nu(2) = 1,\ \nu(4) = 3,\ \nu(8) = 7, \text{ and } \nu(16\tilde{b}) = \nu(\tilde{b}) + 8. \end{aligned}$$

In Theorem 3.1.4, we used Theorem 1.4.2 to control the possible eigenvalue structures of a Riemannian Jordan Osserman algebraic curvature tensor. We may apply the following Lemma to the case $(p, q) = (0, \ell)$ to see that estimates of Theorem 3.1.4 are sharp.

3.2.4 Lemma. *Let V be a vector space of signature (p,q), where $p \equiv 0 \mod \ell$ and $q \equiv 0 \mod \ell$. Let real numbers λ_i and multiplicities μ_i be given for $0 \le i \le s$ so that $\mu_0 \ge \mu_1 \ge ... \ge \mu_s$, so that $\mu_1 + ... + \mu_s \le \nu(\ell)$, and so that $\mu_0 + ... + \mu_s = p+q-1$. Then there exists a Jordan Osserman algebraic curvature tensor R on V so that the reduced Jacobi operator $\tilde{\mathcal{J}}_R(x) := \mathcal{J}_R|_{x^\perp}$ is diagonalizable and has eigenvalues $(x,x)\lambda_i$ of multiplicity μ_i on $x^\perp$ for $0 \le i \le s$ and for every $x \in S^\pm(V)$.*

Proof. We generalize the argument given in [71] from the Riemannian to the higher signature setting to prove Lemma 3.2.4. Let $\{c_0, ..., c_{\nu(\ell)}\}$ be real constants to be determined presently. By Lemma 1.4.3, there exists a collection of skew-symmetric matrices $\{\Phi_1, ..., \Phi_{\nu(\ell)}\}$ on $\mathbb{R}^\ell$ satisfying the *Clifford commutation relations*:

$$\Phi_i \Phi_j + \Phi_j \Phi_i = -2\delta_{ij} \operatorname{Id}.$$

Since $p = \tilde{p}\ell$ and $q = \tilde{q}\ell$, we may choose an isometry between V and $\mathbb{R}^{(0,\ell)} \otimes \mathbb{R}^{(\tilde{p},\tilde{q})}$. Let $\phi_i := \Phi_i \otimes \operatorname{Id} \in \mathfrak{so}(V)$. Define

$$R := c_0 R_{\operatorname{Id}} + \textstyle\sum_{1 \le i \le \nu(\ell)} c_i R_{\phi_i}.$$

Let $x \in S^\pm(V)$. By Lemma 1.4.5, $\{x, \phi_1 x, ..., \phi_\nu x\}$ is an orthonormal set. Thus, we may use Display (3.2.1.a) to see that:

$$\begin{aligned} \mathcal{J}_R(x)\phi_i x &= (x,x)(c_0 + 3c_i)\phi_i x \text{ for } 1 \le i \le \nu(\ell), \text{ and} \\ \mathcal{J}_R(x)y &= (x,x)c_0 y \text{ for } y \perp \{x, \phi_1 x, ..., \phi_{\nu(\ell)}\}. \end{aligned}$$

The desired result now follows by setting $c_0 = \lambda_0$ and by choosing the remaining constants c_i appropriately. □

We can use a similar construction to construct Jordan Osserman algebraic curvature tensors with arbitrary Jordan normal form. We first introduce some additional notation.

3.2.5 Definition. Let V be a vector space of signature (p,q) and let W be a vector space of signature (r,s), where $r \le p$ and $s \le q$. By choosing a suitable orthonormal bases for V and W, we can find an isometry between V and $W \oplus \mathbb{R}^{(p-r,q-s)}$. Let J be a linear map of W. We define the *stabilization* $J \oplus 0$ by setting

$$(J \oplus 0)(w \oplus x) = Jw \oplus 0.$$

If J is a self-adjoint linear map of W, then $J \oplus 0$ is a self-adjoint linear map of V; the stabilization $J \oplus 0$ is uniquely defined up to conjugation.

3.2.6 Definition. Let V and W be vector spaces of signatures (p,q) and (r,s), respectively. We say that V *admits a* $\operatorname{Cliff}(W)$ *module structure* if there exist elements $\phi_i \in \mathfrak{so}(V)$ for $1 \le i \le r+s$ so that

(1) $\phi_i\phi_j + \phi_j\phi_i = 0$ for $i \ne j$.
(2) $\phi_i^2 = \operatorname{Id}$ for r values of i.
(3) $\phi_i^2 = -\operatorname{Id}$ for s values of i.

3.2.7 Lemma. *Let V and W be vector spaces of signatures (p,q) and (r,s), respectively. Assume that V admits a* $\operatorname{Cliff}(W)$ *module structure. Let J be a self-adjoint linear map of W. Then there exists a Jordan Osserman algebraic curvature tensor R on V so that $\mathcal{J}_R(x)$ is conjugate to $(x,x)J \oplus 0$ for every $x \in S^{\pm}(V)$.*

Proof. Let $\mathcal{B} := \{e_1, ..., e_{r+s}\}$ be an orthonormal basis for W. Choose a collection of linear maps $\{\phi_1, ..., \phi_{r+s}\} \subset \mathfrak{so}(V)$ so that $\phi_i^2 = -\varepsilon_i \operatorname{Id}$, where $\varepsilon_i := (e_i, e_i) = \pm 1$. Expand

$$Je_i = \textstyle\sum_j J_{ij} e_j.$$

Let $(x,x) \in S^{\pm}(V)$ and let $f_i := \phi_i x \in V$. If $i \ne j$, then $(f_i, f_j) = 0$ by Lemma 1.4.5. We also have:

$$(f_i, f_i) = (\phi_i x, \phi_i x) = -(\phi_i^2 x, x) = (x,x)\varepsilon_i.$$

Thus, if x is spacelike, then the map $e_i \to f_i$ is an isometric embedding of W in V while if x is timelike, this is a para-isometric embedding. Let c_i and $c_{ij} = c_{ji}$ be constants to be determined presently. Let

$$R := \textstyle\sum_i c_i R_{\phi_i} + \frac{1}{2} \sum_{i \ne j} c_{ij} R_{\phi_i + \phi_j}$$

be an algebraic curvature tensor on V. Let $\mathcal{J}_i$ be the Jacobi operator defined by R_{ϕ_i} and let $\mathcal{J}_{ij}$ be the Jacobi operator defined by $R_{\phi_i+\phi_j}$. Let

$$\pi(x) := \operatorname{span}\{f_1, ..., f_{r+s}\} \in \widetilde{\operatorname{Gr}}_{r+s}(V).$$

Then range $\mathcal{J}_R(x) \subset \pi(x)$ and $\mathcal{J}_R(x) = 0$ on $U := \pi(x)^{\perp}$. We expand

$$\mathcal{J}_R(x) f_i = \textstyle\sum_j \tilde{J}_{ij} f_j.$$

Let i, j, and k be distinct indices. We use Display (3.2.1.a) to see:

$$\mathcal{J}_i f_k = \begin{cases} 3(x,x)\varepsilon_i f_i & \text{if } k = i, \\ 0 & \text{if } k \ne i, \end{cases}$$

$$\mathcal{J}_{ij} f_k = \begin{cases} 3(x,x)\varepsilon_k (f_i + f_j) & \text{if } k = i, j, \\ 0 & \text{if } k \ne i, j, \end{cases} \tag{3.2.7.a}$$

$$\tilde{J}_{ij} = \begin{cases} 3(x,x)\varepsilon_i (c_i + \sum_{k \ne i} c_{ik}) & \text{if } i = j, \\ 3(x,x)\varepsilon_i c_{ij} & \text{if } i \ne j. \end{cases}$$

We solve the equations $\tilde{J}_{ij} = (x,x)J_{ij}$ for $1 \le i,j \le r+s$ to see that if $i \ne j$, then:

$$c_{ij} := \tfrac{1}{3}\varepsilon_i J_{ij} \text{ and } c_i := \tfrac{1}{3}\varepsilon_i(J_{ii} - \textstyle\sum_{j\ne i} J_{ij}).$$

Since J is self-adjoint we have that $\varepsilon_j J_{ij} = \varepsilon_i J_{ji}$ so $\frac{1}{3}\varepsilon_i J_{ij} = \frac{1}{3}\varepsilon_j J_{ji}$ and hence, since $\varepsilon_i = \pm 1$, $c_{ij} = c_{ji}$ as required. It is now clear that $\mathcal{J}_R(x)$ is conjugate to $(x,x)J \oplus 0$. □

Assertion (1) of the following Lemma shows that the Jordan normal form of a timelike or spacelike Jordan Osserman algebraic curvature tensor can be arbitrarily complicated if we are willing to stabilize and greatly increase the dimension by suitable powers of 2; Assertion (2) deals with the 4 dimensional balanced setting:

3.2.8 Theorem.

(1) *Let J be an arbitrary map of a vector space W of dimension m. There exists $\ell = \ell(m)$ and an algebraic curvature tensor R on $\mathbb{R}^{(2^\ell,2^\ell)}$ so that $\mathcal{J}_R(x)$ is conjugate to $\pm J \oplus 0$ if $x \in S^\pm(V)$.*

(2) *Let J be a self adjoint linear transformation of $\mathbb{R}^{(2,1)}$. There exists a Jordan Osserman algebraic curvature tensor R on $\mathbb{R}^{(2,2)}$ so that $\mathcal{J}_R(x)$ is conjugate to $\pm J \oplus 0$ if $x \in S^\pm(\mathbb{R}^{(2,2)})$.*

Proof. Let J be an arbitrary linear transformation of W. We use Lemma 1.2.9 to choose an innerproduct on W so that J is self-adjoint with respect to this inner product. Let $\{e_1, ..., e_m\}$ be an orthonormal basis for W. Let $V = \mathbb{R}^{(2^\ell,2^\ell)}$ for suitably chosen ℓ. We apply Lemma 1.4.5 to choose $\{\phi_1, ..., \phi_m\} \subset \mathfrak{so}(V)$ so that $\phi_i^2 = -(e_i,e_i)\,\mathrm{Id}$ and so that $\phi_i\phi_j + \phi_j\phi_i = 0$ for $i \ne j$. Assertion (1) now follows from Lemma 3.2.7.

We may also use Lemma 1.4.5 to find three skew-symmetric linear transformations of $\mathbb{R}^{(2,2)}$ which satisfy the commutation relations:

$$\phi_1^2 = \mathrm{Id},\ \phi_2^2 = \mathrm{Id},\ \phi_3^2 = -\mathrm{Id}, \text{ and } \phi_i\phi_j + \phi_j\phi_i = 0 \text{ for } i \ne j.$$

The Assertion (2) follows as well from Lemma 3.2.7.

3.3 Examples of higher order Osserman tensors

In this section, we will present two different families of algebraic curvature tensors which are Jordan Osserman of type (r,s) for certain admissible values of (r,s). In Theorem 3.3.1, we present examples Gilkey, Stanilov, and Videv [85] of algebraic curvature tensors which are Jordan Osserman of type (r,s) for $r+s=1$ or $r+s=m-1$ but which are not k Osserman for $2 \le k \le m-2$. In Theorem 3.3.2, we exhibit examples due to Stavrov [146] of algebraic curvature tensors which k Osserman for all values of k and which are Jordan Osserman of type (r,s) for some but not all values of (r,s); the Jacobi operators of these algebraic curvature tensors are not diagonalizable.

3.3.1 Theorem. *Let V be a vector space of signature (p,q) and dimension m. Let ϕ be a skew-adjoint map of V with $\phi^2 = -\mathrm{Id}$.*

(1) *The algebraic curvature tensors R_{Id} and R_ϕ are spacelike and timelike Jordan Osserman for all admissible pairs (r,s).*

(2) *Let c_0 and c_1 be non-zero constants. Let $R := c_0 R_{\mathrm{Id}} + c_1 R_\phi$. Then:*
   2a) *R is Jordan Osserman of type (r,s) if and only if $r+s=1$ or $r+s=m-1$.*
   2b) *R is not k Osserman if $2 \le k \le m-2$.*

Proof. Let $v \in S^\pm(V)$. If $y \perp v$, then

$$\text{(3.3.1.a)} \qquad \begin{aligned} \mathcal{J}_{R_{\mathrm{Id}}}(v)(y) &= (v,v)y, \text{ and} \\ \mathcal{J}_{R_\phi}(v)y &= 3(y,\phi v)\phi v. \end{aligned}$$

Let $\{e_1,\ldots,e_{r+s}\}$ be an orthonormal basis for $\sigma \in \mathrm{Gr}_{r,s}(V)$. By definition, $\mathcal{J}_R(\sigma) = \sum_i (e_i,e_i)\mathcal{J}_R(e_i)$. Note that $\mathcal{J}_R(x)x = 0$ and that $\{\phi e_1,\ldots,\phi e_{r+s}\}$ is an orthonormal basis for $\phi\sigma$ with $(e_i,e_i) = (\phi e_i,\phi e_i)$. Therefore, we have:

$$\text{(3.3.1.b)} \qquad \begin{aligned} \mathcal{J}_{R_{\mathrm{Id}}}(\sigma)(y) &= \begin{cases} (r+s-1)y & \text{if } y \in \sigma, \\ (r+s)y & \text{if } y \perp \sigma, \end{cases} \\ \mathcal{J}_{R_\phi}(\sigma)(y) &= \begin{cases} 3y & \text{if } y \in \phi\sigma, \\ 0 & \text{if } y \perp \phi\sigma. \end{cases} \end{aligned}$$

Thus, the operators $\mathcal{J}_{R_{\mathrm{Id}}}(\sigma)$ and $\mathcal{J}_{R_\phi}$ are diagonalizable and the eigenvalues are determined by $r+s$; Assertion (1) now follows.

Let $R := c_0 R_{\mathrm{Id}} + c_1 R_\phi$. Let σ be a non-degenerate line. Since ϕ is skew-adjoint, $\sigma \perp \phi\sigma$. Thus, we may use Display (3.3.1.a) to see:

$$\mathcal{J}_R(\sigma)y = \begin{cases} 0 & \text{if } y \in \sigma, \\ (c_0+3c_1)y & \text{if } y \in \phi\sigma, \\ c_0 y & \text{if } y \perp \{\sigma,\phi\sigma\}. \end{cases}$$

It now follows that R is Jordan Osserman of types $(1,0)$ and $(0,1)$; dually by Lemma 1.10.3, we have R is Jordan Osserman of types $(p-1,q)$ and $(p,q-1)$. Assertion (2a) now follows.

We complete the proof by showing that R is not k Osserman when we have $2 \leq k \leq m-2$. Since $\phi^2 = -\operatorname{Id}$, ϕ is an isometry which gives V an almost complex structure. We necessarily have that p, q, and m are even. Choose an orthonormal basis $\{e_1, ..., e_m\}$ for v so that

$$\phi e_{2i-1} = e_{2i} \text{ and } \phi e_{2i} = -e_{2i-1} \text{ for } 1 \leq i \leq \tfrac{1}{2}m.$$

Suppose first that $k = 2\ell$ is even. Set:

$$\begin{aligned} \sigma_1 &:= \operatorname{span}\{e_1, ..., e_{2\ell}\}, \text{ and} \\ \sigma_2 &:= \operatorname{span}\{e_1, ..., e_{2\ell-1}, e_{m-1}\}. \end{aligned}$$

We have $2 \leq k = 2\ell \leq m-2$. Since $c_0 \neq 0$ and $c_1 \neq 0$, we show that the eigenvalues and eigenvalue multiplicities of $\mathcal{J}_R(\sigma_1)$ differ by computing:

$$\mathcal{J}_R(\sigma_1)(e_i) = \begin{cases} \{(k-1)c_0 + 3c_1\}e_i & \text{if } i \leq 2\ell, \\ kc_0 e_i & \text{if } 2\ell < i, \end{cases}$$

$$\mathcal{J}_R(\sigma_2)(e_i) = \begin{cases} \{(k-1)c_0 + 3c_1\}e_i & \text{if } i \leq 2\ell - 2, \\ (kc_0 + 3c_1)e_i & \text{if } i = 2\ell, m, \\ (k-1)c_0 e_i & \text{if } i = 2\ell - 1, m-1, \\ kc_0 e_i & \text{if } 2\ell < i < m-1. \end{cases}$$

Suppose now that $k = 2\ell + 1$ is odd. Set:

$$\begin{aligned} \sigma_1 &:= \operatorname{span}\{e_1, ..., e_{2\ell-1}, e_{2\ell}, e_{m-1}\}, \text{ and} \\ \sigma_2 &:= \operatorname{span}\{e_1, ..., e_{2\ell-1}, e_{2\ell+1}, e_{m-1}\}. \end{aligned}$$

We have $3 \leq k = 2\ell + 1 < m - 1$. We show that the eigenvalue multiplicities of $\mathcal{J}_R(\sigma_1)$ and $\mathcal{J}_R(\sigma_2)$ differ and thus that R is not k Osserman by computing:

$$\mathcal{J}_R(\sigma_1)(e_i) = \begin{cases} ((k-1)c_0 + 3c_1)e_i & \text{if } 1 \leq i \leq 2\ell, \\ kc_0 e_i & \text{if } 2\ell < i < m-1, \\ (k-1)c_0 e_i & \text{if } i = m-1, \\ (kc_0 + 3c_1)e_i & \text{if } i = m, \end{cases}$$

$$\mathcal{J}_R(\sigma_2)(e_i) = \begin{cases} ((k-1)c_0 + 3c_1)e_i & \text{if } 1 \leq i \leq 2\ell - 2, \\ kc_0 e_i & \text{if } 2\ell + 2 < i < m-1, \\ (k-1)c_0 e_i & \text{if } i = 2\ell - 1, 2\ell + 1, m-1, \\ (kc_0 + 3c_1)e_i & \text{if } i = 2\ell, 2\ell + 2, m. \quad \square \end{cases}$$

We remark that Lemma 3.3.1 continues to hold true if we replace the condition $\phi^2 = -\operatorname{Id}$ by $\phi^2 = \operatorname{Id}$. We omit details in the interests of brevity.

In Chapter 3, we shall use primarily the curvature tensors R_ϕ which are defined by a skew-symmetric map ϕ. It is, however, possible to construct useful examples by considering tensors of the form R_ϕ, where ϕ is symmetric. Let $\{e_1^-, ..., e_p^-, e_1^+, ..., e_q^+\}$ be a normalized orthonormal basis for V. Let a be a positive integer with $a \le \min(p, q)$. As in the proof of Lemma 2.2.5, we define

$$\begin{aligned} &\Phi_a e_i^\pm = \pm(e_i^+ + e_i^-) \text{ for } i \le a,\ \Phi_a e_i^\pm = 0 \text{ for } i > a, \\ (3.3.1.c)\qquad &R_a(x,y)z := (\Phi_a y, z)\Phi_a x - (\Phi_a x, z)\Phi_a y, \text{ and} \\ &\mathcal{J}_a(x)y := (\Phi_a x, x)\Phi_a y - (\Phi_a y, x)\Phi_a x. \end{aligned}$$

Since Φ_a is self-adjoint, we use Lemma 1.8.1 to see R is an algebraic curvature tensor. If $\Phi_a y = c\Phi_a x$, then $\mathcal{J}_a(x)y = 0$. Thus, $\mathcal{J}_a(\sigma) = 0$ if $a = 1$ so we shall assume $a \ge 2$. We use Lemma 1.10.3 to see that if R is Jordan Osserman of type (r, s), then R is Jordan Osserman of type $(p - r, q - s)$. The following result is due to Stavrov [146].

3.3.2 Theorem. *Let V be a vector space of signature (p, q) and dimension m. Let R_a and $\mathcal{J}_a$ be as in* Display (3.3.1.c), *where $2 \le a \le \min(p, q)$.*

(1) *Let $1 \le k \le m - 1$. Then R_a is k Osserman.*

(2) *Let $2 \le r \le p$ and $2 \le s \le q$. R_a is Jordan Osserman of type*
  2a) *$(1, 0)$ or $(p - 1, q)$ if and only if $p = a$.*
  2b) *$(0, 1)$ or $(p, q - 1)$ if and only if $q = a$.*
  2c) *$(r, 0)$ or $(p - r, q)$ if and only if $p - a + 2 \le r$.*
  2d) *$(0, s)$ or $(p, q - s)$ if and only if $q - a + 2 \le s$.*

(3) *If $1 \le r \le p - 1$ and if $1 \le s \le q - 1$, then R_a is not Jordan Osserman of type (r, s).*

3.3.3 Remark. The algebraic curvature tensor R_q is spacelike but not timelike Jordan Osserman if $2 \le a = q < p$. This gives a slight strengthening of Lemma 3.2.2 (7) by removing the hypothesis that q is even.

Before beginning the proof of Theorem 3.3.2, we must first derive a series of technical results.

3.3.4 Lemma. *Let V be a vector space of signature (p, q). Let R_a and $\mathcal{J}_a$ be as in* Display (3.3.1.c), *where $2 \le a \le \min(p, q)$. Let σ be a spacelike or timelike subspace of dimension k.*

(1) *Let $x \in V$. Then $(\mathcal{J}_a(\sigma)x, x) = 0$ if and only if $\Phi_a x$ and $\Phi_a v$ are colinear for all $v \in \sigma$.*

(2) *Set* $\ell := \dim \Phi_a\sigma$.
  2a) *If* $\ell = 0$, *then* $\operatorname{rank}\{\mathcal{J}_a(\sigma)\} = 0$.
  2b) *If* $\ell = 1$, *then* $\operatorname{rank}\{\mathcal{J}_a(\sigma)\} = a - 1$.
  2c) *If* $\ell \geq 2$, *then* $\operatorname{rank}\{\mathcal{J}_a(\sigma)\} = a$.

Proof. We define a new inner product $\langle\cdot,\cdot\rangle$ on V similar to the one constructed in the proof of Lemma 1.2.13 by setting:

$$\langle e_i^+, e_j^+\rangle = \delta_{ij},\ \langle e_i^+, e_j^-\rangle = 0, \text{ and } \langle e_i^-, e_j^-\rangle = \delta_{ij}.$$

Let ε and δ be choices of sign. We compute:

$$\text{if } i \neq j \text{ or } i > a, \text{ then } (\Phi_a e_i^\varepsilon, e_j^\delta) = 0 \text{ and } \langle \Phi_a e_i^\varepsilon, \Phi_a e_j^\delta\rangle = 0,$$

$$\text{if } i \leq a, \text{ then } (\Phi_a e_i^\varepsilon, e_i^\delta) = \varepsilon\delta \cdot 1 \text{ and } \langle \Phi_a e_i^\varepsilon, \Phi_a e_i^\delta\rangle = \varepsilon\delta \cdot 2.$$

This shows that

$$(\Phi_a v, w) = \tfrac{1}{2}\langle \Phi_a v, \Phi_a w\rangle \text{ for any } v, w \in V. \tag{3.3.4.a}$$

We now prove Assertion (1). Let $\varepsilon = +1$ if σ is spacelike and $\varepsilon = -1$ if σ is timelike. Let the index ν range from 1 to k and index an orthonormal basis $\{w_1, ..., w_k\}$ for σ. We use Display (3.3.1.c) and Equation (3.3.4.a) to compute:

$$\begin{aligned}&4\varepsilon(\mathcal{J}_a(\sigma)x, x)\\ =&4\textstyle\sum_\nu\{(\Phi_a w_\nu, w_\nu)(\Phi_a x, x) - (\Phi_a x, w_\nu)(\Phi_a w_\nu, x)\}\\ =&\textstyle\sum_\nu\{\langle\Phi_a w_\nu, \Phi_a w_\nu\rangle\langle\Phi_a x, \Phi_a x\rangle - \langle\Phi_a x, \Phi_a w_\nu\rangle\langle\Phi_a w_\nu, \Phi_a x\rangle\}.\end{aligned}$$

The inner product $\langle\cdot,\cdot\rangle$ is positive definite. Thus,

$$\langle\Phi_a w_\nu, \Phi_a w_\nu\rangle\langle\Phi_a x, \Phi_a x\rangle - \langle\Phi_a x, \Phi_a w_\nu\rangle\langle\Phi_a w_\nu, \Phi_a x\rangle \geq 0$$

for $1 \leq \nu \leq k$ by the Cauchy-Schwarz inequality. Furthermore, equality occurs if and only if $\Phi_a w_\nu$ and $\Phi_a x$ are colinear. Assertion (1) now follows.

To prove Assertion (2a), we suppose $\ell = 0$. Then $\Phi_a(w_\nu) = 0$ for all k and $\mathcal{J}_a(\sigma) = 0$. This implies that $\operatorname{rank}\{\mathcal{J}_a(\sigma)\} = 0$; Assertion (2a) now follows.

To prove Assertion (2b), we suppose $\ell = 1$. Choose an orthonormal basis $\{w_1, ..., w_k\}$ for σ so that $\Phi_a w_1 \neq 0$. Then there exist constants c_ν so $\Phi_a w_\nu = c_\nu \Phi_a w_1$ for $1 \leq \nu \leq k$. We compute:

$$\begin{aligned}2\varepsilon\mathcal{J}_a(\sigma)x =&2\textstyle\sum_\nu\{(\Phi_a w_\nu, w_\nu)\Phi_a x - (\Phi_a x, w_\nu)\Phi_a w_\nu\}\\ =&\textstyle\sum_\nu\{\langle\Phi_a w_\nu, \Phi_a w_\nu\rangle\Phi_a x - \langle\Phi_a x, \Phi_a w_\nu\rangle\Phi_a w_\nu\}\\ =&\textstyle\sum_\nu c_\nu^2\{\langle\Phi_a w_1, \Phi_a w_1\rangle\Phi_a x - \langle\Phi_a x, w_1\rangle\Phi_a w_1\}\\ =&2\textstyle\sum_\nu c_\nu^2\mathcal{J}_a(w_1)x.\end{aligned}$$

Let ϱ be the natural projection from range$\{\Phi_a\}$ to the quotient space

$$W := \text{range}\{\Phi_a\}/\operatorname{span}\{\Phi_a w_1\}.$$

We will show that range$\{\mathcal{J}_a(w_1)\}$ has dimension $a-1$ by showing that ϱ is an isomorphism from the range of $\mathcal{J}_a(w_1)$ to the quotient space W. Note that $(w_1, \Phi_a w_1) = \langle \Phi_a w_1, \Phi_a w_1\rangle \neq 0$.

(1) If $\varrho\{\mathcal{J}_a(w_1)y\} = 0$, then $\mathcal{J}_a(w_1)y$ is a multiple of $\Phi_a w_1$. Thus, by Display (3.3.1.c), $\Phi_a y$ and $\Phi_a w_1$ are colinear. Since $\Phi_a w_1 \neq 0$, there exists a constant c so $\Phi_a y = c\Phi_a w_1$. Since $\Phi_a(y - cw_1) = 0$ and since $\mathcal{J}_a(w_1)w_1 = 0$, we show that the map ρ is injective from range$\{\mathcal{J}_a(w_1)\}$ to W by computing:

$$\mathcal{J}_a(w_1)y = \mathcal{J}_a(w_1)(y - cw_1) + c\mathcal{J}_a(w_1)w_1 = 0.$$

(2) We show that ϱ is surjective by observing:

$$\varrho\{\mathcal{J}_a(w_1)y\} = (w_1, \Phi_a w_1)\cdot \varrho\{\Phi_a y\}.$$

We have shown that ϱ is a bijective map from range$\{\mathcal{J}_a(w_1)\}$ to W; this establishes Assertion (2b) since:

$$\text{rank}\{\mathcal{J}_a(\sigma)\} = \text{rank}\{\mathcal{J}_a(w_1)\} = \text{rank}\{\Phi_a\} - 1 = a - 1.$$

To prove Assertion (2c), we suppose $\ell \geq 2$. Since $\{\Phi_a w_1, ..., \Phi_a w_k\}$ is not a colinear set, we may use Assertion (1) to see that if $\mathcal{J}_a(\sigma)x = 0$, then $(\mathcal{J}_a(\sigma)x, x) = 0$ and hence $\Phi_a x$ must be zero. Consequently, $\mathcal{J}_a(\sigma)x = 0$ implies $\Phi_a x = 0$ and

$$\text{rank}\{\mathcal{J}_a(\sigma)\} = \text{rank}\{\Phi_a\} = a. \quad \square$$

Proof of Theorem 3.3.2. Since $\Phi_a^2 = 0$, $(\Phi_a u, \Phi_a v) = (\Phi_a^2 u, v) = 0$ so the range of Φ_a is totally isotropic. Thus, $\mathcal{J}_a(\sigma)^2 = 0$ for any non-degenerate k plane σ and Assertion (1) holds.

We now prove Assertion (2a) of Theorem 3.3.2. By duality, we have R_a is Jordan Osserman of type $(1,0)$ if and only if R_a is Jordan Osserman of type $(p-1,q)$. Thus, we need only study the case $(r,s) = (1,0)$. Since $\Phi_a e_1^{\pm} \neq 0$, we apply Lemma 3.3.4 to see that

$$\text{rank}\{\mathcal{J}_a(\text{span}(e_1^{\pm}))\} = a - 1 > 0.$$

Thus, $\mathcal{J}_a$ is Jordan Osserman of type $(1,0)$ if and only if $\Phi_a v \neq 0$ for every $v \in S^-(V)$. If $a < p$, then $\Phi_a e_p^- = 0$ and thus R_a is not Jordan Osserman of type $(1,0)$. Suppose $a = p$. We compute:

$$\begin{aligned} v &:= c_1^- e_1^- + ... + c_p^- e_p^- + c_1^+ e_1^+ + ... + c_q^+ e_q^+, \\ \Phi_a v &= (c_1^+ - c_1^-)(e_1^- + e_1^+) + ... + (c_a^+ - c_a^-)(e_a^- + e_a^+). \end{aligned} \tag{3.3.4.b}$$

Thus, if $\Phi_a v = 0$, then $c_i^+ = c_i^-$ for $i \leq a$. Since $a = p$, this implies that we have $(v,v) = \sum_{i>p}(c_i^+)^2 \geq 0$ so v is not timelike. Thus, $\Phi_a v \neq 0$ for v timelike and hence R_a is Jordan Osserman of type $(1,0)$. This proves Assertion (2a); Assertion (2b) follows similarly.

Next, we prove Assertion (2c). Let $2 \leq r \leq p$. Again, by duality, R_a is Jordan Osserman of type $(r,0)$ if and only if R_a is Jordan Osserman of type $(p-r,q)$. Thus, it suffices to study the case $(r,0)$. Suppose that $2 \leq r \leq p - a + 1$. We define:

$$\begin{aligned} \tau &:= \begin{cases} \{0\} & \text{if } r = 2, \\ \operatorname{span}\{e_{a+2}^-, ..., e_{a+r-1}^-\} & \text{if } r \geq 3, \end{cases} \\ \sigma_1 &:= \operatorname{span}\{e_1^-, e_2^-\} \oplus \tau, \\ \sigma_2 &:= \operatorname{span}\{e_1^-, e_{a+1}^-\} \oplus \tau. \end{aligned}$$

Then $\dim \Phi_a \sigma_1 = 2$ and $\dim \Phi_a \sigma_2 = 1$. We may therefore apply Lemma 3.3.4 to see $\operatorname{rank}\{\mathcal{J}_a(\sigma_1)\} = a$ while $\operatorname{rank}\{\mathcal{J}_a(\sigma_2)\} = a - 1$; consequently, R_a is not Jordan Osserman of type $(r,0)$.

Next, suppose that $p - a + 2 \leq r \leq p$. Let σ be a timelike plane of dimension r. If $\dim \Phi_a \sigma \geq 2$, then $\operatorname{rank}\{\mathcal{J}_a(\sigma)\} = a$ by Lemma 3.3.4. Thus, it suffices to show

$$\dim \Phi_a \sigma \geq 2 \text{ on } \operatorname{Gr}_{r,0}(V).$$

Suppose the contrary. Let $\sigma_0 := \sigma \cap \ker \Phi_a$. If $0 \neq v \in \sigma_0$, then we use Equation (3.3.4.b) to see $c_i^+(v) = c_i^-(v)$ for $1 \leq i \leq a$ and thus

$$0 > (v,v) = -\sum_{i>a}(c_i^-)^2 + \sum_{i>a}(c_i^+)^2.$$

This implies $c_i^-(v) \neq 0$ for some $i > a$. Thus, the linear functionals c_i^- for $a < i \leq p$ separate points of σ_0. Since there are $p-a$ such linear functionals, we have $\dim \sigma_0 \leq p - a$ so

$$r = \dim \sigma = \dim\{\ker \Phi_a \cap \sigma\} + \dim \Phi_a \sigma \leq (p-a) + 1 < r$$

which is false. This completes the proof of Assertion (2c); the proof of Assertion (2d) is similar.

Finally, we establish Assertion (3). Let $1 \le r \le p-1$ and let $1 \le s \le q-1$. We shall suppose that $r \le s$; the case $s \le r$ is similar. Let $\theta \in \mathbb{R}$. We define:

$$\tau := \begin{cases} \{0\} & \text{if } r = 1 \text{ and } s = 1, \\ \operatorname{span}\{e_3^+, ..., e_{s+1}^+\} & \text{if } r = 1 \text{ and } 2 \le s, \\ \operatorname{span}\{e_3^-, ..., e_{r+1}^-, e_3^+, ..., e_{s+1}^+\} & \text{if } 2 \le r. \end{cases}$$

If $\theta \in \mathbb{R}$ is a real parameter, we define:

$$\sigma(\theta) := \operatorname{span}\{\cosh\theta \cdot e_1^+ + \sinh\theta \cdot e_2^-, \cosh\theta \cdot e_1^- - \sinh\theta \cdot e_2^+\}.$$

We have $\tau \oplus \sigma(\theta) \in \mathrm{Gr}_{r,s}(V)$. We will show that the rank of $\mathcal{J}_a(\sigma(\theta) \oplus \tau)$ drops for certain values of θ. Thus, $\mathcal{J}_a$ does not have constant rank on $\mathrm{Gr}_{r,s}(V)$ so R_a is not Jordan Osserman of type (r, s).

Let $1 \le i \le a$. We compute:

$$\begin{aligned} \mathcal{J}_a(e_i^+)v =& (\Phi_a e_i^+, e_i^+)\Phi_a v - (\Phi_a e_i^+, v)\Phi_a e_i^+ = \Phi_a v - (\Phi_a e_i^+, v)\Phi_a e_i^+ \\ \mathcal{J}_a(e_i^-)v =& (\Phi_a e_i^-, e_i^-)\Phi_a v - (\Phi_a e_i^-, v)\Phi_a e_i^- = \Phi_a v - (\Phi_a e_i^-, v)\Phi_a e_i^- . \end{aligned}$$

Since $\Phi_a e_i^+ = -\Phi_a e_i^-$, we see that $\mathcal{J}_a(e_i^+) = \mathcal{J}_a(e_i^-)$; we denote this common value by $\mathcal{J}_{a,i}$. This shows that:

$$\mathcal{J}_{a,i}(e_j^\pm) = \begin{cases} \Phi_a e_j^\pm & \text{if } i \le a \text{ and } i \ne j, \\ 0 & \text{if } a < i \text{ or } i = j. \end{cases}$$

Let $\varepsilon := \min(s+1, a) - \min(r+1, a)$. We compute:

$$\text{(3.3.4.c)} \qquad \mathcal{J}_a(\tau)e_j^\pm = \begin{cases} \pm\varepsilon\Phi_a e_j^+ & \text{if } 1 \le j \le r+1, \\ \pm(\varepsilon - 1)\Phi_a e_j^+ & \text{if } r+1 < j \le s+1, \\ \pm\varepsilon\Phi_a e_j^+ & \text{if } s+1 < j. \end{cases}$$

We make a hyperbolic boost and compute:

$$\begin{aligned} &\mathcal{J}_a(\cosh\theta \cdot e_1^\pm \pm \sinh\theta \cdot e_2^\mp)y \\ &= \cosh^2\theta \cdot \mathcal{J}_{a,1}y + \sinh^2\theta \cdot \mathcal{J}_{a,2}y \\ &\mp \cosh\theta \cdot \sinh\theta\{(\Phi_a e_1^\pm, y)\Phi_a e_2^\mp + (\Phi_a e_2^\mp, y)\Phi_a e_1^\pm\} \\ &= \cosh^2\theta \cdot \mathcal{J}_{a,1}y + \sinh^2\theta \cdot \mathcal{J}_{a,2}y \\ &\pm \cosh\theta \cdot \sinh\theta\{(\Phi_a e_1^+, y)\Phi_a e_2^+ + (\Phi_a e_2^+, y)\Phi_a e_1^+\}. \end{aligned}$$

Consequently, only the cross terms survive and we have

$$\begin{aligned}&\mathcal{J}_a(\sigma(\theta))y\\ =&\mathcal{J}_a(\cosh\theta\cdot e_1^+ + \sinh\theta\cdot e_2^-)y - \mathcal{J}_a(\cosh\theta\cdot e_1^- - \sinh\theta\cdot e_2^+)y\\ =&\sinh(2\theta)\{(\Phi_a e_1^+, y)\Phi_a e_2^+ + (\Phi_a e_2^+, y)\Phi_a e_1^+\}.\end{aligned}$$

This shows that:

$$\text{(3.3.4.d)}\qquad \mathcal{J}_a(\sigma(\theta))e_j^\pm = \begin{cases} \pm\sinh 2\theta\cdot\Phi_a e_2^+ & \text{if } j=1,\\ \pm\sinh 2\theta\cdot\Phi_a e_1^+ & \text{if } j=2,\\ 0 & \text{if } 3\le j.\end{cases}$$

We use Displays (3.3.4.c) and (3.3.4.d) to compute:

$$\mathcal{J}_a(\sigma(\theta)\oplus\tau)e_j^\pm = \begin{cases} \pm\{\varepsilon\Phi_a e_1^+ + \sinh 2\theta\cdot\Phi_a e_2^+\} & \text{if } j=1,\\ \pm\{\sinh 2\theta\cdot\Phi_a e_1^+ + \varepsilon\Phi_a e_2^+\} & \text{if } j=2,\\ \pm\varepsilon\Phi_a e_j^+ & \text{if } 3\le j\le r+1,\\ \pm(\varepsilon-1)\Phi_a e_j^+ & \text{if } r+1<j\le s+1,\\ \pm\varepsilon\Phi_a e_j^+ & \text{if } s+1<j.\end{cases}$$

Consequently, if $\varepsilon^2 = \sinh^2 2\theta$, then the rank drops. Since $\sinh^2 2\theta$ defines a surjective map from $\mathbb{R}^+$ to $\mathbb{R}^+$, we can find θ so $\sinh 2\theta = \pm\varepsilon$. Thus, R_a is not Jordan Osserman of type (r, s). □

3.4 Rakić duality

We begin our proof of Theorem 3.1.2 with a technical observation concerning the Jacobi operator.

3.4.1 Lemma. *Let R be an algebraic curvature tensor on a Riemannian vector space V. Let ρ_τ and ρ_σ be orthogonal projection on subspaces τ and σ of V. Then* $\operatorname{Tr}\{\rho_\tau\mathcal{J}_R(\sigma)\} = \operatorname{Tr}\{\rho_\sigma\mathcal{J}_R(\tau)\}$.

Proof. Let $\{v_1, ..., v_k\}$ and $\{w_1, ..., w_\ell\}$ be orthonormal bases for $\tau\in\operatorname{Gr}_k(V)$ and $\sigma\in\operatorname{Gr}_\ell(V)$, respectively. We prove Lemma 3.4.1 by computing:

$$\begin{aligned}\operatorname{Tr}\{\rho_\tau\mathcal{J}_R(\sigma)\} &= \textstyle\sum_i(\mathcal{J}_R(\sigma)v_i, v_i) = \sum_{i,j}R(v_i, w_j, w_j, v_i)\\ &= \textstyle\sum_{i,j}R(w_j, v_i, v_i, w_j) = \sum_j(\mathcal{J}_R(\tau)w_j, w_j)\\ &= \operatorname{Tr}\{\rho_\sigma\mathcal{J}_R(\tau)\}.\quad\square\end{aligned}$$

The remainder of this section is devoted to the proof of Theorem 3.1.2. We shall first prove Assertion (2a) and then use Assertion (2a) to derive the remaining assertions. We establish the following notational conventions.

Let R be an algebraic curvature tensor on a Riemannian vector space V of dimension m. Let τ be a linear subspace of V. Let $\lambda_1(\tau) \leq ... \leq \lambda_m(\tau)$ be the eigenvalues of $\mathcal{J}_R(\tau)$, where each eigenvalue is repeated according to the multiplicity. Let

$$\Lambda_i(\tau) := \lambda_1(\tau) + ... + \lambda_i(\tau).$$

Let $1 \leq k \leq m-1$. We suppose that R is k Osserman. Thus, the functions $\lambda_i(\cdot)$ and $\Lambda_i(\cdot)$ are constant on $\mathrm{Gr}_k(V)$.

Let $\tau \in \mathrm{Gr}_k(V)$ and let $v \in S(V)$. To prove Assertion (2a), we must show that if $\mathcal{J}_R(\tau)v = \lambda_j(\tau)v$ for some j, then $\mathcal{J}_R(v)\tau \subset \tau$. We choose an orthonormal basis $\mathcal{B} = \{v_1, ..., v_m\}$ for V so that

$$\mathcal{J}(\tau)v_i = \lambda_i(\tau)v_i \text{ for } 1 \leq i \leq m \text{ and so } v_j = v.$$

For $1 \leq i \leq m$, let $\pi_i := \mathrm{span}\{v_1, ..., v_i\} \in \mathrm{Gr}_i(V)$. Let $\tilde{\tau} \in \mathrm{Gr}_k(V)$. We use Lemma 1.3.3 and Lemma 3.4.1 to see that:

$$\text{(3.4.1.a)} \qquad \Lambda_i(\tau) = \Lambda_i(\tilde{\tau}) \leq \mathrm{Tr}\{\rho_{\pi_i}\mathcal{J}_R(\tilde{\tau})\} = \mathrm{Tr}\{\rho_{\tilde{\tau}}\mathcal{J}_R(\pi_i)\}.$$

We apply Lemma 1.3.3 to Display (3.4.1.a) to observe:

$$\text{(3.4.1.b)} \qquad \Lambda_i(\tau) \leq \min_{\tilde{\tau} \in \mathrm{Gr}_k(V)} \mathrm{Tr}\{\rho_{\tilde{\tau}}\mathcal{J}_R(\pi_i)\} = \Lambda_k(\pi_i).$$

We now use Lemma 1.3.3, Lemma 3.4.1, and the definition of π_i to estimate:

$$\text{(3.4.1.c)} \qquad \Lambda_k(\pi_i) \leq \mathrm{Tr}\{\rho_\tau\mathcal{J}_R(\pi_i)\} = \mathrm{Tr}\{\rho_{\pi_i}\mathcal{J}_R(\tau)\} = \Lambda_i(\tau).$$

Consequently, we have equality in Equations (3.4.1.b) and (3.4.1.c) so

$$\Lambda_k(\pi_i) = \mathrm{Tr}\{\rho_\tau\mathcal{J}_R(\pi_i)\}.$$

Hence, by Lemma 1.3.3, we have:

$$\text{(3.4.1.d)} \qquad \mathcal{J}_R(\pi_i)\tau \subset \tau.$$

We set $i = 1$ in Display (3.4.1.d) to see $\mathcal{J}_R(v_1)\tau \subset \tau$. For $i > 1$, we note

$$\mathcal{J}_R(v_i) = \mathcal{J}_R(\pi_i) - \mathcal{J}_R(\pi_{i-1})$$

and complete the proof of Assertion (2a) by using Equation (3.4.1.d) to see:

$$\mathcal{J}_R(v_i)\tau \subset \mathcal{J}_R(\pi_i)\tau \ + \mathcal{J}_R(\pi_{i-1})\tau \subset \tau.$$

To prove Assertion (1) of Theorem 3.1.2, we suppose that $k = 1$ and that $\mathcal{J}_R(v_1)v_2 = \lambda v_2$. We may then apply Assertion (2a) to see $\mathcal{J}_R(v_2)v_1 = \tilde{\lambda} v_1$. We apply complete the proof of Assertion (1) by checking that $\lambda = \tilde{\lambda}$:

$$\begin{aligned}\tilde{\lambda} =&(\mathcal{J}_R(v_2)v_1, v_1) = R(v_1, v_2, v_2, v_1) = R(v_2, v_1, v_1, v_2)\\ =&(\mathcal{J}_R(v_1)v_2, v_2) = \lambda.\end{aligned}$$

To prove Assertion (2b), we assume R is k Osserman. Let $\tau \in \mathrm{Gr}_k(V)$ and let σ be an arbitrary subspace of V. We suppose that $\mathcal{J}_R(\tau)\sigma \subset \sigma$. We must show that $\mathcal{J}_R(\sigma)\tau \subset \tau$.

Since $\mathcal{J}_R(\tau)$ is a self-adjoint map which preserves σ, we can find an orthonormal basis $\{v_1, ..., v_p\}$ for σ so that $\mathcal{J}_R(\tau)v_i = \mu_i v_i$. We then may apply Assertion (2a) to see $\mathcal{J}_R(v_i)\tau \subset \tau$. Assertion (2b) now follows since

$$\mathcal{J}_R(\sigma) = \textstyle\sum_{1\le i\le p} \mathcal{J}_R(v_i). \quad \square$$

3.5 The Osserman conjecture

Let V be a Riemannian vector space V of dimension m; we shall use the fact that the metric is positive definite repeatedly. Let R be an Osserman algebraic curvature tensor on V such that the reduced Jacobi operator $\tilde{\mathcal{J}}_R(x)$ has exactly two eigenvalues λ_0 and λ_1 on $S(V)$, where λ_1 has multiplicity 1. It then follows that λ_0 has multiplicity $m - 2$. If $x \in S(V)$, then set

$$\begin{aligned}&E_i(x) := \{y \in x^\perp : \mathcal{J}_R(x)y = \lambda_i y\};\\ &V = \mathrm{span}\{x\} \oplus E_0(x) \oplus E_1(x).\end{aligned}$$

Theorem 3.1.6 will follow from the following Lemma.

3.5.1 Lemma. *Let R be an Osserman algebraic curvature tensor on a Riemannian vector space V of dimension $m \ge 3$. Suppose that $\tilde{\mathcal{J}}_R$ has two distinct eigenvalues $\{\lambda_0, \lambda_1\}$, where λ_1 has multiplicity 1. Then:*

(1) *Let $\{x, y\}$ be an orthonormal set. Then $\mathcal{J}_R(x)y = \lambda_i y$ if and only if $R(x, y, y, x) = \lambda_i$.*

(2) *There exists a smooth map* $\phi : S(V) \to S(V)$ *so that* $\phi x \in E_1(x)$ *for all* $x \in S(V)$. *We have* $\phi(\cos\theta x + \sin\theta\phi x) = \cos\theta\phi x - \sin\theta x$ *for any* $\theta \in [0, 2\pi]$. *In particular,* $\phi^2 x = -x$ *and* $\phi(-x) = -\phi x$.

(3) *Let* $\{x, y\}$ *be an orthonormal set. Let* $y \in E_0(x)$. *Then:*
 3a) $\{x, \phi x, y, \phi y\}$ *is an orthonormal set.*
 3b) $\{y, \phi y\} \subset E_0(x) \cap E_0(\phi x)$.
 3c) $\{x, \phi x\} \subset E_0(y) \cap E_0(\phi y)$.

(4) *Let* $\{x, y, z\}$ *be an orthonormal set. Assume that* $y, z \in E_0(x)$. *Then* $R(x, y)z + R(x, z)y = 0$.

(5) *Let* $\{x, y, z\}$ *be an orthonormal set. Assume that* $y, z \in E_0(x)$, *that* $x, z \in E_0(y)$, *and that* $x, y \in E_0(z)$. *Then* $R(x, y)z = 0$.

(6) *Let* $\{x, y\}$ *be an orthonormal set with* $y \in E_0(x)$. *Then we have that* $\phi(\cos\theta \cdot x + \sin\theta \cdot y) = \cos\theta \cdot \phi x + \sin\theta \cdot \phi y$ *for any* $\theta \in [0, 2\pi]$.

(7) *Extend* ϕ *from* $S(V)$ *to* V *by defining* $\phi(rx) = r\phi x$ *for* $r \in [0, \infty)$ *and* $x \in S(V)$. *Then* $\phi \in \mathfrak{so}(V)$ *and* $\phi^2 = -\operatorname{Id}$.

(8) *We have* $R = \lambda_0 R_{\mathrm{Id}} + \frac{1}{3}(\lambda_1 - \lambda_0) R_\phi$.

Proof. We follow the discussion in Gilkey, Swann, and Vanhecke [87]. Let $\{x, y\}$ be an orthonormal set. Since λ_i are extremal eigenvalues of $\tilde{\mathcal{J}}_R(x)$ on $x^\perp$, Assertion (1) follows from Lemma 1.3.2 and from the following chain of equivalences:

$$\mathcal{J}_R(x)y = \lambda_i y \iff (\mathcal{J}_R(x)y, y) = \lambda_i \iff R(x, y, y, x) = \lambda_i.$$

Since λ_1 has multiplicity 1, E_1 is a real line bundle over $S^{m-1} = S(V)$. Since $m \geq 3$, S^{m-1} is simply connected. Thus, $H^1(S^{m-1}; \mathbb{Z}_2) = 0$, so we may use Lemma 4.3.11 to see that every real line bundle over S^{m-1} is trivial. Let ϕ be a smooth unit section to E_1. Let $x \in S(V)$. For $\theta \in [0, 2\pi]$, we define orthogonal unit vectors:

$$x_\theta := \cos\theta x + \sin\theta\phi x \text{ and } y_\theta := -\sin\theta x + \cos\theta\phi x.$$

Since $\tilde{\mathcal{J}}_R(x)\phi x = \lambda_1 \phi x$, we have $R(x, \phi x, \phi x, x) = \lambda_1$ so the sectional curvature of the 2 plane $\tau := \operatorname{span}\{x, \phi x\}$ is λ_1. By Lemma 1.6.4, the sectional curvature is independent of the orthonormal basis for τ. Therefore,

$$R(x_\theta, y_\theta, y_\theta, x_\theta) = \lambda_1 \text{ for all } \theta \in [0, 2\pi].$$

Thus, by Assertion (1), $\mathcal{J}_R(x_\theta)y_\theta = \lambda_1 y_\theta$ for all $\theta \in [0, 2\pi]$. Consequently,

$$y_\theta = \varepsilon_\theta \phi x_\theta, \text{ where } \varepsilon_\theta = \pm 1.$$

Since y_θ and ϕx_θ are smooth functions of the parameter θ, the sign ε_θ is constant. Since $y_0 = \phi x = \phi x_0$, $\varepsilon_0 = 1$ so:

$$y_\theta = \phi x_\theta \text{ for all } \theta.$$

We complete the proof of Assertion (2) by computing:

$$\phi^2 x = \phi x_{\frac{\pi}{2}} = y_{\frac{\pi}{2}} = -x \text{ and } \phi(-x) = \phi x_\pi = y_\pi = -\phi x.$$

To prove Assertion (3), we suppose that $\{x, y\}$ is an orthonormal set with $y \in E_0(x)$. We have $E_1(x) = \text{span}\{\phi x\}$ and dually, by Theorem 3.1.2, $E_1(\phi x) = \text{span}\{x\}$. Thus,

$$E_0(x) = \text{span}\{x, \phi x\}^\perp = E_0(\phi x). \tag{3.5.1.a}$$

We check the six orthogonality relations needed to prove Assertion (3a):

(1) We have that $x \perp \phi x$ since $\phi x \in E_1(x)$ and $x \perp E_1(x)$.
(2) By assumption, $x \perp y$.
(3) Since $y \in E_0(x)$, we dually have $x \in E_0(y)$. Since $E_0(y) \perp E_1(y)$, since $x \in E_0(y)$, and since $\phi y \in E_1(y)$, we have $x \perp \phi y$.
(4) Since $E_1(x) \perp E_0(x)$, since $y \in E_0(x)$, and since $\phi x \in E_1(x)$, we have $\phi x \perp y$.
(5) Since $\phi x \perp y$, since $y \perp x$, and since dually x spans $E_1(\phi x)$, we have $y \in E_0(\phi x)$. Thus, dually, $\phi x \in E_0(y)$. Since $\phi y \in E_1(y)$ and since $E_0(y) \perp E_1(y)$, we have $\phi x \perp \phi y$.
(6) We have that $y \perp \phi y$ since $\phi y \in E_1(y)$ and $y \perp E_1(y)$.

Assertion (3b) now follows from Assertion (3a) and Equation (3.5.1.a). Since by Theorem 3.1.2, $y \in E_0(x)$ implies $x \in E_0(y)$, we see that the roles of x and y are symmetric and thus we may complete the proof of Assertion (3) by noting that Assertion (3c) follows from Assertion (3b).

Let $\{x, y, z\}$ be an orthonormal set, where $y, z \in E_0(x)$. Define:

$$w_\theta := \cos\theta y + \sin\theta z \in E_0(x) \text{ for } \theta \in [0, 2\pi].$$

Since $w_\theta \in E_0(x)$, we have dually by Theorem 3.1.2 that $x \in E_0(w_\theta)$. We prove Assertion (4) by computing:

$$\begin{aligned}\lambda_0 x =& R(x, w_\theta)w_\theta \\ =& \cos^2\theta R(x,y)y + \sin^2\theta R(x,z)z + \sin\theta\cos\theta\{R(x,z)y + R(x,y)z\} \\ =& \lambda_0 x + \sin\theta\cos\theta\{R(x,z)y + R(x,y)z\}.\end{aligned}$$

Let $\{x, y, z\}$ be an orthonormal set. Suppose that $y, z \in E_0(x)$, that $x, y \in E_0(z)$, and that $x, z \in E_0(y)$. Then the roles of $\{x, y, z\}$ are symmetric. We may therefore use Assertion (4) and the Bianchi identities to establish Assertion (5) by computing:

$$R(x,y)z = -R(y,z)x - R(z,x)y = R(y,x)z + R(y,x)z = -2R(x,y)z.$$

We now prove Assertion (6). Let $\{x, y\}$ be an orthonormal set with $y \in E_0(x)$. Dually then we also have $x \in E_0(y)$. We set $a_\theta := \cos\theta x + \sin\theta y$ and compute:

$$\begin{aligned}\mathcal{J}_R(a_\theta)\phi x =& \cos^2\theta R(\phi x, x)x + \sin^2\theta R(\phi x, y)y \\ & \sin\theta\cos\theta\{R(\phi x, y)x + R(\phi x, x)y\}, \text{ and} \\ \mathcal{J}_R(a_\theta)\phi y =& \cos^2\theta R(\phi y, x)x + \sin^2\theta R(\phi y, y)y \\ & + \sin\theta\cos\theta\{R(\phi y, y)x + R(\phi y, x)y\}.\end{aligned}$$

Since $\phi x \in E_1(x)$, $\phi x \in E_0(y)$, $\phi y \in E_0(x)$, and $\phi y \in E_1(y)$, we have:

$$\begin{aligned}\mathcal{J}_R(a_\theta)\phi x =& (\cos^2\theta\lambda_1 + \sin^2\theta\lambda_0)\phi x \\ & + \cos\theta\sin\theta\{R(\phi x, x)y + R(\phi x, y)x\}, \text{ and} \\ \mathcal{J}_R(a_\theta)\phi y =& (\sin^2\theta\lambda_1 + \cos^2\theta\lambda_0)\phi y \\ & + \cos\theta\sin\theta\{R(\phi y, x)y + R(\phi y, y)x\}.\end{aligned} \tag{3.5.1.b}$$

Let $\tau := \text{span}\{x, \phi x, y, \phi y\}$. Suppose that $z \perp \tau$. Then $x, z \in E_0(y)$ and $y, z \in E_0(x)$. Thus, dually, $x, y \in E_0(z)$, so by Assertion (5), $R(x, z)y = 0$ and $R(y, z)x = 0$. This implies that:

$$\begin{aligned}& R(\cdot, y, x, z) = -(R(x,z)y, \cdot) = 0,\ R(\cdot, x, y, z) = -(R(y,z)x, \cdot) = 0 \text{ so} \\ & R(\phi x, x)y \in \tau,\ R(\phi x, y)x \in \tau,\ R(\phi y, x)y \in \tau, \text{ and } R(\phi y, y)x \in \tau.\end{aligned}$$

We show that $R(\phi x, y)x$ is a multiple of ϕy by computing:

$$\begin{aligned}& R(\phi x, y, x, x) = 0, \\ & R(\phi x, y, x, y) = -R(x, y, y, \phi x) = -\lambda_0(x, \phi x) = 0, \text{ and} \\ & R(\phi x, y, x, \phi x) = R(x, \phi x, \phi x, y) = \lambda_1(x, y) = 0.\end{aligned}$$

Similarly, $R(\phi x, x)y$ is a multiple of ϕy, $R(\phi y, x)y$ is a multiple of ϕx, and $R(\phi y, y)x$ is a multiple of ϕx. Therefore, there exist constants c and $\tilde{c}$ so that we may express Display (3.5.1.b) in the form:

$$\begin{aligned}\mathcal{J}_R(a_\theta)\phi x =& (\cos^2\theta\lambda_1 + \sin^2\theta\lambda_0)\phi x - \cos\theta\sin\theta c\phi y, \text{ and} \\ \mathcal{J}_R(a_\theta)\phi y =& -\cos\theta\sin\theta\tilde{c}\phi x + (\sin^2\theta\lambda_1 + \cos^2\theta\lambda_0)\phi y.\end{aligned} \tag{3.5.1.c}$$

Since $\mathcal{J}_R(a_\theta)$ is self-adjoint, $c = \tilde{c}$. Let $\sigma := \operatorname{span}\{\phi x, \phi y\}$. Since $a_\theta \perp \sigma$, σ is in the domain of $\tilde{\mathcal{J}}_R(a_\theta)$. Since $\mathcal{J}_R(a_\theta)$ preserves σ, $\tilde{\mathcal{J}}_R(a_\theta)$ has two eigenvalues $\mu_0(\theta)$ and $\mu_1(\theta)$ on σ. Since these eigenvalues vary continuously and since R is Osserman, these eigenvalues are constant. Taking $\theta = 0$ we see that we may take $\mu_0 = \lambda_0$ and $\mu_1 = \lambda_1$.

We take $\theta = \frac{\pi}{4}$ and compute:

$$\mathcal{J}_R(a_{\frac{\pi}{4}})(\phi x + \phi y) = \tfrac{1}{2}(\lambda_1 + \lambda_0 - c)(\phi x + \phi y).$$

Consequently, one of the two alternatives holds:

$$\begin{aligned}
\lambda_0 &= \tfrac{1}{2}(\lambda_1 + \lambda_0 - c) \text{ and } c = \lambda_1 - \lambda_0 \text{ or} \\
\lambda_1 &= \tfrac{1}{2}(\lambda_1 + \lambda_0 - c) \text{ and } c = \lambda_0 - \lambda_1.
\end{aligned}$$

Let $b_\theta := \cos\theta\phi x + \varepsilon \sin\theta\phi y$, where $\varepsilon = \pm 1$ is chosen so that $\varepsilon c = \lambda_0 - \lambda_1$. We use Equation (3.5.1.c) to compute:

$$\begin{aligned}
\mathcal{J}_R(a_\theta)b_\theta =& (\cos^2\theta\lambda_1 + \sin^2\theta\lambda_0 - c\varepsilon\sin^2\theta)\cos\theta\phi x \\
&+ (\sin^2\theta\lambda_1 + \cos^2\theta\lambda_0 - c\varepsilon\cos^2\theta)\varepsilon\sin\theta\phi y
\end{aligned}$$

We then have $b_\theta \in E_1(a_\theta)$ since

$$\begin{aligned}
&\cos^2\theta\lambda_1 + \sin^2\theta\lambda_0 - c\varepsilon\sin^2\theta = \lambda_1, \text{ and} \\
&\sin^2\theta\lambda_1 + \cos^2\theta\lambda_0 - c\varepsilon\cos^2\theta = \lambda_1.
\end{aligned}$$

Thus, there exists $\delta_\theta = \pm 1$ so that $b_\theta = \delta_\theta \phi a_\theta$, i.e.

$$\phi(\cos\theta x + \sin\theta y) = \delta_\theta(\cos\theta\phi x + \varepsilon\sin\theta\phi y). \tag{3.5.1.d}$$

By continuity, δ_θ is independent of θ. We show $\delta_0 = \varepsilon = 1$ and complete the proof of Assertion (6) by computing:

$$\begin{aligned}
\theta = 0: \ \phi x &= \delta_0\phi x \text{ so } \delta_0 = 1 \\
\theta = \tfrac{\pi}{2}: \ \phi y &= \varepsilon\delta_0\phi y \text{ so } \varepsilon = 1.
\end{aligned}$$

We now prove Assertion (7). We extend ϕ radially to all of V by setting

$$\phi(rv) = r\phi(v) \text{ for } r \geq 0 \text{ and } v \in S(V).$$

As $\phi(-x) = -\phi x$ by Assertion (2), we have $\phi(rx) = r\phi x$ for all $r \in \mathbb{R}$. Let $v_i \in S(V)$. We may Assertions (2) and (6) to see:

$$\phi(av_1 + bv_2) = a\phi v_1 + b\phi v_2 \text{ if } v_2 \in E_0(v_1) \text{ or if } v_2 \in E_1(v_1).$$

Let $u_1, u_2 \in V$. If $u_1 = 0$, then $\phi(u_1 + u_2) = \phi u_1 + \phi u_2$ so we suppose $u_1 \neq 0$ and let $u_1 = av_1$, where v_1 is a unit vector. We may then decompose $u_2 = bv_1 + c\phi v_1 + dv_2$, where $\{v_1, \phi v_1, v_2\}$ is an orthonormal set. We then have $\{v_1, \phi v_1\} \subset E_0(v_2)$ and thus

$$\begin{aligned}\phi(u_1 + u_2) =&\phi(av_1 + bv_1 + c\phi v_1 + dv_2) = \phi(av_1 + bv_1 + c\phi v_1) + d\phi v_2\\ =&(a+b)\phi(v_1) + c\phi\phi v_1 + d\phi v_2 = a\phi v_1 + \phi(bv_1 + c\phi v_1) + d\phi v_2\\ =&a\phi v_1 + \phi(bv_1 + c\phi v_1 + dv_2) = \phi u_1 + \phi u_2.\end{aligned}$$

This shows that ϕ is linear. By Assertion (2), $\phi^2 = -\operatorname{Id}$. If $x \in S(V)$, then $x \perp \phi x$. We rescale to see that $v \perp \phi v$ for any $v \in V$. We show $\phi \in \mathfrak{so}(V)$ and complete the proof of Assertion (7) by computing:

$$\begin{aligned}&(u_1, \phi u_1) = 0,\ (u_2, \phi u_2) = 0,\\ &(u_1 + u_2, \phi(u_1 + u_2)) = 0 \text{ so}\\ &(u_1, \phi u_2) + (u_2, \phi u_1) = 0.\end{aligned}$$

We complete the proof by establishing Assertion (8). Let

$$\tilde{R} := \lambda_0 R_{\mathrm{Id}} + \tfrac{1}{3}(\lambda_1 - \lambda_0) R_\phi.$$

We use Display (3.2.1.a) to see that

$$\begin{aligned}&\mathcal{J}_{\tilde{R}}(x)\phi x = (\lambda_0 + 3 \cdot \tfrac{1}{3}(\lambda_1 - \lambda_0))\phi x = \lambda_1 \phi x,\\ &\mathcal{J}_{\tilde{R}}(x)y = \lambda_0 y \text{ if } y \perp \{x, \phi x\}.\end{aligned}$$

Thus, $\mathcal{J}_{\tilde{R}-R}(x) = 0$ for all $x \in S(V)$. Therefore, $\tilde{R} - R$ has zero sectional curvature. We apply Lemma 1.6.4 to conclude $\tilde{R} - R = 0$. □

3.6 Space forms and (para-)complex space forms

Let V be a vector space of signature (p, q). We suppose given $\phi \in \mathfrak{so}(V)$ with $\phi^2 = \pm \operatorname{Id}$. We shall be interested in curvature tensors of the form $R = c_0 R_{\mathrm{Id}} + c_1 R_\phi$. If $p + q > 2$, then the constants c_0 and c_1 are uniquely determined by this identity; if $p + q = 2$, then $R = (c_0 - 3\varepsilon c_1)R_{\mathrm{Id}}$ has constant sectional curvature.

Let (M, g) be a connected pseudo-Riemannian manifold of signature (p, q) and dimension $m = p + q$. We say that (M, g) is a *space form* if for every point P of M, there is a real number $\kappa(P)$ so ${}^gR_P = \kappa(P) R_{\mathrm{Id}}$. Suppose that (M, g) is not a space form. We also suppose that for every point P of

M, there exist real numbers $c_0(P)$ and $c_1(P)$ and there exists $\phi \in \mathfrak{so}(TM)$ so that ${}^gR_P = c_0(P)R_{\text{Id}} + c_1(P)R_{\phi(P)}$. If $\phi^2 = -\operatorname{Id}$, then (M,g) is said to be a *complex space form*, while if $\phi^2 = \operatorname{Id}$, then (M,g) is said to be a *para-complex space form*. Let M be contractible. By Lemmas 1.14.1 and 1.15.1:

(1) If $m \neq 2$, then (M,g) is a space form if and only if ${}^gR = \kappa R_{\text{Id}}$, where κ is constant.

(2) If $m \neq 2, 4$, then (M,g) is a complex space form if and only if ${}^gR = c_0(R_{\text{Id}} + R_\phi)$, where c_0 is constant and where $\phi \in \mathfrak{so}(TM)$ satisfies $\phi^2 = -\operatorname{Id}$ and $\nabla\phi = 0$.

(3) If $m \neq 2, 4$, then (M,g) is a para-complex space form if and only if ${}^gR = c_0(R_{\text{Id}} - R_\phi)$, where c_0 is constant and where $\phi \in \mathfrak{so}(TM)$ satisfies $\phi^2 = +\operatorname{Id}$ and $\nabla\phi = 0$.

In this section, we shall use results from Section 1.14 to complete the classification of the local isometry type of space forms in Theorem 3.6.1 if $m \neq 2$, of complex space forms in Theorem 3.6.5 if $m \neq 2, 4$, and of para-complex space forms in Theorem 3.6.7 if $m \neq 2, 4$.

Let $g_{p,q}$ be the flat metric on $\mathbb{R}^{(p,q)}$ and let $g^{\pm}_{p,q}$ be the induced metric on the *pseudo-spheres*:

$$S^+_{p,q} := \{v \in \mathbb{R}^{(p,q+1)} : (v,v) = 1\} \text{ and}$$
$$S^-_{p,q} := \{v \in \mathbb{R}^{(p+1,q)} : (v,v) = -1\}.$$

We note that $(S^+_{p,q}, g^+_{p,q}) = (S^-_{q,p}, -g^-_{q,p})$.

3.6.1 Theorem. *Let (M,g) be a space form of dimension $m \neq 2$, signature (p,q), and sectional curvature κ.*

(1) *If $\kappa = r^{-2} > 0$, then (M,g) and $(S^+_{p,q}, r^2 g^+_{p,q})$ are locally isometric.*

(2) *If $\kappa = 0$, then (M,g) and $(\mathbb{R}^{(p,q)}, g_{p,q})$ are locally isometric.*

(3) *If $\kappa = -r^{-2} < 0$, then (M,g) and $(S^-_{p,q}, r^2 g^-_{p,q})$ are locally isometric.*

Proof. Let (M,g) and $(\tilde{M},\tilde{g})$ be two connected pseudo-Riemannian manifolds of signature (p,q) and constant sectional curvature κ. By Lemma 1.14.2, (M,g) and $(\tilde{M},\tilde{g})$ are locally isometric. We have that $(\mathbb{R}^{(p,q)}, g_{p,q})$ is a pseudo-Riemannian manifold of signature (p,q) and zero sectional curvature. In Lemma 2.6.1, we showed that the pseudo-spheres $(S^{\pm}_{p,q}, g^{\pm}_{p,q})$ are homogeneous pseudo-Riemannian manifolds of signature (p,q) with constant sectional curvature ± 1. Thus, $(S^{\pm}_{p,q}, r^2 g^{\pm}_{p,q})$ has constant sectional curvature $\pm r^{-2}$. Theorem 3.6.1 now follows. $\square$

It is worth making this explicit in the Riemannian setting. The space

$$(S^q, r^2 g_q) := (S^+_{0,q}, r^2 g^+_{0,q})$$

is the sphere of radius r in $\mathbb{R}^{q+1}$ with the standard metric of constant sectional curvature r^{-2}. The space $S^-_{0,q}$ has two connected components. Let H^q be one of these components with the metric $r^2 g^-_{0,q}$. We use Theorem 4.1.1 to see that H^q is contractible. Since $(H^q, r^2 g_{0,q})$ is homogeneous, it is complete. Since it has a metric of constant sectional curvature -1, it can be identified with hyperbolic space.

Before discussing (para-)complex space forms, we must establish some additional technical results. We say that a submanifold N of a pseudo-Riemannian manifold (M, g_M) is *non-degenerate* if the restriction of the metric g_M to the submanifold N is a non-degenerate metric on N. Let i be the inclusion of N in M, let $g_N := i^* g_M$ be the restriction of the metric g_M to N, let R_M be the curvature tensor of (M, g_M), let $i^* R_M$ be the restriction of R_M to N, and let R_N be the curvature tensor of (N, g_N). In general, $i^* R_M \neq R_N$.

A smooth map $\psi : (M, g) \to (M, g)$ is said to be an *isometry* if the associated linear map $\psi_*(P) : T_P M \to T_{\psi P} M$ is an isometry for any point $P \in M$; necessarily ψ_* is injective so ψ is a local diffeomorphism.

3.6.2 Lemma. *Let (N, g_N) be a non-degenerate submanifold of a pseudo-Riemannian manifold (M, g_M). If there exists an isometry ψ of M so that N is a connected component of the fixed point set of ψ, then $i^* R_M = R_N$.*

Proof. Fix a point $P \in N$. Then ψ_* induces an isometry of $T_P M$. Let

$$V := \{v \in T_P M : \psi_* v = v\}.$$

The exponential map $\exp_P$ discussed in Section 1.11 is a diffeomorphism from a small convex neighborhood $\mathcal{O}$ of $0 \in T_P M$ to a small neighborhood $\mathcal{U} := \exp_P(\mathcal{O})$ of $P \in M$. Choose a smaller convex open set $\mathcal{O}_1$ so that $0 \in \mathcal{O}_1 \subset \mathcal{O}$ and so that:

$$\psi \exp_P(\mathcal{O}_1) \subset \mathcal{U} \cap \psi^{-1}\mathcal{U}. \tag{3.6.2.a}$$

Let γ be a geodesic in M with $\gamma(0) = P$ and $\gamma'(0) \in V$. Since ψ is an isometry, $\psi\gamma$ is also a geodesic of M. We have:

$$\psi\gamma(0) = P \text{ and } (\psi\gamma)'(0) = \psi_*(\gamma'(0)) = \gamma'(0).$$

Thus, the uniqueness theorem for ordinary differential equations implies that $\psi\gamma = \gamma$ so

$$\exp_P(V \cap \mathcal{O}) \subset N \text{ and } V \subset T_P N.$$

Conversely, if $Q \in \exp_P(\mathcal{O}_1) \cap N$, then there exists a unique geodesic $\gamma : [0,1] \to \mathcal{O}_1$ so $\gamma(0) = P$ and $\gamma(1) = Q$. We use Equation (3.6.2.a) to see

$$\psi\gamma \subset \psi \exp_P(\mathcal{O}_1) \subset \psi(\psi^{-1}\mathcal{U}) = \mathcal{U}.$$

Therefore, γ and $\psi\gamma$ are both geodesics mapping $[0,1]$ to $\mathcal{U}$ which start at P and end at Q. Since $\exp_P$ is a diffeomorphism from $\mathcal{O}$ to $\mathcal{U}$, we must have $\gamma = \psi\gamma$. Thus, $\gamma'(0) \in V$. Consequently, $\exp_P(V \cap \mathcal{O}_1)$ is an open neighborhood of P in N. This shows

$$\dim V = \dim N \text{ so } V = T_P N.$$

Let $m := \dim(M)$ and let $n := \dim(N)$. Since the metric on N is assumed to be non-degenerate, we can find an orthonormal basis $\{e_1, ..., e_m\}$ for $T_P M$ so that $\{e_1, ..., e_n\}$ is a basis for $T_P N = V$. Let

$$\tilde{\mathcal{O}}_1 := \{\xi = (\xi_1, ..., \xi_m) \in \mathbb{R}^m : \xi_1 e_1 + ... + \xi_m e_m \in \mathcal{O}_1\}.$$

We have exponential coordinates on M, centered at P, which are given by:

$$\xi \to \exp_P(\xi_1 e_1 + ... + \xi_m e_m) \text{ for } \xi \in \tilde{\mathcal{O}}_1.$$

Since $\exp_P(\mathcal{O}_1 \cap V) \subset N$, these coordinates induce coordinates on N so:

$$N \cap \exp_P(\mathcal{O}_1) = \{\xi : \xi_{n+1} = ... = \xi_m = 0\}.$$

Let $\partial_i := \frac{\partial}{\partial \xi_i}$. By Lemma 1.11.4, $\partial_k g_{ij}(P) = 0$. Let $1 \le i, j, k, l \le m$ and let $1 \le a, b, c, d \le n$. We use Lemma 1.11.1 to complete the proof by computing:

$$\begin{aligned} {}^M R_{ijkl}(P) &= \tfrac{1}{2}\{\partial_i\partial_k g_{jl} + \partial_j\partial_l g_{ik} - \partial_i\partial_l g_{jk} - \partial_j\partial_k g_{il}\}(P), \\ {}^N R_{abcd}(P) &= \tfrac{1}{2}\{\partial_a\partial_c g_{bd} + \partial_b\partial_d g_{ac} - \partial_a\partial_d g_{bc} - \partial_b\partial_c g_{ad}\}(P). \qquad \square \end{aligned}$$

Let V be a vector space of signature (p,q). Suppose that $\phi \in \mathfrak{so}(V)$ satisfies $\phi^2 = \pm\text{Id}$. We say that $S = \{e_1^+, \phi e_1^+, ..., e_k^+, \phi e_k^+\}$ is a spacelike (para-)complex k frame if S is an orthonormal set and if the vectors $\{e_1^+, ..., e_k^+\}$ are spacelike. Let $U(V,\phi)$ be the *(para-)unitary group* of all isometries of V which commute with ϕ. We can characterize the algebraic curvature tensors which have the form $c_0 R_{\text{Id}} + c_1 R_\phi$ as the fixed points of the action of $U(V,\phi)$ on the set of algebraic curvature tensors on V:

3.6.3 Lemma. *Let R be an algebraic curvature tensor on a vector space V of signature (p,q), where $m=p+q\neq 2,4$. Let $\phi\in\mathfrak{so}(V)$ satisfy $\phi^2=\pm\operatorname{Id}$.*

(1) *$U(V,\phi)$ acts transitively on spacelike (para-)complex k frames.*
(2) *If $\psi^*R=R$ for every $\psi\in U(V,\phi)$, then there are constants b_i so:*
  2a) *$R(\phi x,x)x=b_0(x,x)\phi x$ for any $x\in V$.*
  2b) *$R(w,x,x,w)=b_1(x,x)(w,w)$ for any $x\in V$ and for any $w\in V$ with $w\perp\{x,\phi x\}$.*
(3) *The following conditions are equivalent:*
  3a) *$\psi^*R=R$ for all $\psi\in U(V,\phi)$.*
  3b) *There exist constants c_0 and c_1 so $R=c_0R_{\operatorname{Id}}+c_1R_\phi$.*

Proof. The proof of Assertion (1) is essentially the same as the proof given for Lemma 1.15.3 (1). Let S and $\tilde S$ be spacelike (para-)complex k frames. Suppose first that $\phi^2=-\operatorname{Id}$; thus ϕ is an isometry and p and q are both even. Extend S and $\tilde S$ to orthonormal bases for V of the form:

$$\mathcal{B}:=\{e_1^-,\phi e_1^-,...,e_{\frac12 p}^-,\phi e_{\frac12 p}^-,e_1^+,\phi e_1^+,...,e_{\frac12 q}^+,\phi e_{\frac12 q}^+\},\text{ and}$$
$$\tilde{\mathcal{B}}:=\{\tilde e_1^-,\phi\tilde e_1^-,...,\tilde e_{\frac12 p}^-,\phi\tilde e_{\frac12 p}^-,\tilde e_1^+,\phi\tilde e_1^+,...,\tilde e_{\frac12 q}^+,\phi\tilde e_{\frac12 q}^+\}.$$

Here the vectors with the superscript '$-$' are timelike and the vectors with the superscript '+' are spacelike. Define $\psi\in U(V,\phi)$ so $\psi\mathcal{B}=\tilde{\mathcal{B}}$ by setting:

$$\psi(e_i^-)=\tilde e_i^-\text{ and }\psi(\phi e_i^-)=\phi\tilde e_i^-\text{ for }1\le i\le\tfrac12 p;$$
$$\psi(e_j^+)=\tilde e_j^+\text{ and }\psi(\phi e_j^+)=\phi\tilde e_j^+\text{ for }1\le j\le\tfrac12 q.$$

Suppose next that $\phi^2=\operatorname{Id}$. Then ϕ is a para-isometry and $p=q$. We can extend S and $\tilde S$ to orthonormal bases for V of the form:

$$\mathcal{B}:=\{e_1^+,\phi e_1^+,...,e_p^+,\phi e_p^+\},\text{ and}$$
$$\tilde{\mathcal{B}}:=\{\tilde e_1^+,\phi\tilde e_1^+,...,\tilde e_p^+,\phi\tilde e_p^+\};$$

here the vectors $\{e_1^+,...,e_p^+\}$ and $\{\tilde e_1^+,...,\tilde e_p^+\}$ are spacelike, while the vectors $\{\phi e_1^+,...,\phi e_p^+\}$ and $\{\phi\tilde e_1^+,...,\phi\tilde e_p^+\}$ are timelike. We define $\psi\in U(V,\phi)$ so $\psi\mathcal{B}=\tilde{\mathcal{B}}$ and complete the proof of Assertion (1) by setting:

$$\psi(e_i)=\tilde e_i\text{ and }\psi(\phi e_i)=\phi\tilde e_i\text{ for }1\le i\le p.$$

By changing the sign on the innerproduct if necessary, we may assume without loss of generality in the proof of Assertions (2) and (3) that $p\le q$.

If $\phi^2 = -\operatorname{Id}$, then p and q both are even. If $\phi^2 = \operatorname{Id}$, then $p = q$ so m is even. Thus, in either case, m is even and hence $m \geq 6$. This implies $q \geq 3$.

Suppose that $\psi^* R = R$ for all $\psi \in U(V, \phi)$. We first prove Assertion (2a) for the special case, where $x \in S^+(V)$. As $\operatorname{span}\{x, \phi x\}$ is non-degenerate plane which is invariant under ϕ, we may define $\psi \in U(V, \phi)$ by:

$$\psi(w) = \begin{cases} w & \text{if } w \in \operatorname{span}\{x, \phi x\}, \\ -w & \text{if } w \perp \operatorname{span}\{x, \phi x\}. \end{cases}$$

If $w \perp \operatorname{span}\{x, \phi x\}$, then

$$\begin{aligned} R(\phi x, x, x, w) =& \psi^* R(\phi x, x, x, w) = R(\phi x, x, x, -w) \\ =& -R(\phi x, x, x, w) = 0. \end{aligned}$$

This shows that $R(\phi x, x)x \in \operatorname{span}\{x, \phi x\}$. As $R(\phi x, x, x, x) = 0$, there exists a real number $b_0(x)$ so that

$$R(\phi x, x)x = b_0(x)\phi x \text{ if } x \in S^+(V).$$

Let $\phi^2 = \varepsilon \operatorname{Id}$, where $\varepsilon = \pm 1$. Let $\tilde{x}$ be another spacelike unit vector. We use Assertion (1) to find ψ in $U(V, \phi)$ so $\psi x = \tilde{x}$. Since $\psi^* R = R$, we have:

$$\begin{aligned} -\varepsilon b_0(x) &= R(\phi x, x, x, \phi x) = \psi^* R(\phi x, x, x, \phi x) \\ &= R(\phi \tilde{x}, \tilde{x}, \tilde{x}, \phi \tilde{x}) = -\varepsilon b_0(\tilde{x}). \end{aligned}$$

This shows that $b_0(x) = b_0$ is constant. Because $R(\phi x, x)x$ is a cubic function of x, we can rescale this identity to see that:

$$R(\phi x, x)x = b_0(x, x)\phi x \text{ if } x \text{ is spacelike.}$$

Let $x \in V$ be arbitrary. Since $q \geq 3$, there exists a spacelike vector y_0. We set $x_t := x + t y_0$. As x_t is spacelike for large values of t,

$$R(\phi x_t, x_t)x_t = b_0(x_t, x_t)\phi x_t \text{ for } t >> 0.$$

Since this is a cubic identity in t, it continues to hold when $t = 0$; Assertion (2a) now follows.

Let $w \perp \operatorname{span}\{x, \phi x\}$. We first suppose that w and x are unit spacelike vectors. We can apply Assertion (1) to the spacelike (para-)complex 2 frames $S := \{x, \phi x, w, \phi w\}$ and $\tilde{S} := \{\tilde{x}, \phi \tilde{x}, \tilde{w}, \phi \tilde{w}\}$ to find $\psi \in U(V, \phi)$ so

$\psi x = \tilde{x}$ and $\psi w = \tilde{w}$. Since $\psi^* R = R$, we see $R(x,w,w,x) = R(\tilde{x},\tilde{w},\tilde{w},\tilde{x})$. Let b_1 be this common value. Again, we can rescale to see that this implies

$$R(x,w,w,x) = b_1(x,x)(w,w) \text{ for } w,x \text{ spacelike and } w \perp \{x,\phi x\}.$$

Suppose next that x is spacelike but w is arbitrary with $w \perp \operatorname{span}\{x,\phi x\}$. Since $q \geq 3$, we can choose z_0 spacelike so that $z_0 \perp \operatorname{span}\{x,\phi x\}$. Set $w_t := w + tz_0$. Since w_t is spacelike for large values of t,

$$R(x,w_t,w_t,x) = b_1(x,x)(w_t,w_t) \text{ for } t >> 0.$$

Since this is a quadratic identity in t, it continues to hold for $t = 0$. Thus,

$$R(x,w,w,x) = b_1(x,x)(w,w) \text{ if } x \text{ is spacelike and } w \perp \{x,\phi x\}.$$

Now let x be arbitrary and let $w \perp \{x,\phi x\}$ be arbitrary. Since $q \geq 3$, we can choose y_0 spacelike so $w \perp \{y_0,\phi y_0\}$. Let $x_t := x + ty_0$. Since x_t is spacelike for large values of t,

$$R(x_t,w,w,x_t) = b_1(x_t,x_t)(w,w) \text{ for } t >> 0.$$

As this identity is quadratic in t, it continues to hold for $t = 0$. This completes the proof of Assertion (2b).

To prove the first implication of Assertion (3), we suppose $\psi^* R = R$ for every $\psi \in U(V,\phi)$. Let x be a non-degenerate vector, let $\delta := (x,x)$, and let $\pi := \operatorname{span}\{x,\phi x\} \in \widetilde{\operatorname{Gr}}_2(V)$. Let w be a non-degenerate vector in $\pi^\perp$. Our first task is to show $\mathcal{J}_R(x)w = b_1(x,x)w$. We polarize. Let $\tilde{w} \in \pi^\perp$ with $\tilde{w} \perp w$. Let $w_t := w + t\tilde{w}$. We use Assertion (2b) and compute:

$$0 = R(w_t,x,x,w_t) - b_1(x,x)(w_t,w_t) = 2tR(w,x,x,\tilde{w}).$$

Since $R(w,x,x,\phi x) = R(w,x,x,\tilde{w}) = R(w,x,x,x) = 0$, we can express $R(w,x)x = b_1(x,x)w$ for any non-degenerate vector $w \in \pi^\perp$; by continuity this relationship holds for all $w \in \pi^\perp$. Let $\tilde{R} = b_1 R_{\mathrm{Id}} + \frac{1}{3}\varepsilon(b_1 - b_0)R_\phi$. If $y \perp x$, we have $\mathcal{J}_{R_\phi}(x)y = 3(y,\phi x)\phi x$ and $\mathcal{J}_{R_{\mathrm{Id}}}(x)y = (x,x)y$. Consequently:

$$\begin{aligned}
\mathcal{J}_R(x)x &= \mathcal{J}_{\tilde{R}}(x)x = 0,\\
\mathcal{J}_R(x)\phi x &= \mathcal{J}_{\tilde{R}}(x)\phi x = b_1(x,x)\phi x + \varepsilon(b_1 - b_0)(\phi x,\phi x)\phi x = (x,x)b_0\phi x,\\
\mathcal{J}_R(x)w &= \mathcal{J}_{\tilde{R}}(x)w = (x,x)b_1 w \text{ for } w \perp \operatorname{span}\{x,\phi x\}.
\end{aligned}$$

Let $\bar{R} = R - \tilde{R}$. Then $\mathcal{J}_{\bar{R}}(x) = 0$ for all non-degenerate x and hence by continuity $\mathcal{J}_{\bar{R}}(x) = 0$ for all x. This implies $\bar{R}$ has constant sectional curvature 0; we may then use Lemma 1.6.4 to see that $\bar{R} = 0$ and, consequently, R has the form described in Assertion (3b).

Conversely, suppose that $R = c_0 R_{\text{Id}} + c_1 R_\phi$. We show that $\psi^* R = R$ for all $\psi \in U(V, \phi)$ and complete the proof by computing:

$$\begin{aligned}
\psi^* R_{\text{Id}}(x, y, z, w) =& (\psi y, \psi z)(\psi x, \psi w) - (\psi x, \psi z)(\psi y, \psi w) \\
=& (y, z)(x, w) - (x, z)(y, w) = R_{\text{Id}}(x, y, z, w) \\
\psi^* R_\phi(x, y, z, w) =& (\phi\psi y, \psi z)(\phi\psi x, \psi w) - (\phi\psi x, \psi z)(\phi\psi y, \psi w) \\
& -2(\phi\psi x, \psi y)(\phi\psi z, \psi w) \\
=& (\phi y, z)(\phi x, w) - (\phi x, z)(\phi y, w) - 2(\phi x, y)(\phi z, w) \\
=& R_\phi(x, y, z, w). \quad \square
\end{aligned}$$

Let (E, g_E) and (X, g_X) be pseudo-Riemannian manifolds. We suppose given a smooth surjective map $\pi : E \to X$. We say that π is a *submersion* if π_* is a surjective map from $T_e E$ to $T_{\pi e} X$ for each $e \in E$. Let g_E and g_X be pseudo-Riemannian metrics on E and on X. Let

$$\mathcal{V} := \ker \pi_* \text{ and } \mathcal{H} := \mathcal{V}^\perp$$

be the vertical and the horizontal distributions. We assume the restriction of the metric to $\mathcal{V}$ is non-degenerate and thus $TE = \mathcal{V} \oplus \mathcal{H}$. We say that a submersion π is a *pseudo-Riemannian submersion* if $\pi_* : \mathcal{H}_e \to T_{\pi e} X$ is an isometry for each $e \in E$.

Let V be a complex vector space, where $\dim_{\mathbb{C}}(V) = p+q+1$. We assume the underlying real vector space is equipped with a metric of signature $(2p, 2q+2)$ and that the almost complex structure Φ is skew-adjoint. Let $\mathbb{CP}_+[V]$ be the space of complex spacelike lines in V. Let

$$\mathbb{CP}_+^{(p,q)} := \mathbb{CP}_+[\mathbb{C}^{(p,q+1)}] \text{ and } \mathbb{CP}^n := \mathbb{CP}_+^{(0,n)}.$$

The circle acts on $S^+(V)$ by complex multiplication; we identify the orbit space with $\mathbb{CP}_+[V]$ to define the *Hopf fibration*

$$\pi : S^+(V) \to \mathbb{CP}_+[V]$$

which associates to a spacelike unit vector v the complex line through v; $\pi(v) := \text{span}\{v, \Phi v\}$. The vertical and horizontal distributions of the fibration are given by:

$$\mathcal{V}_v := \text{span}\{\Phi v\} = \ker \pi_* \text{ and } \mathcal{H}_v := \text{span}\{v, \Phi v\}^\perp.$$

The metric on the horizontal distribution $\mathcal{H}$ has signature $(2p, 2q)$. Since S^1 acts by isometries on $\mathcal{H}$, there is a canonical pseudo-Riemannian metric

g_{FS} of signature $(2p, 2q)$ on $\mathbb{CP}_+[V]$. This metric is called the *Fubini-Study metric*, and with this metric, the Hopf fibration described above is a pseudo-Riemannian submersion. The map Φ preserves the horizontal distributions. Since Φ also commutes with the circle action, Φ descends to define a pseudo-Riemannian almost complex structure ϕ on $\mathbb{CP}_+[V]$.

We could also work with timelike complex lines to define $(\mathbb{CP}_-^{(p,q)}, g_{FS})$. We then have:

$$(\mathbb{CP}_+^{(p,q)}, g_{FS}) = (\mathbb{CP}_-^{(q,p)}, -g_{FS}).$$

Let V_1 and V_2 be complex vector spaces. The natural embeddings

$$i_1 : v_1 \to v_1 \oplus 0 \text{ and } i_2 : v_2 \to 0 \oplus v_2$$

extend to isometric embeddings of $S^+(V_i)$ in $S^+(V_1 \oplus V_2)$ which are equivariant with respect to the natural S^1 action. These embeddings therefore descend to isometric embeddings

$$[i_1] : \mathbb{CP}_+[V_1] \to \mathbb{CP}_+[V_1 \oplus V_2] \text{ and } [i_2] : \mathbb{CP}_+[V_2] \to \mathbb{CP}_+[V_1 \oplus V_2].$$

3.6.4 Lemma. *Let V be a complex vector space of signature $(2p, 2q+2)$. Let ϕ be the induced almost complex structure and let g_{FS} be the Fubini-Study metric on $\mathbb{CP}_+[V]$.*

(1) *Suppose that $V := V_1 \oplus V_2$. Let R and R_1 be the curvature tensors of $(\mathbb{CP}_+[V], g_{FS})$ and $(\mathbb{CP}_+[V_1], g_{FS})$. Then $[i_1]^*(R) = R_1$.*
(2) *$(\mathbb{CP}^1, g_{FS})$ is isometric to $(S^2, \frac{1}{4}g_2)$.*
(3) *$(\mathbb{CP}^2, g_{FS})$ does not have constant sectional curvature.*
(4) *$(\mathbb{CP}_+[V], g_{FS})$ is a homogeneous space.*
(5) *If $p + q \geq 2$, then $(\mathbb{CP}_+[V], g_{FS})$ is a complex space form with curvature tensor $R = R_{\mathrm{Id}} + R_\phi$.*

Proof. Let $V = V_1 \oplus V_2$. We define $\psi \in U(V \oplus W, \phi_1 \oplus \phi_2)$ by setting $\psi(v_1 \oplus v_2) = (v_1 \oplus -v_2)$. Assertion (1) follows from Lemma 3.6.2 since:

$$\mathrm{Fix}(\psi) = \mathbb{CP}_+[V_1] \cup \mathbb{CP}_+[V_2].$$

To prove Assertion (2), we give a very concrete description of the Hopf fibration $\pi : S^1 \to S^3 \to \mathbb{CP}^1$ that is of interest in its own right. Let $(x_1, x_2, x_3, x_4) \in S^3 \subset \mathbb{R}^4$. We define $\pi : \mathbb{R}^4 \to \mathbb{R}^3$ by setting:

$$\begin{aligned} &\pi(x_1, x_2, x_3, x_4) = (\xi_1(\vec{x}), \xi_2(\vec{x}), \xi_3(\vec{x})) \text{ where} \\ &\xi_1(x_1, x_2, x_3) := 2(x_1x_3 + x_2x_4) \\ &\xi_2(x_1, x_2, x_3) := 2(x_2x_3 - x_1x_4), \text{ and} \\ &\xi_3(x_1, x_2, x_3) := x_1^2 + x_2^2 - x_3^2 - x_4^2. \end{aligned} \tag{3.6.4.a}$$

We may then compute that

$$\begin{aligned}
&\xi_1^2 + \xi_2^2 + \xi_3^2 \\
=&4x_1^2x_3^2 + 4x_2^2x_4^2 + 8x_1x_2x_3x_4 + 4x_2^2x_3^2 + 4x_1^2x_4^2 - 8x_1x_2x_3x_4 \\
&+x_1^4 + x_2^4 + x_3^4 + x_4^4 + 2x_1^2x_2^2 + 2x_3^2x_4^2 \\
&-2x_1^2x_3^2 - 2x_1^2x_4^2 - 2x_2^2x_3^2 - 2x_2^2x_4^2 \\
=&(x_1^2 + x_2^2 + x_3^2 + x_4^2)^2
\end{aligned}$$

and thus π induces a smooth map from S^3 to S^2. We let S^1 act on $S^3 = S(\mathbb{C}^2)$ by complex multiplication. Let $\Re$ and $\Im$ denote the real and imaginary parts of a complex number. Let $(z_0, z_1) \in S^3$. We expand $z_0 = x_1 + \sqrt{-1}x_2$ and $z_1 = x_3 + \sqrt{-1}x_4$. We then have that:

$$\pi(z_0, z_1) := (2\Re(z_0\bar{z}_1), 2\Im(z_0\bar{z}_1), |z_0|^2 - |z_1|^2).$$

It is immediate from this definition that $\pi(z_0, z_1) = \pi(w_0, w_1)$ if and only if $(z_0, z_1) = \lambda(w_0, w_1)$ for $\lambda \in S^1$. Thus, $\mathbb{CP}^1 = S^2$ and Display (3.6.4.a) describes the Hopf fibration from S^3 to $\mathbb{CP}^1$.

Let $\{\omega_{ij} := dx_i \circ dx_j\}_{1\le i\le j\le 4}$ be a basis for the space of symmetric 2 tensors on $\mathbb{R}^4$. By Display (3.6.4.a),

$$\begin{aligned}
\tfrac{1}{4}\pi^*(d\xi_1 \circ d\xi_1) =& x_3^2\omega_{11} + x_4^2\omega_{22} + x_1^2\omega_{33} + x_2^2\omega_{44} \\
&+2x_3x_4\omega_{12} + 2x_1x_3\omega_{13} + 2x_2x_3\omega_{14} \\
&+2x_1x_4\omega_{23} + 2x_2x_4\omega_{24} + 2x_1x_2\omega_{34}, \\
\tfrac{1}{4}\pi^*(d\xi_2 \circ d\xi_2) =& x_4^2\omega_{11} + x_3^2\omega_{22} + x_2^2\omega_{33} + x_1^2\omega_{44} \\
&-2x_3x_4\omega_{12} - 2x_2x_4\omega_{13} + 2x_1x_4\omega_{14} \\
&+2x_2x_3\omega_{23} - 2x_1x_3\omega_{24} - 2x_1x_2\omega_{34}, \\
\tfrac{1}{4}\pi^*(d\xi_3 \circ d\xi_3) =& x_1^2\omega_{11} + x_2^2\omega_{22} + x_3^2\omega_{33} + x_4^2\omega_{44} \\
&+2x_1x_2\omega_{12} - 2x_1x_3\omega_{13} - 2x_1x_4\omega_{14} \\
&-2x_2x_3\omega_{23} - 2x_2x_4\omega_{24} + 2x_3x_4\omega_{34}.
\end{aligned}$$

The flow for the fiber circles is given by $e^{it} \cdot (z, w)$. Let θ be the induced vector field and let Θ be the dual 1 form. Then

$$\begin{aligned}
\theta =& (-x_2, x_1, -x_4, x_3), \\
\Theta =& -x_2dx_1 + x_1dx_2 - x_4dx_3 + x_3dx_4, \\
\Theta \circ \Theta =& x_2^2\omega_{11} + x_1^2\omega_{22} + x_4^2\omega_{33} + x_3^2\omega_{44} \\
&-2x_1x_2\omega_{12} + 2x_2x_4\omega_{13} - 2x_2x_3\omega_{14} \\
&-2x_1x_4\omega_{23} + 2x_1x_3\omega_{24} - 2x_3x_4\omega_{34}.
\end{aligned}$$

We have $\mathcal{V}^* = \text{span}\{\Theta\}$ and $\mathcal{H} = \Theta^\perp$. We compute:

$$\begin{aligned}&\tfrac{1}{4}\pi^*(d\xi_1 \circ d\xi_1 + d\xi_2 \circ d\xi_2 + d\xi_3 \circ d\xi_3) + \Theta \circ \Theta \\ =&(x_1^2 + x_2^2 + x_3^2 + x_4^2)(\omega_{11} + \omega_{22} + \omega_{33} + \omega_{44}).\end{aligned}$$

Since the canonical metric on S^2 is given by $d\xi_1^2 + d\xi_2^2 + d\xi_3^2$, we have

$$ds^2_{S^3} = \tfrac{1}{4}\pi^*\{ds^2_{S^2}\} + \Theta \circ \Theta.$$

Thus, $\frac{1}{4}\pi^* ds^2_{S^2}$ is the induced metric on the horizontal subspace; this completes the proof of Assertion (2).

We prove Assertion (3) by showing there exist 2 planes in $T\mathbb{CP}^2$ with sectional curvatures $+1$ and $+4$. Decompose $\mathbb{C}^3 = \mathbb{C}^2 \oplus \mathbb{C}$; by Assertion (1), the curvature tensor of $\mathbb{CP}^1$ is the restriction of the curvature tensor of $\mathbb{CP}^2$. Since the sectional curvature of S^2 with the metric $\frac{1}{4}g_2$ has constant sectional curvature $+4$, by Assertion (2) there exists a 2 plane π in $T\mathbb{CP}^2$ with $\kappa(\pi) = +4$.

To construct π with sectional curvature $+1$, we let

$$\psi(z_1, z_2, z_3) = (\bar{z}_1, \bar{z}_2, \bar{z}_3)$$

be complex conjugation on $\mathbb{C}^3$. This map is an isometry of $\mathbb{C}^3$. Since $\psi(\lambda\vec{z}) = \bar{\lambda}\vec{z}$, ψ preserves the vertical circles and hence preserves the horizontal and vertical subspaces. Thus, ψ induces an isometry $[\psi]$ of $\mathbb{CP}^2$. The fixed point set of ψ is $S^2 \subset \mathbb{R}^3$, the fixed point set of $[\psi]$ is the projective space $\mathbb{RP}^2$, and the map from S^2 to $\mathbb{RP}^2$ is the usual double covering. Thus, the sub-manifold

$$\mathbb{RP}^2 \subset \mathbb{CP}^2$$

has constant sectional curvature $+1$. Since $\mathbb{RP}^2$ is the fixed point set of the isometry $[\psi]$, we apply Lemma 3.6.2 to see the curvature tensor of $\mathbb{RP}^2$ is the restriction of the curvature tensor of $\mathbb{CP}^2$. Thus, there exist 2 planes with sectional curvature $+1$ in $T\mathbb{CP}^2$ and Assertion (3) follows.

Since the elements of $U(V, \Phi)$ commute with the circle action, an element $\psi \in U(V, \Phi)$ defines an isometric action $[\psi]$ on $\mathbb{CP}_+[V]$ which commutes with ϕ. By Lemma 3.6.3, $U(V, \Phi)$ acts transitively on $S^+(V)$. Consequently, $\mathbb{CP}_+[V]$ is a homogeneous space; Assertion (4) now follows.

Fix $x \in S^+(V)$ and let $[x] := \text{span}\{x, \Phi x\} \in \mathbb{CP}_+[V]$. We then have

$$U(\mathcal{H}_x, \Phi) = U(T_{[x]}\mathbb{CP}_+[V], \phi).$$

We may decompose $V = \text{span}\{x, \phi x\} \oplus \mathcal{H}_x$. The map $\psi \to \text{Id} \oplus \psi$ embeds $U(\mathcal{H}_x, \Phi|_{\mathcal{H}})$ in $U(V, \Phi)$. As $[\text{Id} \oplus \psi]$ is an isometry of $\mathbb{CP}_+[V]$, we have that $\psi^* R = [\text{Id} \oplus \psi]^* R = R$ so

$$\psi^* R = R \text{ for all } [\psi] \in U(\mathcal{H}_x, \Phi) = U(T_{[x]}\mathbb{CP}(V)).$$

Consequently, we may apply Lemma 3.6.3 to see

$$R_{\pi P} = \lambda_0([x]) R_{\text{Id}} + \lambda_1([x]) R_\phi.$$

Since $\mathbb{CP}_+[V]$ is a homogeneous space, $\lambda_0 = \lambda_0(V)$ and $\lambda_1 = \lambda_1(V)$ are constant. We use Assertion (1) to see that $\lambda_i(V) = \lambda_i(V \oplus W) = \lambda_i(W)$; thus the λ_i are universal constants. By Assertion (3), $\lambda_1 \neq 0$. By Assertion (2), $\mathbb{CP}^1 = S^2$ has constant sectional curvature 4. Thus, $\lambda_0 + 3\lambda_1 = 4\lambda_0 = 4$. This implies $\lambda_0 = \lambda_1 = 1$ and completes the proof of Assertion (5). □

We make the following observation which plays an important role in various 'pinching' theorems in the Riemannian setting. Let $n \geq 2$. Fix $P \in \mathbb{CP}^n$ and let $\{x, \phi x, w, \phi w\}$ be an orthonormal set contained in $T_P\mathbb{CP}^n$. The proof of Assertion (2) shows there is an isometric totally geodesic embedding of $(S^2, \frac{1}{4}g_2)$ in $\mathbb{CP}^n$ whose tangent plane at P is spanned by $\{x, \phi x\}$. Similarly, the proof of Assertion (3) shows that there is an isometric totally geodesic embedding of $(\mathbb{RP}^2, g_2)$ in $\mathbb{CP}^n$ whose tangent plane at P is spanned by $\{x, w\}$.

If $c \neq 0$, then $c^{-1} g_{FS}$ defines a non-degenerate pseudo-Riemannian metric on $(\mathbb{CP}^{(p,q)})$ with

$$^{g_c}R = c\{R_{\text{Id}} + R_{\phi_{\mathbb{CP}}}\}.$$

If $c > 0$, the metric has signature $(2p, 2q)$ while if $c < 0$, then the metric has signature $(2q, 2p)$.

The following classification theorem is now an immediate consequence of Lemmas 1.15.1, 1.15.3 and 3.6.4.

3.6.5 Theorem. *Let (M, g_M) be a complex space form of signature $(2p, 2q)$ and dimension $m = 2p + 2q \neq 2, 4$. Let $^gR = c_0(R_{\text{Id}} + R_\phi)$.*

(1) *If $c_0 = r^{-2} > 0$, (M, g_M) is locally isometric to $(\mathbb{CP}^{(p,q)}, r^2 g_{FS})$.*

(2) *If $c_0 = -r^{-2} < 0$, (M, g_M) is locally isometric to $(\mathbb{CP}^{(q,p)}, -r^2 g_{FS})$.*

(3) *The isometries in Assertions (1) and (2) can be chosen to intertwine the two almost complex structures involved.*

We make this quite specific in the Riemannian setting. Let $c > 0$. Then $(\mathbb{CP}^n, c^{-1} g_{FS})$ is a compact homogeneous Riemannian complex space form whose curvature tensor is given by $c(R_{\mathrm{Id}}+R_\phi)$. We use Theorem 4.1.1 to see that the negative curvature dual $(\mathbb{CP}^{(n,0)}, -c^{-1}g_{FS})$ is a non-compact homogeneous contractible Riemannian complex space form whose curvature tensor is given by $-c(R_{\mathrm{Id}}+R_\phi)$.

We now turn to para-complex space forms. Let $\widetilde{\mathbb{C}} := \mathbb{R}^2$ with the multiplication

$$(x_1+iy_1)(x_2+iy_2) = (x_1x_2+y_1y_2) + i(x_1y_2+y_1x_2).$$

Since the roles of (x_1, y_1) and (x_2, y_2) are symmetric, the multiplication is commutative. Using the commutative law, we check the multiplication is associative by observing that (x_1, y_1), (x_2, y_2), and (x_3, y_3) play symmetric roles in the product:

$$\begin{aligned}&\{(x_1+iy_1)(x_2+iy_2)\}(x_3+iy_3)\\ =&\{(x_1x_2+y_1y_2)x_3+(x_1y_2+y_1x_2)y_3\}\\ &+i\{x_1x_2+y_1y_2)y_3+(x_1y_2+y_1x_2)x_3\}.\end{aligned}$$

If $z = x+iy$, then $\bar{z} := x - iy$ is the para-complex conjugate. We show $\overline{zw} = \bar{z}\bar{w}$ by computing:

$$(x-iy)(u-iv) = (xu+yv) - i(xv+yu).$$

Let $\Re(x+iy) = x$ and $\Im(x+iy) = y$ be the real and imaginary parts of a para-complex number. Let

$$(z, w) := \Re(z\bar{w}) = xu - yv.$$

Since $(z, z) = x^2 - y^2$, we have $\widetilde{\mathbb{C}} = \mathbb{R}^{(1,1)}$ with this inner product. Let

$$\tilde{S}^1 := \{(x, y) \in \mathbb{R} : x^2 - y^2 = 1 \text{ and } x > 0\}$$

be the connected component of $S^+(\widetilde{\mathbb{C}})$ which contains 1. Let $\sigma \in \tilde{S}^1$. We can show that para-complex multiplication by σ is an isometry of $\widetilde{\mathbb{C}}$:

$$(\sigma z, \sigma w) = \Re(\sigma z\overline{\sigma w}) = \Re(\sigma\bar{\sigma}z\bar{w}) = \Re(z\bar{w}) = (z, w).$$

Thus, in particular $\tilde{S}^1$ is a group under para-complex multiplication. We may identify $\tilde{S}^1$ with $\mathbb{R}$ using the parametrization

$$t \to (\cosh t, \sinh t).$$

Let $\Phi z \rightarrow iz$, i.e. $\Phi(x,y) = (y,x)$. We then have $\Phi^2 = \text{Id}$ and $\Phi \in \mathfrak{so}(\widetilde{\mathbb{C}})$.

Let V be a vector space with an inner product of signature (p,q) and let $\phi \in \mathfrak{so}(V)$ satisfy $\phi^2 = \text{Id}$. Since ϕ is a paraisometry which interchanges the roles of spacelike and timelike vectors, necessarily $p = q = n$. We use ϕ to give V a para-complex structure by defining $(x+iy)v = xv + y\phi v$. In contrast to the complex setting, where the signature plays an important role, we use Lemma 1.15.3 to see that (V,ϕ) is determined up to isomorphism by the dimension. We may therefore assume without loss of generality that

$$V = \widetilde{\mathbb{C}}^n = \widetilde{\mathbb{C}} \oplus \ldots \oplus \widetilde{\mathbb{C}} \text{ and } \phi\vec{v} = i\vec{v}.$$

The *generalized Hopf fibration* is then given by:

$$\text{(3.6.5.b)} \qquad \pi : S^+(\widetilde{\mathbb{C}}^{n+1}) \rightarrow \widetilde{\mathbb{CP}}^n := S^+(\widetilde{\mathbb{C}}^{n+1})/\tilde{S}^1.$$

We may identify $\widetilde{\mathbb{CP}}^n$ with the non-degenerate para-complex lines in $\widetilde{\mathbb{C}}^{n+1}$, i.e. those elements of $\text{Gr}_{1,1}(\widetilde{\mathbb{C}}^{n+1})$ which preserved by ϕ. There is a canonical pseudo-Riemannian metric g_{FS} on $\widetilde{\mathbb{CP}}^n$, which is called the *Fubini-Study metric*, so that π is a pseudo-Riemannian submersion. The natural inclusion $i : \widetilde{\mathbb{C}}^{n+1} \subset \widetilde{\mathbb{C}}^{n+2}$ induces a natural inclusion $[i] : \widetilde{\mathbb{CP}}^n \subset \widetilde{\mathbb{CP}}^{n+1}$.

Lemma 3.6.4 generalizes to this setting as:

3.6.6 Lemma.

(1) *Let R_n be the curvature tensor of $(\widetilde{\mathbb{CP}}^n, g_{FS})$. Then $[i]^* R_{n+k} = R_n$.*

(2) *$(\widetilde{\mathbb{CP}}^1, g_{FS})$ is isometric to $(S^{(1,1)}, \frac{1}{4}g_2)$.*

(3) *$(\widetilde{\mathbb{CP}}^2, g_{FS})$ does not have constant sectional curvature.*

(4) *$(\widetilde{\mathbb{CP}}^n, g_{FS})$ is a homogeneous space.*

(5) *If $n \geq 2$, then $\widetilde{\mathbb{CP}}^n$ is a complex space form and $R_n = R_{\text{Id}} - R_\phi$.*

Proof. The proof of Assertions (1), (3), and (5) is essentially the same as that given in the complex case in Lemma 3.6.4 and are therefore omitted in the interests of brevity; in proving Assertion (4) the projective space $(\mathbb{RP}^2, g_2)$ is replaced by $(\mathbb{RP}^{(1,1)}, g_2)$. The proof of Assertion (2), however, requires a direct computation as certain signs change.

We introduce para-complex coordinates $z = x_1 + ix_2$ and $w = x_3 + ix_4$ on $\mathbb{R}^{(2,2)} = \widetilde{\mathbb{C}}^2$ and define

$$\begin{aligned} \pi(z,w) :=& (2\Re(z\bar{w}), 2\Im(z\bar{w}), |z|^2 - |w|^2) = (\xi_1, \xi_2, \xi_3) \text{ where} \\ \xi_1 :=& 2(x_1x_3 - x_2x_4), \ \xi_2 := 2(x_2x_3 - x_1x_4), \text{ and} \\ \xi_3 :=& x_1^2 - x_2^2 - x_3^2 + x_4^2. \end{aligned}$$

We may then compute that

$$\begin{aligned}
&\xi_1^2 - \xi_2^2 + \xi_3^2 \\
=&4x_1^2x_3^2 + 4x_2^2x_4^2 - 8x_1x_2x_3x_4 - 4x_2^2x_3^2 - 4x_1^2x_4^2 + 8x_1x_2x_3x_4 \\
&+x_1^4 + x_2^4 + x_3^4 + x_4^4 - 2x_1^2x_2^2 - 2x_3^2x_4^2 \\
&-2x_1^2x_3^2 + 2x_1^2x_4^2 + 2x_2^2x_3^2 - 2x_2^2x_4^2 \\
=&(x_1^2 - x_2^2 + x_3^2 - x_4^2)^2
\end{aligned}$$

and thus π induces a smooth map from $S^+(R^{(2,2)})$ to $S^+(R^{(1,2)}) = S^{(1,1)}$; consequently, we may identify $\widetilde{\mathbb{CP}^1}$ with $S^{(1,1)}$. Let $\omega_{ij} := dx_i \circ dx_j$. We then have:

$$\begin{aligned}
\tfrac{1}{4}\pi^*(d\xi_1 \circ d\xi_1) =& \; x_3^2\omega_{11} + x_4^2\omega_{22} + x_1^2\omega_{33} + x_2^2\omega_{44} \\
&-2x_3x_4\omega_{12} + 2x_1x_3\omega_{13} - 2x_2x_3\omega_{14} \\
&-2x_1x_4\omega_{23} + 2x_2x_4\omega_{24} - 2x_1x_2\omega_{34}, \\
-\tfrac{1}{4}\pi^*(d\xi_2 \circ d\xi_2) =& - x_4^2\omega_{11} - x_3^2\omega_{22} - x_2^2\omega_{33} - x_1^2\omega_{44} \\
&+2x_3x_4\omega_{12} + 2x_2x_4\omega_{13} - 2x_1x_4\omega_{14} \\
&-2x_2x_3\omega_{23} + 2x_1x_3\omega_{24} + 2x_1x_2\omega_{34}, \\
\tfrac{1}{4}\pi^*(d\xi_3 \circ d\xi_3) =& \; x_1^2\omega_{11} + x_2^2\omega_{22} + x_3^2\omega_{33} + x_4^2\omega_{44} \\
&-2x_1x_2\omega_{12} - 2x_1x_3\omega_{13} + 2x_1x_4\omega_{14} \\
&+2x_2x_3\omega_{23} - 2x_2x_4\omega_{24} - 2x_3x_4\omega_{34}.
\end{aligned}$$

The flow for the fiber circles is given by $e^{it} \cdot (z, w)$. Let θ be the induced vector field and let Θ be the dual 1 form. Since we are raising and lowering indices using the metric tensor, certain signs are introduced. As θ is timelike, we shall compute $-\Theta \circ \Theta$ rather than $\Theta \circ \Theta$:

$$\begin{aligned}
\theta =&(x_2, x_1, x_4, x_3), \\
\Theta =&x_2dx_1 - x_1dx_2 + x_4dx_3 - x_3dx_4, \\
-\Theta \circ \Theta =& - x_2^2\omega_{11} - x_1^2\omega_{22} - x_4^2\omega_{33} - x_3^2\omega_{44} \\
&+2x_1x_2\omega_{12} - 2x_2x_4\omega_{13} + 2x_2x_3\omega_{14} \\
&+2x_1x_4\omega_{23} - 2x_1x_3\omega_{24} + 2x_3x_4\omega_{34}.
\end{aligned}$$

Consequently we have:

$$\begin{aligned}
&\tfrac{1}{4}\pi^*(d\xi_1 \circ d\xi_1 - d\xi_2 \circ d\xi_2 + d\xi_3 \circ d\xi_3) - \Theta \circ \Theta \\
=&(x_1^2 - x_2^2 + x_3^2 - x_4^2)(\omega_{11} - \omega_{22} + \omega_{33} - \omega_{44}).
\end{aligned}$$

This shows that

$$ds^2_{S(2,1)} = \pi^*(\tfrac{1}{4}ds^2_{S(1,1)} - \Theta \circ \Theta)$$

and, consequently, $\widetilde{\mathbb{CP}}^1$ inherits the metric $\frac{1}{4}ds^2_{S(1,1)}$; the sectional curvature is $+4$ and Assertion (2) follows. □

The following result follows immediately from Lemmas 1.15.1, 1.15.3 and 3.6.6:

3.6.7 Theorem. *Let (M, g_M) be a para-complex space form of dimension $m = 2p$. If $m \neq 2, 4$, then (M, g_M) is locally isometric to $(\widetilde{\mathbb{CP}}^n, \lambda_1^{-1} g_{FS})$.*

3.7 The higher order Jacobi operator

Let R be an algebraic curvature tensor on a Riemannian vector space V of dimension m which is k Osserman for $2 \leq k \leq m-2$. We shall show that either R has constant sectional curvature or $R = cR_\phi$, where $\phi \in \mathfrak{so}(V)$ satisfies $\phi^2 = -\operatorname{Id}$ and where c is a suitably chosen constant. This will prove Theorem 3.1.7. Since dually, by Lemma 1.10.3, R is $m-k$ Osserman, we may suppose without loss of generality that $k \leq m-k$, i.e. that $2k \leq m$. We shall apply a number of topological results from Section 4.3 in our discussion.

We introduce the following notational conventions. Let A be a self-adjoint map of a ℓ dimensional Riemannian vector space W. Let $\lambda_1 \leq ... \leq \lambda_\ell$ be the eigenvalues of A, where we repeat the eigenvalues $\lambda_i = \lambda_i(A)$ according to their multiplicity and list them in increasing order. We set:

$$\operatorname{spec}(A) := (\lambda_1, ..., \lambda_\ell) \in \mathbb{R}^\ell.$$

Let $\tau \in \operatorname{Gr}_\ell(V)$. Let $I := \{1 \leq i_1 < ... < i_s \leq m\}$ be a collection of s disjoint indices. Set $|I| = s$ and set:

$$\lambda_i(\tau) := \lambda_i(\mathcal{J}_R(\tau)) \text{ and } \Lambda_I(\tau) = \lambda_{i_1}(\tau) + ... + \lambda_{i_s}(\tau).$$

We say that an orthonormal basis $\{\xi_1, ..., \xi_m\}$ for V *diagonalizes* $\mathcal{J}_R(\tau)$ if

$$\mathcal{J}_R(\tau)\xi_i = \lambda_i(\tau)\xi_i \text{ for } 1 \leq i \leq m.$$

Denote the eigenspace and eigenvalue multiplicities of $\mathcal{J}_R(\tau)$ by:

$$E_\lambda(\tau) := \{y \in V : \mathcal{J}_R(\tau)y = \lambda y\} \text{ and } \mu_\lambda(\tau) := \dim E_\lambda(\tau).$$

We shall need the following measurement of the extent to which eigenvalue multiplicities are large. Set

$$\Xi_\nu(\tau) := \textstyle\sum_\lambda \max\{0, \mu_\lambda(\tau) - \nu\}.$$

As R is k Osserman, λ_i, Λ_I, μ_λ, and Ξ_ν are constant on $\operatorname{Gr}_k(V)$.

We prepare for the proof of Theorem 3.1.7 by establishing a series of technical Lemmas. We begin with:

3.7.1 Lemma. *Let R be a k Osserman algebraic curvature tensor on a Riemannian vector space V of dimension m, where $2 \le k \le m-2$ and $2k \le m$. Let ℓ be a positive integer with $1 \le \ell < k$.*

(1) *Suppose that $\Xi_\ell \ge k-\ell$ on $\mathrm{Gr}_k(V)$.*
   1a) *Let $\tau \in \mathrm{Gr}_\ell(V)$ and let $\pi \in \mathrm{Gr}_k(V)$. There exists τ_1 in $\mathrm{Gr}_{k-\ell}(V)$ so that $\tau_1 \perp \tau$ and so that $\mathcal{J}_R(\pi)\tau_1 \subset \tau_1$.*
   1b) *If $|I| = k$, then the function Λ_I is constant on $\mathrm{Gr}_\ell(V)$.*
   1c) *The algebraic curvature tensor R is ℓ Osserman.*

(2) *If there exists λ with $\mu_\lambda \ge k$ on $\mathrm{Gr}_k(V)$, then R is ℓ Osserman.*

Proof. Assume that $\Xi_\ell \ge k-\ell$ on $\mathrm{Gr}_k(V)$ and that $1 \le \ell < k$. Let $\pi \in \mathrm{Gr}_k(V)$ and let $\tau \in \mathrm{Gr}_\ell(V)$. We estimate:

$$\textstyle\sum_\lambda \dim\{E_\lambda(\pi) \cap \tau^\perp\} \ge \sum_\lambda \max\{0, \mu_\lambda(\pi) - \ell\} = \Xi_\ell(\pi) \ge k - \ell.$$

Thus, we can find an orthonormal set $\{y_1, ..., y_{k-\ell}\}$ of eigenvectors of $\mathcal{J}_R(\pi)$ which are all perpendicular to τ. We set $\tau_1 := \mathrm{span}\{y_1, ..., y_{k-\ell}\}$ to prove Assertion (1a).

To prove Assertion (1b), let $I = \{1 \le i_1 < ... < i_k \le m\}$ be a collection of k distinct indices. Let $\tau \in \mathrm{Gr}_\ell(V)$. Choose an orthonormal basis $\{\xi_1, ..., \xi_m\}$ for V which diagonalizes $\mathcal{J}_R(\tau)$ and let

$$\pi := \mathrm{span}\{\xi_{i_1}, \dots, \xi_{i_k}\} \in \mathrm{Gr}_k(V).$$

We use Assertion (1a) to find $\tau_1 \in \mathrm{Gr}_{k-\ell}(V)$ so that $\tau \perp \tau_1$ and so that $\mathcal{J}_R(\pi)\tau_1 \subset \tau_1$. We may then apply Theorem 3.1.2 to see dually that:

$$\mathcal{J}_R(\tau_1)\pi \subset \pi. \tag{3.7.1.a}$$

Since π is spanned by eigenvectors of $\mathcal{J}_R(\tau)$, we have

$$\mathcal{J}_R(\tau)\pi \subset \pi. \tag{3.7.1.b}$$

Since $\tau \perp \tau_1$, we have $\mathcal{J}_R(\tau \oplus \tau_1) = \mathcal{J}_R(\tau) + \mathcal{J}_R(\tau_1)$. Thus, Equations (3.7.1.a) and (3.7.1.b) imply that $\mathcal{J}_R(\tau \oplus \tau_1)\pi \subset \pi$. Since $\tau \oplus \tau_1 \in \mathrm{Gr}_k(V)$, we use Theorem 3.1.2 once again to see dually that:

$$\mathcal{J}_R(\pi)(\tau \oplus \tau_1) \subset \tau \oplus \tau_1. \tag{3.7.1.c}$$

Since $\mathcal{J}_R(\pi)\tau_1 \subset \tau_1$, since $\tau \perp \tau_1$, and since $\mathcal{J}_R(\pi)$ is self-adjoint, Equation (3.7.1.c) shows:

$$\mathcal{J}_R(\pi)\tau \subset \tau. \tag{3.7.1.d}$$

Let ρ_τ be orthogonal projection on τ. It is immediate from the definition that $\Lambda_I(\tau) = \text{Tr}\{\rho_\pi \mathcal{J}_R(\tau)\}$. By Lemma 3.4.1, we have the trace duality $\text{Tr}\{\rho_\pi \mathcal{J}_R(\tau)\} = \text{Tr}\{\rho_\tau \mathcal{J}_R(\pi)\}$. Thus,

$$\Lambda_I(\tau) = \text{Tr}\{\rho_\tau \mathcal{J}_R(\pi)\}.$$

By Equation (3.7.1.d), $\text{Tr}\{\rho_\tau \mathcal{J}_R(\pi)\}$ is a sum of ℓ eigenvalues of $\mathcal{J}_R(\pi)$. Therefore, there exists J with $|J| = \ell$ so

$$\Lambda_I(\tau) = \Lambda_J(\pi).$$

Since R is k Osserman and $\pi \in \text{Gr}_k(V)$, $\Lambda_J(\pi)$ is independent of π. Since there are a finite number of such collections J, $\Lambda_I(\tau)$ takes values in a finite set. By Lemma 1.3.4, the map $\tau \to \Lambda_I(\tau)$ is continuous. Therefore, $\Lambda_I(\tau)$ is constant and Assertion (1b) follows.

Fix $1 < i < j \leq m$. Since $2 \leq k \leq m-2$, we can choose a set K of $k-1$ distinct indices which does not contain the indexes i and j. Then

$$c_{ij} := \Lambda_{\{i\}\cup K}(\tau) - \Lambda_{\{j\}\cup K}(\tau) = \lambda_i(\tau) - \lambda_j(\tau) \tag{3.7.1.e}$$

is constant on $\text{Gr}_\ell(V)$. By Lemma 1.10.3, R is Einstein so there is a constant c_0 so $\text{Tr}\{\mathcal{J}_R(e)\} = c_0$ for any unit vector e. Let $\{e_1, ..., e_\ell\}$ be an orthonormal basis for τ. Then

$$\text{Tr}\{\mathcal{J}_R(\tau)\} = \textstyle\sum_\nu \text{Tr}\{\mathcal{J}_R(e_\nu)\} = \ell c_0.$$

If the functions $\lambda_i(\cdot)$ are constant on $\text{Gr}_\ell(V)$, then R will be ℓ Osserman. We use Equation (3.7.1.e) to prove Assertion (2c) by computing:

$$\begin{aligned}\textstyle\sum_{1\leq j\leq m} c_{ij} =& m\lambda_i(\tau) - \textstyle\sum_{1\leq j\leq m} \lambda_j(\tau) = m\lambda_i(\tau) - \text{Tr}\{\mathcal{J}_R(\tau)\}\\ =& m\lambda_i(\tau) - \ell c_0.\end{aligned}$$

If there exists λ so $\mu_\lambda(\pi) \geq k$ on $\text{Gr}_k(V)$, then Assertion (2) will follow from Assertion (1) as

$$\Xi_\ell \geq \max\{0, \mu_\lambda(\pi) - \ell\} \geq k - \ell. \quad \square$$

Let $\pi_1, \pi_2 \in \text{Gr}_k(V)$. If $\mathcal{J}_R(\pi)\pi_2 \not\subset \pi_2$, then $\text{spec}(\pi_1, \pi_2)$ is undefined. If $\mathcal{J}_R(\pi_1)\pi_2 \subset \pi_2$, then we define the *relative spectrum* by setting:

$$\text{spec}(\pi_1, \pi_2) := \text{spec}(\mathcal{J}_R(\pi_1)|_{\pi_2}) \in \mathbb{R}^k.$$

Let $S = \text{spec}(\pi_1, \pi_2) \in \mathbb{R}^k$ be such a relative spectrum. Let $\{\mu_1^S, ..., \mu_\nu^S\}$ be the multiplicities with which the distinct eigenvalues of $\mathcal{J}_R(\pi_1)|_{\pi_2}$ appear in S and let $\{\mu_1, ..., \mu_\nu\}$ be the corresponding multiplicities with which these eigenvalues appear in $\mathcal{J}_R(\pi_1)$. We have that:

$$\mu_j^S \leq \mu_j \text{ for } 1 \leq j \leq \nu \text{ and } k = \mu_1^S + ... + \mu_\nu^S.$$

3.7.2 Lemma. *Let R be a k Osserman algebraic curvature tensor on a Riemannian vector space V of dimension m, where $2 \le k \le m-2$ and $2k \le m$. Let S be a relative spectrum.*

(1) *Let $\mathcal{P}_S := \{(\pi_1, \pi_2) \in \mathrm{Gr}_k(V) \times \mathrm{Gr}_k(V) : \mathrm{spec}(\pi_1, \pi_2) = S\}$. Projection $\rho_1(\pi_1, \pi_2) = \pi_1$ defines a fiber bundle $F_S \to \mathcal{P}_S \xrightarrow{\rho_1} \mathrm{Gr}_k(V)$, where $F_S = \mathrm{Gr}_{\mu_1^S}(\mathbb{R}^{\mu_1}) \times ... \times \mathrm{Gr}_{\mu_\nu^S}(\mathbb{R}^{\mu_\nu})$ and where $\mathcal{P}_S$ is connected.*

(2) *The following conditions are equivalent and any of them define the notion of a* **regular** *relative spectrum S.*

2a) *F_S is a single point, i.e. ρ_1 is a diffeomorphism.*

2b) $\dim F_S = 0$, *i.e.* $\dim \mathcal{P}_S = \dim \mathrm{Gr}_k(V)$.

2d) $\mu_1 = \mu_1^S$, ..., $\mu_\nu = \mu_\nu^S$, *i.e.* $k = \mu_1 + ... + \mu_\nu$.

(3) *If there exists a regular relative spectrum and if $1 \le \ell < k$, then R is ℓ Osserman.*

Proof. Assertions (1) and (2) are immediate from the definition.

To prove Assertion (3), we suppose that S is a regular relative spectrum. Let $(\pi_1, \pi_2) \in \mathcal{P}_S$. We use Theorem 3.1.2 to see that $\mathcal{J}_R(\pi_1)\pi_2 \subset \pi_2$ implies $\mathcal{J}_R(\pi_2)\pi_1 \subset \pi_1$. Consequently, $\mathrm{spec}(\pi_2, \pi_1)$ is well defined. By Lemma 1.3.4, the function $(\pi_1, \pi_2) \to \mathrm{spec}(\pi_2, \pi_1)$ is a continuous map from $\mathcal{P}_S$ to $\mathbb{R}^k$. Since R is k Osserman, $\mathrm{spec}(\pi_2, \pi_1)$ takes values in a finite set. Since $\mathcal{P}_S$ is connected, $\mathrm{spec}(\pi_2, \pi_1) := \tilde{S}$ is constant on $\mathcal{P}_S$. Thus, the map which interchanges π_1 and π_2 defines a diffeomorphism from $\mathcal{P}_S$ to $\mathcal{P}_{\tilde{S}}$.

We use Assertion (2b) to see that if S is a regular relative spectrum, then the dual spectrum $\tilde{S} := \mathrm{spec}(\pi_2, \pi_1)$ is a regular relative spectrum. Assertion (2a) then shows that $\Phi : \rho_1 \circ \rho_2^{-1} : \pi_2 \to \pi_1$ is a diffeomorphism of $\mathrm{Gr}_k(V)$.

Let $\{\alpha_1, ..., \alpha_\nu\}$ be the distinct eigenvalues of $\mathcal{J}_R(\pi_1)|_{\pi_2}$. Since S is a regular relative spectrum, we may decompose

$$\text{(3.7.2.a)} \qquad \pi_2 = E_{\alpha_1}(\pi_1) \oplus ... \oplus E_{\alpha_\nu}(\pi_1) = E_{\alpha_1}(\Phi\pi_2) \oplus ... \oplus E_{\alpha_\nu}(\Phi\pi_2).$$

Let γ_k be the classifying k plane bundle over $\mathrm{Gr}_k(V)$ defined by:

$$\gamma_k := \{(\pi, v) \in \mathrm{Gr}_k(V) \times V : v \in \pi\}.$$

Then Equation (3.7.2.a) defines a decomposition

$$\gamma_k = \Phi^* E_{\alpha_1} \oplus ... \oplus \Phi^* E_{\alpha_\nu}^S,$$

where we pull back the eigen-bundles using the diffeomorphism Φ. Theorem 4.3.13 then shows this decomposition must be trivial since $2k \le m$ implies $k \le m-k$. Consequently, $\nu = 1$ so the eigenvalue α_1 has multiplicity k. We can now use Lemma 3.7.1 to see R is ℓ Osserman. □

We can now prove that R is Osserman.

3.7.3 Lemma. *Let R be a k Osserman algebraic curvature tensor on a Riemannian vector space V of dimension m, where $2 \leq k \leq m - 2$ and $2k \leq m$. Then R is Osserman.*

Proof. We suppose that R is not Osserman and argue for a contradiction. Let a_i be the number of eigenvalues of $\mathcal{J}_R(\tau)$ which have multiplicity i and let $\Xi_\nu = \Xi_\nu(\tau)$ for any $\tau \in \mathrm{Gr}_k(V)$. Then:

$$m = a_1 + 2a_2 + 3a_3 + ... + ma_m \text{ and } \Xi_\nu = \textstyle\sum_{n>\nu} a_n(n - \nu).$$

Let $\tilde{a}_i$ be non-negative integers with $\tilde{a}_i \leq a_i$. If $k = \tilde{a}_1 + 2\tilde{a}_2 + \dots$, then R admits a regular relative spectrum and R is Osserman by Lemma 3.7.2. Thus, by assumption, this does not happen. Consequently, if $\{\mu(\alpha_1), ..., \mu(\alpha_i)\}$ are the multiplicities of distinct eigenvalues of $\mathcal{J}_R$ on $\mathrm{Gr}_k(V)$, then we have

$$\mu(\alpha_1) + ... + \mu(\alpha_i) \neq k.$$

By Lemma 3.7.1, R is Osserman if some eigenvalue has multiplicity at least k. Thus, no eigenvalue has multiplicity which is at least k so $a_i = 0$ for $i \geq k$. This means that

$$m = a_1 + 2a_2 ... + (k - 1)a_{k-1}.$$

We apply Lemma 3.7.1 to see that since R is not Osserman, then

$$\Xi_1 = a_2 + 2a_3 + ... + (k - 2)a_{k-1} < k - 1. \tag{3.7.3.a}$$

If $a_1 = 0$, then every eigenvalue has multiplicity at least 2. Thus,

$$m = 2a_2 + 3a_3 \cdots \geq 2(a_2 + a_3 + \dots).$$

Therefore, $a_2 + \cdots + a_{k-1} \leq \frac{m}{2}$ and we may estimate:

$$\begin{aligned}\Xi_1 =& a_2 + 2a_3 + 3a_4 + \dots \\ =& (2a_2 + 3a_3 + \dots) - (a_2 + a_3 + \dots) \\ =& m - (a_2 + a_3 + \dots) \geq m - \tfrac{m}{2} \geq k\end{aligned}$$

which contradicts Display (3.7.3.a). Thus, $a_1 \neq 0$. On the other hand, if $a_1 \geq k$, then R has a regular relative spectrum which is false. Thus,

$$1 \leq a_1 \leq k - 1.$$

As $m = a_1 + 2a_2 + \cdots \geq 2k$ and as $a_1 \leq k-1$, there exists $q \geq 2$ so

$$\text{(3.7.3.b)} \qquad a_1 + \cdots + (q-1)a_{q-1} \leq k \leq a_1 + \cdots + (q-1)a_{q-1} + qa_q.$$

If we have equality in either inequality of Display (3.7.3.b), then R has a regular relative spectrum which is false. Thus, $1 \leq a_q$.

Choose an integer b_q with $1 \leq b_q \leq a_q$ so that:

$$\text{(3.7.3.c)} \; a_1 + \cdots + (q-1)a_{q-1} + q(b_q - 1) \leq k \leq a_1 + \cdots + (q-1)a_q + qb_q.$$

Since R does not have a regular relative spectrum, we have strict inequalities in Display (3.7.3.c). Choose the integer b_1 so that:

$$b_1 + 2a_2 + \cdots + (q-1)a_{q-1} + qb_q = k.$$

Since we must have strict inequalities in Display (3.7.3.c), $a_1 - q < b_1 < a_1$. Since R does not have a regular relative spectrum, $b_1 < 0$ and therefore $a_1 - q < b_1 \leq -1$. This shows that $a_1 \leq q - 2$; since $1 \leq a_1$, we have $q \geq 3$.

We compute that:

$$\begin{aligned} a_1 + qa_q &= a_1 + (q-2)a_q + 2a_q \\ &\geq a_1 + a_1 a_q + 2a_q \geq 2a_1 + 2a_q \\ m &= a_1 + 2a_2 + 3a_3 + \cdots + (k-1)a_{k-1} \\ &\geq 2a_1 + 2a_2 + 2a_3 + \cdots + 2a_{k-1}. \end{aligned}$$

Thus, $a_1 + a_2 \cdots \leq \frac{m}{2}$. This provides the final contradiction by showing:

$$\Xi_1 = a_2 + 2a_3 + \cdots \geq m - a_1 - a_2 - \cdots \geq m - \tfrac{m}{2} \geq k. \quad \square$$

We must now define a relative spectrum which analogous to that used in the proof of Lemma 3.7.2. Let λ be an eigenvalue of $\mathcal{J}_R(\cdot)$ on $\mathrm{Gr}_k(V)$ which has multiplicity μ. Let S_λ be the associated unit sphere bundle:

$$S_\lambda := \{(\pi, x) \in \mathrm{Gr}_k(V) \times S(V) : \mathcal{J}_R(\pi)x = \lambda x\}.$$

Let $(\pi, x) \in S_\lambda$. Since $\mathcal{J}_R(\pi)x = \lambda x$, $\mathcal{J}_R(x)\pi \subset \pi$ by Theorem 3.1.2. Let $\mathrm{spec}(x, \pi) \in \mathbb{R}^k$ be the spectrum of $\mathcal{J}_R(x)|_\pi$. The proof of the following Lemma involves some non-trivial computations in characteristic classes.

3.7.4 Lemma. *Let R be a k Osserman algebraic curvature tensor on a Riemannian vector space V of dimension m, where $2 \le k \le m-2$ and $2k \le m$. Let λ be an eigenvalue of $\mathcal{J}_R$ on $\mathrm{Gr}_k(V)$ which has multiplicity μ.*

(1) *The function* $\mathrm{spec}(x,\pi)$ *is constant on* S_λ.

(2) *Let* $(\pi, x) \in S_\lambda$.
  (2a) *If $\mu \ne m-k$, if $(k,m) \ne (2,7)$, and if $(k,m) \ne (3,8)$, then 0 is an eigenvalue of $\mathcal{J}_R(x)|_\pi$.*
  (2b) *If $\mu \ne k$, then all the eigenvalues of $\mathcal{J}_R(x)|_\pi$ are equal.*

Proof. As the fiber of S_λ is not connected when $\mu = 1$, we introduce the associated projective bundle

$$\mathbb{P}_\lambda := \{(\pi,\sigma) \in \mathrm{Gr}_k(V) \times \mathbb{RP}(V) : \mathcal{J}_R(\pi)|_\sigma = \lambda \,\mathrm{Id}\}$$

in the proof of Assertion (1). If $(\pi,\sigma) \in \mathbb{P}_\lambda$, then let $\mathrm{spec}(\sigma,\pi) \in \mathbb{R}^k$ be the spectrum of $\mathcal{J}_R(\sigma)|_\pi$. By Lemma 1.3.4, the map $(\pi,\lambda) \to \mathrm{spec}(\sigma,\pi)$ is continuous on $\mathbb{P}_\lambda$. By Lemma 3.7.3, R is Osserman. Therefore, $\mathrm{spec}(\sigma,\pi)$ can take on at most a finite number of values. Since the fiber $\mathbb{RP}^{\mu-1}$ and the base $\mathrm{Gr}_k(V)$ are connected, the total space $\mathbb{P}_\lambda$ is connected. Thus, $\mathrm{spec}(\sigma,\pi)$ is constant on $\mathbb{P}_\lambda$.

The natural map $(\pi,x) \to (\pi, \mathrm{span}\{x\})$ defines a covering projection

$$\mathbb{Z}_2 \to S_\lambda \to \mathbb{P}_\lambda.$$

Assertion (1) now follows since $\mathrm{spec}(x,\pi) = \mathrm{spec}(\mathrm{span}\{x\},\pi)$. This shows that to prove Assertion (2a) it suffices to show there **exists** $(\pi,x) \in S_\lambda$ so that $0 \in \mathrm{spec}(x,\pi)$. Similarly, to prove Assertion (2b), it suffices to show there **exists** $(\pi,x) \in S_\lambda$ so $\mathcal{J}_R(x)|_\pi$ is a multiple of the identity.

Suppose that $\mu \ne m-k$, that $(k,m) \ne (2,7)$, and that $(k,m) \ne (3,8)$. We suppose Assertion (1) fails and argue for a contradiction; this means 0 is not an eigenvalue of $\mathcal{J}_R(x)|_\pi$.

Suppose that there exists $\pi \in \mathrm{Gr}_k(V)$ so that $\pi \cap E_\lambda(\pi) \ne \{0\}$. Then we may choose a unit vector $x \in \pi \cap E_\lambda(\pi)$. Since $\mathcal{J}_R(x)x = 0$, 0 is an eigenvalue of $\mathcal{J}_R(x)$ on π contrary to the assumption we have made. Thus,

$$\pi \cap E_\lambda(\pi) = \{0\} \text{ for all } \pi \in \mathrm{Gr}_k(V).$$

Let $\gamma_k^\perp$ be the orthogonal complement of the classifying k plane bundle γ_k over $\mathrm{Gr}_k(V)$:

$$\gamma_k^\perp := \{(\tau, y) : \tau \in \mathrm{Gr}_k(V) \text{ and } y \perp \tau\}.$$

Let $\rho_{\pi^\perp} = (1 - \rho_\pi)$ be orthogonal projection on $\pi^\perp$. Since $\ker(\rho_{\pi^\perp}) = \pi$, $\rho_{\pi^\perp} : E_\lambda(\pi) \to \gamma_k^\perp$ is injective. Thus,

$$\dim E_\lambda(\pi) = \mu \leq m - k = \dim \gamma_k^\perp .$$

Since we have assumed $\mu \neq m-k$, $\rho_{\pi^\perp} E_\lambda$ is a proper sub-bundle of $\gamma_k^\perp$. Since $2k \leq m$, since $(k, m) \neq (2, 7)$, and since $(k, m) \neq (3, 8)$, this contradicts Theorem 4.3.13 and establishes Assertion (2a).

To prove Assertion (2b), we suppose that $\mu \neq k$. Let $(\pi, x) \in S_\lambda$. Let $\{\alpha_1, ..., \alpha_\ell\}$ be the distinct eigenvalues of $\mathcal{J}_R(x)|_\pi$ on π. We suppose $\ell \geq 2$ and argue for a contradiction. Let $x_i := w_i(\gamma_k) \in H^i(\mathrm{Gr}_k(V); \mathbb{Z}_2)$. By Theorem 4.3.12, the cohomology groups

(3.7.4.a) $$H^i(\mathrm{Gr}_k(V); \mathbb{Z}_2) = \mathbb{Z}_2[x_1, ..., x_i]_i \text{ for } i \leq m - k$$

can be identified with the vector space of all graded homogeneous polynomials of degree i in these variables; since $k \leq m - k$, there are no non-trivial relations for $i \leq k$.

Let $\rho_1(\pi, \sigma) = \pi$ be the natural projection from $\mathbb{P}_\lambda$ to $\mathrm{Gr}_k(V)$. Let x be the first Stiefel-Whitney class of the canonical line bundle over $\mathbb{P}_\lambda$. We apply Theorem 4.3.9 to see ρ_1^* embeds $H^*(\mathrm{Gr}_k(V); \mathbb{Z}_2)$ in $H^*(\mathbb{P}_\lambda; \mathbb{Z}_2)$ and that $H^*(\mathbb{P}_\lambda; \mathbb{Z}_2)$ is the free $H^*(\mathrm{Gr}_k(V); \mathbb{Z}_2)$ module on generators $\{1, x, ..., x^{\mu-1}\}$. Thus, we may use Equation (3.7.4.a) to see that:

(3.7.4.b) $$\begin{aligned} &H^k(\mathbb{P}_\lambda; \mathbb{Z}_2) = \mathbb{Z}_2[x, x_1, ..., x_k]_k && \text{if } k < \mu, \\ &H^k(\mathbb{P}_\lambda; \mathbb{Z}_2) = \oplus_{i=0}^{\mu-1} \mathbb{Z}_2[x_1, ..., x_i]_i x^{k-i} && \text{if } \mu < k. \end{aligned}$$

We consider the following vector bundles over $\mathbb{P}_\lambda$:

$$\begin{aligned} &\Gamma_k := \{(\pi, x, y) \in \mathbb{P}_\lambda \times V : y \in \pi\} \\ &\Gamma_{i;1} := \{(\pi, x, y) \in \mathbb{P}_\lambda \times V : y \in \pi \text{ and } \mathcal{J}_R(x)y = \alpha_i y\} \text{ for } 1 \leq i \leq \ell. \end{aligned}$$

Since $\Gamma_k = \rho_1^*(\gamma_k)$, we use the decomposition $\Gamma_k = \oplus_{i=1}^\ell \Gamma_{i;1}$ to see

$$\rho_1^*(x_k) = w_k(\Gamma_k) = \textstyle\prod_{i=1}^\ell w_{\dim(\Gamma_{i;1})}.$$

Since $\dim(\Gamma_{i;1}) < k$, there is a polynomial χ_i so that we may express $w_{\dim(\Gamma_{i;a})} = \chi_i(x, x_1, ..., x_{k-1})$. Consequently, we have

(3.7.4.c) $$x_k = \textstyle\prod_{1 \leq i \leq \ell} \chi_i(x, x_1, ..., x_{k-1}) = \chi(x, x_1, ..., x_{k-1}).$$

If $k < \mu$, then Equation (3.7.4.c) contradicts Display (3.7.4.b) since it implies the existence of a non-trivial relationship. Since we have assumed $k \neq \mu$, we must have $\mu < k$. We use Theorem 4.3.9 to see that the defining relation of the algebra $H^*(\mathbb{P}_\lambda; \mathbb{Z}_2)$ is

$$x^\nu = \textstyle\sum_{0<j\leq\mu} \rho_1^*(x_j)x^{\mu-j}.$$

Consequently, we may use this relationship to assume the polynomial χ in Equation (3.7.4.c) involves only powers of x to powers less than k. In particular, we do not introduce an x_k. Consequently, we again get a non-trivial relationship which is impossible. □

We continue our investigations with the following Lemma.

3.7.5 Lemma. *Let R be a k Osserman algebraic curvature tensor on a Riemannian vector space V of dimension m, where $2 \leq k \leq m-2$ and $2k \leq m$.*

(1) *Let $x \in S(V)$. Let $I = \{i_1, ..., i_k\}$ be given, where we have the relation $1 \leq i_1 < ... < i_k \leq m$. There exists π in $\mathrm{Gr}_k(V)$ such that $\mathrm{spec}(x,\pi) = (\lambda_{i_1}(x), ..., \lambda_{i_k}(x))$ and such that $\mathcal{J}_R(\pi)x = \Lambda_I(x)x$.*

(2) *R is 2 Osserman.*

(3) *$\mathcal{J}_R$ does not have two unequal and non-zero eigenvalues on $\mathrm{Gr}_1(V)$.*

Proof. By Lemma 3.7.3, R is Osserman. Let $x \in S(V)$. Let $\{\xi_1, ..., \xi_m\}$ be an orthonormal basis for V so that $\mathcal{J}_R(x)\xi_i = \lambda_i$ for $1 \leq i \leq m$. Let

$$\pi := \mathrm{span}\{\xi_{i_1}, \dots, \xi_{i_k}\}.$$

Then by definition, $\mathrm{spec}(x,\pi) = (\lambda_{i_1}, ..., \lambda_{i_k})$. Since $\mathcal{J}_R(x)\xi_i = \lambda_i\xi_i$, we have dually by Theorem 3.1.2 that $\mathcal{J}_R(\xi_i)x = \lambda_i x$. We complete the proof of Assertion (1) by computing:

$$\mathcal{J}_R(\pi)x = \mathcal{J}_R(\xi_{i_1})x + \cdots + \mathcal{J}_R(\xi_{i_k})x = \lambda_{i_1} + ... + \lambda_{i_k} = \Lambda_I(x).$$

If R has constant sectional curvature, then Assertions (2) and (3) are immediate. We therefore suppose R does not have constant sectional curvature. Consequently, the reduced Jacobi operator $\tilde{\mathcal{J}}_R$ has at least two distinct eigenvalues $a \neq b$ on $\mathrm{Gr}_1(V)$. Let $\lambda := a + b + \dots$, where we choose the remaining $k-2$ eigenvalues of $\mathcal{J}_R$ arbitrarily. We use Assertion (1) to choose π with $(\pi, x) \in S_\lambda$ so a and b are eigenvalues of $\mathcal{J}_R(x)|_\pi$. Since $a \neq b$, Lemma 3.7.4 shows that $\mu_\lambda(\pi) = k$. Thus, by Lemma 3.7.1, R is i Osserman for any $1 \leq i \leq k$. In particular R is 2 Osserman. This completes the proof of Assertion (2).

Suppose that m is odd. Since R is Osserman, R has constant sectional curvature by Theorem 3.1.4 and Theorem 3.1.6. We suppose therefore that m is even.

We set $k = 2$. Since m is even, $(k, m) \neq (2, 7)$ and $(k, m) \neq (3, 8)$. Thus, we may apply Lemma 3.7.4. Suppose Assertion (3) fails and that we have two non-equal and non zero eigenvalues $0 \neq a$, $0 \neq b$, and $a \neq b$ of $\mathcal{J}_R$ on $\mathrm{Gr}_1(V)$. Fix a unit vector x and let $\{x, \xi_a, \xi_b\}$ be an orthonormal set of associated eigenvectors of $\mathcal{J}_R(x)$ corresponding to the eigenvalues $\{0, a, b\}$ so

$$\mathcal{J}_R(x)x = 0,\ \mathcal{J}_R(x)\xi_a = a\xi_a,\ \text{and}\ \mathcal{J}_R(x)\xi_b = b\xi_b.$$

Let

$$\pi_a := \mathrm{span}\{x, \xi_a\},\ \pi_b := \mathrm{span}\{x, \xi_b\},\ \text{and}\ \pi_{a+b} := \mathrm{span}\{\xi_a, \xi_b\}.$$

Then dually by Theorem 3.1.2, we have:

$$\begin{aligned} &\mathcal{J}_R(x)x = 0,\ \ \mathcal{J}_R(\xi_a)x = ax,\ \text{and}\ \mathcal{J}_R(\xi_b)x = bx\ \text{so} \\ &\mathcal{J}_R(\pi_a)x = ax,\ \mathcal{J}_R(\pi_b)x = bx,\ \text{and}\ \mathcal{J}_R(\pi_{a+b})x = (a+b)x. \end{aligned}$$

The eigenvalues $\{a, b, a+b\}$ of $\mathcal{J}_R$ on $\mathrm{Gr}_2(V)$ are distinct. The eigenvalues of $\mathcal{J}_R(x)$ restricted to π_a, to π_b, and to π_{a+b} are $\{0, a\}$, $\{0, b\}$, and $\{a, b\}$, respectively. These eigenvalues are distinct so by Lemma 3.7.4, we have:

$$\mu_a(\pi) = 2,\ \mu_b(\pi) = 2,\ \text{and}\ \mu_{a+b}(\pi) = 2$$

for any $\pi \in \mathrm{Gr}_2(V)$. Consequently, $m \geq 6$. On the other hand, since the eigenvalues of $\mathcal{J}_R(x)$ on $\pi_{a,b}$ are $\{a, b\}$ and since 0 does not belong to this set, $\mu_{a+b}(\pi) = m - 2$ by Lemma 3.7.4. Thus, $m - 2 = 2$ so $m = 4$. This contradiction completes the proof of Assertion (3). □

Proof of Theorem 3.1.7. Let R be a k Osserman algebraic curvature tensor on a Riemannian vector space V of dimension m, where $2 \leq k \leq m - 2$. By Lemma 3.7.3, R is Osserman. The desired result follows from Theorem 3.1.4 and from Theorem 3.1.6 if m is odd, so we suppose m even. We use Lemma 3.7.5 to see that we may suppose $k = 2$. Thus, $(k, m) \neq (2, 7)$ and $(k, m) \neq (3, 8)$. Thus, we may apply Lemma 3.7.4.

We use Lemma 3.7.5 to see that $\mathcal{J}_R$ has two distinct eigenvalues $\{0, a\}$ with multiplicities $\{\mu_0, \mu_a\}$ on $\mathrm{Gr}_1(V)$. If $\mu_0 = 1$ or $\mu_a = 1$, then $\mathcal{J}_R$ and hence $\tilde{\mathcal{J}}_R$ has only two eigenvalues; furthermore, one of these eigenvalues must has multiplicity at most 1. Thus, by Theorem 3.1.6, $R = c_0 R_{\mathrm{Id}} + c_1 R_\phi$, where ϕ is an almost complex structure on V. Since R is k Osserman, we use Theorem 3.3.1 to see $c_0 = 0$ or $c_1 = 0$.

We therefore assume $\mu_0 \geq 2$ and $\mu_a \geq 2$ and argue for a contradiction. Let x be a unit vector and let $\{\xi_1, \xi_2, \psi_1, \psi_2\}$ be an orthonormal set so that

$$\mathcal{J}_R(x)\xi_1 = 0,\ \mathcal{J}_R(x)\xi_2 = 0,\ \mathcal{J}_R(x)\psi_1 = a\psi_1, \text{ and } \mathcal{J}_R(x)\psi_2 = a\psi_2.$$

We then have dually by Theorem 3.1.2 that

$$\mathcal{J}_R(\xi_1)x = 0,\ \mathcal{J}_R(\xi_2)x = 0,\ \mathcal{J}_R(\psi_1)x = ax, \text{ and } \mathcal{J}_R(\psi_2)x = ax.$$

Let

$$\begin{array}{lll} \pi_0 := \operatorname{span}\{\xi_1, \xi_2\}, & \pi_a := \operatorname{span}\{\xi_1, \psi_1\}, & \pi_{2a} := \operatorname{span}\{\psi_1, \psi_2\}; \\ \mathcal{J}_R(\pi_0)x = 0, & \mathcal{J}_R(\pi_a)x = ax, & \mathcal{J}_R(\pi_{2a})x = 2ax. \end{array}$$

Let π be any 2 plane. Since the eigenvalue structure of $\mathcal{J}_R(x)$ on π_{2a} does not include the eigenvalue 0, we use Lemma 3.7.4 to see $\mu_{2a}(\pi) = m - 2$. Since the eigenvalues of $\mathcal{J}_R(x)$ on π_a are unequal, we have $\mu_a(\pi) = 2$. Since $\mathcal{J}_R(\pi_0)x = 0$, $\mu_0(\pi) \geq 1$. We complete the proof by deriving the following contradiction:

$$m \geq \mu_0(\pi) + \mu_a(\pi) + \mu_{2a}(\pi) \geq m + 1. \quad \square$$

3.8 The Szabó operator

Let V be a vector space of signature (p, q). A 5 tensor $\nabla R \in \otimes^5 V^*$ is said to be an *algebraic covariant derivative curvature tensor* if it satisfies the symmetries of Equations (1.5.4.b), (1.5.4.c), and (1.5.4.d). The *Szabó* operator is the self-adjoint operator on V defined by the relation:

$$(\mathcal{S}_{\nabla R}(x)y, z) = \nabla R(y, x, x, z; x).$$

Before beginning the proof of Theorem 3.1.9, we must first establish the following vanishing result:

3.8.1 Lemma. *Let V be a vector space of signature (p, q). Let ∇R be an algebraic covariant derivative curvature tensor on V. If $\mathcal{S}_{\nabla R}$ vanishes identically, then ∇R vanishes identically.*

Proof. Since we are in the purely algebraic setting, we can not use the argument using Jacobi fields given by Besse [16, p65]. Instead, we generalize the argument given by Vanhecke and Willmore [156] from the Riemannian setting to the pseudo-Riemannian setting.

As $\mathcal{S}_{\nabla R} = 0$,

$$\nabla R(x,y,y,w;y) = 0 \text{ for all } w,x,y. \tag{3.8.1.a}$$

We polarize. Set $y(t) := y + tx$ and expand Equation (3.8.1.a) in terms of powers of t. We set the term which is linear in t to zero and then use the curvature symmetries to compute:

$$\begin{aligned} 0 =& \nabla R(x,x,y,w;y) + \nabla R(x,y,x,w;y) + \nabla R(x,y,y,w;x) \\ =& 0 + \nabla R(x,y,x,w;y) - \nabla R(x,y,w,x;y) - \nabla R(x,y,x,y;w) \\ =& -2\nabla R(x,y,w,x;y) + \nabla R(x,y,y,x;w). \end{aligned} \tag{3.8.1.b}$$

Set $w = x$ in Equation (3.8.1.a) to see:

$$\nabla R(x,y,y,x;y) = 0 \text{ for all } x,y. \tag{3.8.1.c}$$

Again, we polarize setting $y(t) := y + tw$ and expanding (3.8.1.c) in terms of powers of t. We set linear term in t to zero to see:

$$0 = 2\nabla R(x,y,w,x;y) + \nabla R(x,y,y,x;w) \text{ for all } w,x,y. \tag{3.8.1.d}$$

We add Equations (3.8.1.b) and (3.8.1.d) to obtain:

$$0 = \nabla R(y,x,x,y;w) \text{ for all } x,y,w. \tag{3.8.1.e}$$

We polarize. Set $y(t) := y + tz$ and expand Equation (3.8.1.e) in terms of powers of t. The term which is linear in t then yields the identity:

$$0 = \nabla R(y,x,x,z;w) \text{ for all } x,y,z,w \in V. \tag{3.8.1.f}$$

We polarize. Set $x(t) = x + tv$ and expand Equation (3.8.1.f) in terms of powers of t. The term which is linear in t then yields the identity:

$$0 = \nabla R(y,x,v,z;w) + \nabla R(y,v,x,z;w) \text{ for all } v,w,x,y,z \in V. \tag{3.8.1.g}$$

We then use Equation (3.8.1.g) and the curvature symmetries to complete the proof by computing:

$$\begin{aligned} 0 =& \nabla R(y,x,v,z;w) + \nabla R(y,v,z,x;w) + \nabla R(y,z,x,v;w) \\ =& \nabla R(y,x,v,z;w) - \nabla R(y,v,x,z;w) + \nabla R(y,z,x,v;w) \\ =& \nabla R(y,x,v,z;w) + \nabla R(y,x,v,z;w) - \nabla R(y,x,z,v;w) \\ =& 3\nabla R(y,z,v,z;w) \text{ for all } v,w,x,y,z \in V. \quad \square \end{aligned}$$

Proof of Theorem 3.1.9. Let $\mathcal{S}_{\nabla R}$ be the Szabó operator defined by an algebraic covariant derivative curvature tensor on a Riemannian or on a Lorentzian vector space V. We suppose that $\mathcal{S}_{\nabla R}(\cdot)$ has constant eigenvalues on $S^{\pm}(V)$. We wish to show that $\nabla R = 0$. By Lemma 3.8.1, it suffices to show $\mathcal{S}_{\nabla R} = 0$.

First, we suppose that we are in the Riemannian setting. Since $\mathcal{S}_{\nabla R}$ has constant eigenvalues, it has has constant rank on $S(V)$. Since $\mathcal{S}_{\nabla R}(x)$ is self-adjoint, since $\mathcal{S}_{\nabla R}(x)x = 0$, and since $\mathcal{S}_{\nabla R}(-x) = -\mathcal{S}_{\nabla R}(x)$, we may use Theorem 4.1.2 to see that $\mathcal{S}_{\nabla R}(x) = 0$ for all x.

Thus, we may suppose that we are in the Lorentzian setting. We follow the treatment of Gilkey and Stavrov [86]. Let $\mathcal{B} := \{e_0, e_1, ..., e_q\}$ be an orthonormal basis for V, where e_0 is timelike and e_i is spacelike for $i > 0$. Let θ be a real parameter. We make a hyperbolic boost to define a new orthonormal basis $\mathcal{B}^\theta$ by setting:

$$e_i^\theta := \begin{cases} \cosh\theta \cdot e_0 + \sinh\theta \cdot e_1 & \text{if } i = 0, \\ \sinh\theta \cdot e_0 + \cosh\theta \cdot e_1 & \text{if } i = 1, \\ e_i & \text{if } i \geq 2. \end{cases}$$

Since e_0^θ is timelike and since ∇R is Szabó, there exists a constant C, which is independent of the parameter θ, so that for any $\theta \in \mathbb{R}$ we have:

$$C = \operatorname{Tr}\{\mathcal{S}_{\nabla R}(e_0^\theta)^2\}.$$

We may express $\cosh\theta = \frac{1}{2}(e^\theta + e^{-\theta})$ and $\sinh\theta = \frac{1}{2}(e^\theta - e^{-\theta})$. Thus, there are real numbers $a_{ij,\nu}$ so that:

$$\nabla R(e_i^\theta, e_0^\theta, e_0^\theta, e_j^\theta; e_0^\theta) = \textstyle\sum_{-5\leq\nu\leq5} a_{ij,\nu} e^{\nu\theta}. \tag{3.8.2.a}$$

Consequently, we may compute that:

$$\begin{aligned} e^{-10\theta}C =& e^{-10\theta}\operatorname{Tr}\{\mathcal{S}_{\nabla R}(e_0^\theta)^2\} \\ =& e^{-10\theta}\textstyle\sum_{1\leq i,j\leq q} \nabla R(e_i^\theta, e_0^\theta, e_0^\theta, e_j^\theta; e_0^\theta)^2 \\ =& \textstyle\sum_{1\leq i,j\leq q}\{a_{ij,5}\}^2 + O(e^{-\theta}). \end{aligned}$$

We take the limit as $\theta \to \infty$ to see $\sum_{ij}\{a_{ij,5}\}^2 = 0$ and hence $a_{ij,5} = 0$ for all i, j; we can show similarly that $a_{ij,-5} = 0$. An analogous argument then shows $a_{ij,\nu} = 0$ for $\nu \neq 0$. Consequently, we use Equation (3.8.2.a) to see

$$\nabla R(e_2, e_0^\theta, e_0^\theta, e_2; e_0^\theta) = a_{22,0}.$$

On the other hand, since there are three terms involving θ, the powers of e^θ which appear are odd. Thus, we see $a_{22,0} = 0$ and hence

$$\nabla R(e_2, e_0, e_0, e_2; e_0) = 0.$$

Similarly, we conclude $\nabla R(e_i, e_0, e_0, e_i; e_0) = 0$ for any $i \geq 1$. We polarize to see $\nabla R(e_i, e_0, e_0, e_j; e_0) = 0$ for any i, j; the vanishing being automatic if $i = 0$ or $j = 0$. Consequently, $\mathcal{S}_{\nabla R}(e_0) = 0$. As e_0 was an arbitrary unit timelike vector, $\mathcal{S}_{\nabla R}(\cdot) = 0$ on $S^-(V)$. Rescaling and analytic continuation then imply $\mathcal{S}_{\nabla R}(\cdot) = 0$ on V. $\square$

Let $F : M \to W$ be an embedding of a manifold M as a non-degenerate hypersurface in a vector space of signature (r, s). Let ν be a unit normal along M so that $(\nu, \nu) = \varepsilon = \pm 1$. If L is the second fundamental form and if S is the associated shape operator, then by Lemma 1.12.3, $R = \varepsilon R_S$, i.e.

$$\begin{aligned} &R(x,y)z = \varepsilon\{(Sy,z)Sx - (Sx,z)Sy\} \text{ and} \\ &R(x,y,z,w) = \varepsilon\{L(y,z)L(x,w) - L(x,z)L(y,w)\}. \end{aligned} \tag{3.8.3.a}$$

In Lemma 1.12.3, we showed that the ∇S and hence ∇L were totally symmetric. We covariantly differentiate Equation (3.8.3.a) to see:

$$\begin{aligned} \nabla R(x,y,z,w;v) :=& \varepsilon\{\nabla L(v,y,z)L(x,w) - \nabla L(v,x,z)L(y,w) \\ &+ L(y,z)\nabla L(v,x,w) - L(x,z)\nabla L(v,y,w)\}. \end{aligned}$$

We use this observation to motivate the following construction we shall employ in the proof of Theorem 3.1.10.

3.8.4 Lemma. *Let V be a vector space of signature (p, q). Let $\tilde{L}$ be a completely symmetric trilinear form and L be a symmetric bilinear form on V. Define*

$$\begin{aligned} \tilde{R}(x,y,z,w;v) :=& \tilde{L}(v,y,z)L(x,w) - \tilde{L}(v,x,z)L(y,w) \\ &+ \tilde{L}(v,x,w)L(y,z) - \tilde{L}(v,y,w)L(x,z). \end{aligned}$$

Then $\tilde{R}$ is an algebraic covariant derivative algebraic curvature tensor.

Proof. It is immediate that $\tilde{R}(x,y,z,w;v) = -\tilde{R}(y,x,z,w;v)$. Thus, we may interchange the roles of x and z and of y and w to show that the remaining symmetry of Equation (1.5.4.b) is satisfied:

$$\begin{aligned} \tilde{R}(z,w,x,y;v) =& \tilde{L}(v,w,x)L(z,y) - \tilde{L}(v,z,x)L(w,y) \\ &+ \tilde{L}(v,z,y)L(w,x) - \tilde{L}(v,w,y)L(z,x) \\ =& \tilde{R}(x,y,z,w;v). \end{aligned}$$

We check the first Bianchi identity of Equation (1.5.4.c):

$$\begin{aligned}
&\tilde{R}(x,y,z,w;v) + R(x,z,w,y;v) + R(x,w,y,z;v)\\
=&\tilde{L}(v,y,z)L(x,w) + \tilde{L}(v,z,w)L(x,y) + \tilde{L}(v,w,y)L(x,z)\\
-&\tilde{L}(v,x,z)L(y,w) - \tilde{L}(v,x,w)L(z,y) - \tilde{L}(v,x,y)L(z,w)\\
+&\tilde{L}(v,x,w)L(y,z) + \tilde{L}(v,x,y)L(z,w) + \tilde{L}(v,x,z)L(w,y)\\
-&\tilde{L}(v,y,w)L(x,z) - \tilde{L}(v,w,z)L(x,y) - \tilde{L}(v,z,y)L(x,w)\\
=&0.
\end{aligned}$$

We check the second Bianchi identity of Equation (1.5.4.d) satisfied and complete the proof of Lemma 3.8.4 by computing:

$$\begin{aligned}
&\tilde{R}(x,y,z,w;v) + \tilde{R}(x,y,w,v;z) + \tilde{R}(x,y,v,z;w)\\
=&\tilde{L}(v,y,z)L(x,w) + \tilde{L}(z,y,w)L(x,v) + \tilde{L}(w,y,v)L(x,z)\\
-&\tilde{L}(v,x,z)L(y,w) - \tilde{L}(z,x,w)L(y,v) - \tilde{L}(w,x,v)L(y,z)\\
+&\tilde{L}(v,x,w)L(y,z) + \tilde{L}(z,x,v)L(y,w) + \tilde{L}(w,x,z)L(y,v)\\
-&\tilde{L}(v,y,w)L(x,z) - \tilde{L}(z,y,v)L(x,w) - \tilde{L}(w,y,z)L(x,v)\\
=&0 \quad \square.
\end{aligned}$$

Proof of Theorem 3.1.10. Let V be a vector space of signature (p,q), where $p \geq 2$ and $q \geq 2$. To show that Theorem 3.1.9 fails in this setting, we must show that there exists a non-trivial algebraic covariant derivative curvature tensor on V so that $\mathcal{S}_{\nabla R}(\cdot)$ has constant eigenvalues on $S^{\pm}(V)$ and so that $\mathcal{S}_{\nabla R}$ does not vanish identically.

Let $\tilde{L}$ and L be a completely symmetric trilinear and bilinear forms on V. Let $\tilde{\phi}_x$ and ϕ be the associated self-adjoint operators defined by:

$$\tilde{L}(x,y,z) = (\phi_x y, z) \text{ and } L(x,y) = (\phi x, y).$$

We use Lemma 3.8.4 to define an algebraic covariant derivative tensor ∇R with associated Szabó tensor and operator given by:

$$\begin{aligned}
(\mathcal{S}_{\nabla R}(x)y,w) =&\tilde{L}(x,x,x)L(y,w) - \tilde{L}(x,y,x)L(x,w)\\
&+\tilde{L}(x,y,w)L(x,x) - \tilde{L}(x,x,w)L(y,x), \text{ and}\\
\mathcal{S}_{\nabla R}(x)y =&\tilde{L}(x,x,x)\phi y - \tilde{L}(y,x,x)\phi x + L(x,x)\tilde{\phi}_x y - L(y,x)\tilde{\phi}_x x.
\end{aligned}$$

Let $\{e_1^-, ..., e_p^-, e_1^+, ..., e_q^+\}$ be a normalized orthonormal basis for V. Let ε, δ, and ϱ be choices of ± 1 signs. Let $1 \le a \le \min(p,q)$. We define:

$$L_a(e_i^\varepsilon, e_j^\varepsilon) = \begin{cases} 1 & \text{if } i = j \le a, \\ 0 & \text{otherwise,} \end{cases}$$

$$L_a(e_i^\varepsilon, e_j^\delta, e_k^\varrho) = \begin{cases} 1 & \text{if } i = j = k \le a, \\ 0 & \text{otherwise,} \end{cases}$$

$$\phi_a e_i^\varepsilon = \begin{cases} e_i^+ - e_i^- & \text{if } i \le a, \\ 0 & \text{if } i > a, \end{cases}$$

$$\phi_{a,e_i^\varepsilon} e_j^\delta = \begin{cases} \phi_a e_i & \text{if } i = j \\ 0 & \text{if } i \ne j. \end{cases}$$

The map ϕ_a is, modulo a suitable change of basis, equivalent to the map Φ_a defined in Equation (3.3.1.c). Let ∇R_a be the associated algebraic covariant derivative curvature tensor and let $\mathcal{S}_a$ be the associated Szabó operator.

Since $\phi_a^2 = 0$, $\mathcal{S}_a(x)^2 = 0$. This shows that ∇R_a is Szabó. On the other hand, we show $\mathcal{S}_a$ does not vanish identically by computing

$$\begin{aligned}(\mathcal{S}_a(e_1^+)e_2^+, e_2^+) =& \nabla R_a(e_2^+, e_1^+, e_1^+, e_2^+; e_1^+) \\ =& \tilde{L}_a(e_1^+, e_1^+, e_1^+) L_a(e_2^+, e_2^+) \\ \ne& 0. \quad \square\end{aligned}$$

Chapter 4

Controlling the Eigenvalue Structure

4.1 Introduction

In this chapter, we use topological methods to establish results needed previously. In Section 4.2, we review the basic facts concerning fiber bundles which we shall need; much of this section is expository in nature so we shall simply state many results and we refer to [53, 117, 139, 147, 159] for details concerning the proofs. We show that a submersion with compact fibers is a fiber bundle and we establish the homotopy property for fiber bundles with compact fibers. We discuss vector bundles, subbundles, quotient bundles, and bundles defined by the image and kernel of vector bundle maps of constant rank. We recall the long exact sequence in homotopy associated with a fiber bundle as well as the Lerray–Hirsch theorem that relates the cohomology of the fiber, the base, and the total space of a fibration under certain circumstances; these will be fundamental tools we will use several times.

One of the main results of Section 4.2 is the determination of the homotopy type of the space forms $S^{\pm}(\mathbb{R}^{(p,q)})$, of the complex space forms $\mathbb{CP}^{(p,q)}$, and the para-complex space forms $\widetilde{\mathbb{CP}}^{p}$ which were defined in Section 3.6. We follow the discussion in Blažić and Vukmirović [24] to prove that:

4.1.1 Theorem.

(1) *$S^{+}(\mathbb{R}^{(p,q)})$ is homotopy equivalent to S^{q-1}.*

(2) *$\mathbb{CP}^{(p,q)}$ is homotopy equivalent to $\mathbb{CP}^{(0,q)}$.*

(3) *$\widetilde{\mathbb{CP}}^{p}$ is homotopy equivalent to S^{p}.*

We conclude Section 4.2 by deriving a result, due to Szabó [150], concerning equivariant vector fields on spheres which was used in Section 3.8 to discuss the Szabó operator:

4.1.2 Theorem.

(1) *Let $\vec{s}$ be a continuous tangent vector field on S^{m-1}. Then there exists $x \in S^{m-1}$ so that $\vec{s}(x) = -\vec{s}(-x)$.*

(2) *Let $A(x)$ be a continuous map from S^{m-1} to the space of self-adjoint linear maps of $\mathbb{R}^m$. Assume that $A(-x) = -A(x)$, that $A(x)x = 0$, and that* $\dim\ker(A)$ *is constant on S^{m-1}. Then $A \equiv 0$.*

Like the material in Section 4.2, the material of Section 4.3 is almost entirely expository in nature; we omit proofs for the most part and refer to [7, 33, 109, 120] for further details. We begin by discussing the Grassmannians; these are the classifying spaces for vector bundles. We give the basic properties of the Stiefel-Whitney classes for real vector bundles and the basic properties of the Chern classes for complex vector bundles; we relate the Chern classes of a complex vector bundle to the Stiefel-Whitney classes of the underlying real vector bundle. We review the splitting principle which reduces computations in characteristic classes for arbitrary vector bundles to bundles which split as the direct sum of line bundles. We recall a result of Borel [31] which expresses the $\mathbb{Z}_2$ cohomology algebra of the real Grassmannians in terms of the Stiefel-Whitney classes of the classifying bundle γ_k. We present results of Glover, Homer, and Stong [89] and of Stong [149] concerning the indecomposability of the bundles γ_k and $\gamma_k^\perp$ over the Grassmannians $\mathrm{Gr}_k(\mathbb{R}^m)$ and $\mathrm{Gr}_k(\mathbb{C}^m)$. We introduce the Steenrod squares Sq^i and the higher power operations $\mathcal{P}^i$. We define the real and complex K theory groups and relate reduced K theory groups to stable equivalence classes of vector bundles. We use a result of Adams [1] concerning the real K theory groups $\mathbb{RP}^n$ to complete the proof of Theorem 2.1.4.

In Section 4.4, we discuss the theory of geometrically symmetric vector bundles. We take cohomology with coefficients in $\mathcal{R} = \mathbb{Z}_2$, let cw be the total Stiefel-Whitney class, and let $\mathbb{F} = \mathbb{R}$ when dealing with real vector bundles; we take coefficients in an arbitrary principle ideal domain $\mathcal{R}$, let cw be the Chern class, and let $\mathbb{F} = \mathbb{C}$ when dealing with complex vector bundles. Let $\mathbb{FP}^{n-1}$ be the projective space of lines in $\mathbb{F}^n$. The cohomology groups $H^*(\mathbb{FP}^{n-1};\mathcal{R})$ are truncated polynomial rings generated by the element $x := cw_1(\gamma_1)$:

$$H^*(\mathbb{FP}^{n-1};\mathcal{R}) = \mathcal{R}[x]/x^n = 0.$$

Let $\mathbf{1}^n := \mathbb{FP}^{n-1} \times \mathbb{F}^n$ be the trivial n plane bundle. Let $V \in \mathrm{Vect}_{\mathbb{F}}^k(\mathbb{FP}^{n-1})$ be a non-trivial k dimensional sub-bundle of $\mathbf{1}^n$ over $\mathbb{FP}^{n-1}$; we shall always assume $0 < k < n$. Let $V^\perp \subset \mathbf{1}^n$ be the complementary bundle; we have $\dim(V^\perp) = n - k$. We say that V is *geometrically symmetric* if $\sigma \subset V(\tau)$ implies $\tau \subset V(\sigma)$ for any $\sigma, \tau \in \mathbb{FP}^{n-1}$. The main result of Section 4.4 is:

4.1.3 Theorem. *Let $n = a2^s$, where a is odd.*

(1) *Let $V \subset \mathbf{1}^n$ be a geometrically symmetric k-plane bundle. Assume that $2k \leq n$. Then $cw(V) = (1+x)^k$ in $H^*(\mathbb{FP}^{n-1};\mathbb{Z}_2)$.*

(2) *Suppose $\mathbf{1}^n = \oplus_i V_i$, where V_i are vector bundles of dimension k_i over $\mathbb{FP}^{n-1}$. Assume at least one of the V_i is geometrically symmetric. Order the dimensions so $k_0 \geq ... \geq k_\ell \geq 1$. Assume $\ell \geq 1$.*

  2a) *We have $k_1 + ... + k_\ell \leq 2^s$.*

  2b) *Suppose that $\mathbb{F} = \mathbb{C}$. If n is odd, then $\ell = 1$ and $k_1 = 1$. If n is even, then $\ell = 1$ and $k_1 \leq 2$ or $\ell = 2$ and $k_1 = k_2 = 1$.*

We now apply Theorem 4.1.3 to complete the proof of three results which were used previously:

Proof of Theorem 2.1.8. Let V be a vector space which is equipped with a positive definite inner product and a Hermitian almost complex structure J. Let R be an *almost complex algebraic curvature tensor* on V, i.e.:

$$R(x, Jx)J = JR(x, Jx) \text{ for all } x \in V$$

(see Definition 1.5.6 for details). Let $\mathbb{CP}(V)$ be the space of all complex lines in V - i.e. J invariant real 2 planes. Let $\pi(x) := \text{span}\{x, Jx\} \in \mathbb{CP}(V)$ be the complex line determined by $x \in S(V)$. The operator $-JR(\pi(\cdot))$ is self-adjoint on $\mathbb{CP}(V)$. Suppose R is *almost complex IP*; this means that the eigenvalues of $-JR(\cdot)$ are constant $\mathbb{CP}(V)$. Let

$$V_\lambda(\pi) := \{y \in V : -JR(\pi)y = \lambda y\}$$

be the associated eigenbundles over $\mathbb{CP}(V)$; since they are J invariant, they are complex. Let $x_i \in S(V)$. We showed in Lemma 2.11.2 that if λ is a minimal or maximal eigenvalue, then:

$$x_2 \in E_\lambda(\pi(x_1)) \Leftrightarrow R(x_1, Jx_1, x_2, Jx_2) = \lambda \Leftrightarrow x_1 \in E_\lambda(\pi(x_2)).$$

Consequently, these vector bundles are geometrically symmetric vector bundles over $\mathbb{CP}(V)$ and the estimates given in Theorem 2.1.8 now follow from Assertion (2b) of Theorem 4.1.3. □

Proof of Theorem 3.1.3. Let V be a Riemannian vector space. Let R be an algebraic curvature tensor on V. Let $\mathcal{J}_R(x) : y \to R(y,x)x$ be the *Jacobi operator*; this operator is self-adjoint and quadratic in x. We suppose R is *Osserman*; i.e. the eigenvalues of the Jacobi operator $\mathcal{J}_R(x)$ are constant on $S(V)$ or, equivalently, since $\mathcal{J}_R(-x) = \mathcal{J}_R(x)$, on the associated projective space $\mathbb{P}(V)$. If $x \in S(V)$, let $\pi(x) \in \mathbb{P}(V)$ be the real line determined by x. Let

$$V_\lambda(\pi) := \{y \in V : \mathcal{J}_R(\pi)y = \lambda y\}$$

be the associated eigenbundles. We use Theorem 3.1.2 (Rakić duality) to see that these eigenbundles are geometrically symmetric. Therefore, Theorem 3.1.3 now follows from Theorem 4.1.3 (1). □

Proof of Theorem 3.1.4 *if* $m \equiv 1$ *mod* 2 *or* $m \equiv 2$ *mod* 4. Let V be a Riemannian vector space. Let R be an Osserman algebraic curvature tensor on V. Let $x \in S(V)$. The *reduced Jacobi operator* $\tilde{\mathcal{J}}_R(x)$ is the restriction of the Jacobi operator $\mathcal{J}_R(x)$ to the subspace $x^\perp$. Let $\{\lambda_s, \mu_s\}$ be the associated eigenvalues and multiplicities of this operator, where we order $\mu_0 \geq ... \geq \mu_\ell \geq 1$; by assumption, these are independent of x. As noted above, Theorem 3.1.2 implies that the eigenbundles V_{λ_ν} are geometrically symmetric. Let $\gamma_1(\pi) := \pi$ be the classifying line bundle over $\mathbb{P}V$. We may decompose:

$$\mathbf{1}^n = \gamma_1 \oplus V_{\lambda_{\mu_0}} \oplus ... \oplus V_{\lambda_{\mu_\ell}}.$$

Thus, Theorem 4.1.3 (2a) implies $1 + \mu_1 + ... + \mu_\ell \leq 2^s$. Therefore, we have that $\mu_1 + ... + \mu_\ell \leq 0$ if n is odd and hence $\ell = 0$. Similarly, we may conclude $\mu_1 + ... + \mu_\ell \leq 1$ if $n \equiv 2$ mod 4 and hence either $\ell = 0$ or $\ell = 1$ and $\mu_1 = 1$. In general, of course, we get the estimate $\mu_1 + ... + \mu_\ell \leq 2^s - 1$ which is much weaker than the estimate provided by Lemma 4.3.14 if $2^s \geq 16$. □

The Lie algebra $\mathfrak{so}(m)$ of the special orthogonal group $\mathrm{SO}(m)$ is the space of skew-symmetric linear transformations of $\mathbb{R}^m$:

$$\mathfrak{so}(m) := \{A \in M_m(\mathbb{R}) : A + A^t = 0\}.$$

Let X be a smooth connected manifold on which $\mathbb{Z}_2 = \{\pm 1\}$ acts smoothly and without fixed points. We say that a continuous map $T : X \to \mathfrak{so}(m)$ is an *odd map of constant rank* r if $T(-x) = -T(x)$ and if $\mathrm{rank}\{T(x)\} = r$ for any $x \in X$. In Section 4.5, we discuss odd maps with values in $\mathfrak{so}(m)$ and prove: the following result:

4.1.4 Theorem.

(1) *Let* $T : S^n \to \mathfrak{so}(n + 1 + \alpha)$ *be an odd map of constant rank* $r > 0$, *where* $n \geq 9$ *and* $0 \leq \alpha \leq n$. *Choose* k *so* $2^k \leq \frac{r}{2} < 2^{k+1}$. *Then:*

1a) $r \leq 2\alpha$.

1b) 2^k *divides at least one element of the set* $\{n+1, ..., n+1+\alpha-\frac{r}{2}\}$.

1c) *If* $\alpha = 2$ *and* $r = 4$, *then* n *is odd.*

(2) *If* $T : S^8 \to \mathfrak{so}(10)$ *is an odd map of constant rank* r, *then* $r \leq 2$.

(3) *Let* $T : \mathrm{Gr}_2^+(\mathbb{R}^m) \to \mathfrak{so}(m)$ *be an odd map of constant rank* r. *If* $m = 5$, *or* $m = 6$, *or* $m = 9$, *then* $r \leq 2$.

(4) *Let* $T : Gr_2^+(\mathbb{R}^m) \to \mathfrak{so}(m)$ *be an odd map so* T *has constant eigenvalues and constant rank* $r > 2$. *Let* $\{\lambda_i, \mu_i\}$ *be the eigenvalues and the associated multiplicities for the self-adjoint operator* T^2, *where* $\lambda_0 = 0$ *and* $\mu_1 \geq ... \geq \mu_l \geq 1$.

4a) *If* $m = 7$, *then* $\vec{\mu} = (1,6)$, $\vec{\mu} = (1,4,2)$, *or* $\vec{\mu} = (3,4)$.

4b) *If* $m = 8$, *then* $\vec{\mu} = (0,8)$, $\vec{\mu} = (0,6,2)$, *or* $\vec{\mu} = (2,6)$.

(5) *Let* $T : \mathrm{Gr}_2^+(\mathbb{R}^m) \to \mathfrak{so}(m+2)$ *be an odd map of constant rank* $r = 4$. *If* $m \geq 10$ *and if* m *is even, then* m *or* $m+2$ *is a power of* 2.

We decompose the proof of Theorem 4.1.4 into a sequence of different Lemmas; Assertion (1) will follow from Lemma 4.5.3, Assertion (2) from Lemma 4.5.5, Assertion (3) from Lemma 4.5.7, and Assertion (4) from Lemma 4.5.8. In Lemmas 4.5.4 and 4.5.9, we show the results of Assertions (1), (2), (3), and (4) are sharp. Assertion (5) will follow from Lemma 4.5.12.

Proof of Theorem 2.1.1. Let R be an algebraic curvature tensor on $\mathbb{R}^{(p,q)}$ which has constant spacelike, mixed, and timelike rank r. Then R defines odd maps from $\mathrm{Gr}_{0,2}^+(\mathbb{R}^{(p,q)})$ and $\mathrm{Gr}_{1,1}^+(\mathbb{R}^{(p,q)})$ to $\mathfrak{so}(p,q)$.

Suppose first $p = 0$. To prove Assertion (1) of Theorem 2.1.1, we must show that $r \leq 2$ if $m = 5$, if $m = 6$, or if $m \geq 9$. We can apply Theorem 4.1.4 directly. Let $n = m - 2$. Let $x \in S^n \subset \mathbb{R}^{n+1}$. Let

$$\mathbb{R}^{n+2} = \mathbb{R}^{n+1} \oplus e_{n+2} \cdot \mathbb{R}.$$

The map $x \to \mathrm{span}\{x, e_{n+2}\}$ defines an inclusion $i : S^n \to \mathrm{Gr}_2^+(\mathbb{R}^m)$. Let R be an algebraic curvature tensor on $\mathbb{R}^m$ so that $R(\pi)$ has constant rank r for any $\pi \in \mathrm{Gr}_2^+(\mathbb{R}^m)$. Then

$$T(x) := R(\mathrm{span}\{x, e_m\})$$

defines an odd map of constant rank r from S^n to $\mathfrak{so}(m)$. Thus, by Assertions (1) and (2) of Theorem 4.1.4, we have that $r \leq 2$ if $n \geq 8$ or, equivalently, if $m \geq 10$. We use Assertion (3) of Theorem 4.1.4 to see that $r \leq 2$ if $m = 5, 6, 9$ and complete the proof of Assertion (1) of Theorem 2.1.1.

Next, suppose $p > 0$. Since the metric is indefinite, we can not use Theorem 4.1.4 directly. We apply Lemma 1.2.13 to find a self-adjoint linear map ψ of $\mathbb{R}^{(p,q)}$ so that

$$(v_1, v_2)_e := (v_1, \psi v_2)$$

is a positive definite inner product on $\mathbb{R}^{(p,q)}$ and so the map $\mathcal{T} \to \psi\mathcal{T}$ defines a linear isomorphism from $\mathfrak{so}(p,q)$ to $\mathfrak{so}(m)$.

Let $\{e_1^-, ..., e_p^-, e_1^+, ..., e_q^+\}$ be an orthonormal basis for $\mathbb{R}^{(p,q)}$, where

$$V^- := \operatorname{span}\{e_1^-, ..., e_p^-\} \text{ and } V^+ := \operatorname{span}\{e_1^+, ..., e_q^+\}$$

are a maximal timelike subspace and a complementary maximal spacelike subspace. If $x \in S(V^+)$, we define

$$T_1(x) := \psi(R(\operatorname{span}\{x, e_1^-\})) \text{ mapping } S(V^+) \text{ to } \mathfrak{so}(p+q).$$

Since R has constant mixed rank r, T_1 an odd map of constant rank r.

Suppose that $p = 1$ and $q - 1 \geq 8$. We apply Theorem 4.1.4 to see that $r \leq 2$; this completes the proof of Assertion (2) of Theorem 2.1.1.

Suppose that $p = 2$ and $q - 1 \geq 9$. We apply Theorem 4.1.4 to see that $r \leq 4$. Furthermore, if $r = 4$, then $q - 1$ is odd and hence q is even. The inclusion i of V^+ in $\mathbb{R}^{(p,q)}$ defines a natural inclusion

$$i : Gr_2^+(\mathbb{R}^q) = \operatorname{Gr}_2^+(V^+) \to \operatorname{Gr}_{0,2}^+(\mathbb{R}^{(p,q)}).$$

We define $T_2(\pi) := \psi \cdot R(i(\pi))$. Then T_2 has constant rank $r = 4$. Thus, we may apply Theorem 4.1.4 (6) to complete the proof of Assertion (3) of Theorem 2.1.1 to see that either q or $q + 2$ are powers of 2. □

4.2 Fiber bundles

The theory of fiber bundles is central to many mathematical developments. We refer to [53, 117, 139, 147, 159] for further details concerning the material of this section as it is expository in nature.

Let E and M be smooth manifolds. We say that a smooth surjective map π from E to M is a *fiber bundle* with fiber F if there is an open cover $\{\mathcal{O}_i\}$ of M and diffeomorphisms ϕ_i from $\pi^{-1}(\mathcal{O}_i)$ to $\mathcal{O}_i \times F$ which intertwine π and projection on the first factor - i.e. we have the following commutative diagram:

$$\begin{array}{ccc} \pi^{-1}(\mathcal{O}_i) & \xrightarrow{\phi_i} & \mathcal{O}_i \times F \\ \pi \searrow & \circ & \swarrow \pi_1 \\ & \mathcal{O}_i & \end{array}.$$

The pair $(\mathcal{O}_i, \phi_i)$ is said to *trivialize the fiber bundle* over $\mathcal{O}_i$. Let $\text{Diff}(F)$ be the group of diffeomorphisms of the fiber F. Over the intersection $\mathcal{O}_i \cap \mathcal{O}_j$, let $\phi_{ij} := \phi_i \phi_j^{-1}$. Then we have a commutative diagram:

$$\begin{array}{ccc} (\mathcal{O}_i \cap \mathcal{O}_i) \times F & \xrightarrow{\phi_{ij}} & (\mathcal{O}_i \cap \mathcal{O}_j) \times F \\ \pi_1 \searrow & \circ & \swarrow \pi_1 \\ & \mathcal{O}_i \cap \mathcal{O}_j & \end{array} .$$

Consequently, the map ϕ_{ij} is *fiber preserving* and we may express:

$$\phi_{ij}(\cdot, f) = (\cdot, \Phi_{ij}(\cdot)f), \text{ where } \Phi_{ij} : \mathcal{O}_i \cap \mathcal{O}_j \to \text{Diff}(F).$$

The $\{\Phi_{ij}\}$ are said to be the *transition diffeomorphisms* of the fiber bundle; they satisfy the *cocycle condition*

$$\text{(4.2.1.a)} \qquad \Phi_{ii} = \text{Id} \text{ and } \Phi_{ij}\Phi_{jk}\Phi_{ki} = \text{Id}$$

on the appropriate domains of definitions. If $F = \mathbb{R}^n$ and if the Φ_{ij} take values in the general linear group $\text{Gl}(n, \mathbb{R})$, then π defines a real *vector bundle*; the notion of a complex vector bundle is defined similarly. Conversely, given an open cover $\mathcal{O}_i$ of M and smooth maps $\Phi_{ij} : \mathcal{O}_i \cap \mathcal{O}_j \to \text{Gl}(n, \mathbb{R})$ satisfying the relations of Display (4.2.1.a), then we can define a smooth real vector bundle V over M by using the Φ_{ij} to glue the local trivializations together:

$$\begin{aligned} &V := \dot{\sqcup}\, \mathcal{O}_i \times \mathbb{R}^n / \sim, \text{ where} \\ &(Q_i, w_i) \sim (Q_j, w_j) \text{ if } Q_i = Q_j = Q \in \mathcal{O}_i \cap \mathcal{O}_j \text{ and } w_i = \Phi_{ij}(Q)w_j. \end{aligned}$$

If the fiber is a Lie group G and if the Φ_{ij} are defined by left multiplication in the group, then we can define an associated *principle bundle* with *structure group* G by using the Φ_{ij} to glue the local trivializations together. Since left and right multiplication commute, we have a natural right fiber preserving G action on E so that $E/G = M$.

We say that two fiber bundles $\pi : E \to M$ and $\tilde{\pi} : \tilde{E} \to M$ are said to be *isomorphic fiber bundles* if there exists a diffeomorphism $\Psi : E \to \tilde{E}$ which intertwines the two projections - i.e. Ψ is *fiber preserving.* We then have a commutative diagram:

$$\begin{array}{ccc} E & \xrightarrow{\Psi} & \tilde{E} \\ \pi \searrow & \circ & \swarrow \tilde{\pi} \\ & M & \end{array} .$$

If V and $\tilde{V}$ are real or complex vector bundles, and if Ψ is linear on the fibers, then V and $\tilde{V}$ are said to be *isomorphic vector bundles.* If P and $\tilde{P}$ are principle bundles with the same structure group G, and if Ψ commutes with the G actions, then P and $\tilde{P}$ are said to be *isomorphic principle bundles.*

We say that a smooth surjective map π from E to M is a *submersion* if $\pi_* : T_eE \to T_{\pi e}M$ is surjective for every point $e \in E$. We may then use the implicit function Theorem to see that the *fibers* $F_P := \pi^{-1}\{P\}$ are smooth submanifolds of E for any $P \in M$. If π is a submersion, then the *vertical subspace* is defined by:

$$\mathcal{V}_e := \ker \pi_* : T_eE \to T_{\pi e}M.$$

Let g_E and g_M be pseudo-Riemannian metrics on E and M. Assume that $g_E|_\mathcal{V}$ is non-degenerate and define the *horizontal subspace* by:

$$\mathcal{H}_e := \mathcal{V}_e^\perp.$$

We say that π is a *pseudo-Riemannian submersion* if $\pi_* : \mathcal{H}_e \to T_{\pi e}M$ is an isometry for all $e \in E$. The *horizontal lift* operator $\mathcal{H}$ applied to a vector field Y on M is the vector field $\mathcal{H}Y$ on E which is characterized by the two properties:

$$\pi_*\{\mathcal{H}Y(e)\} = Y(\pi e) \text{ and } \mathcal{H}Y(e) \perp \mathcal{V}_e.$$

If Φ is the flow for $\mathcal{H}Y$ on E and if ϕ is the flow for Y on M, then π intertwines Φ and ϕ, or more precisely:

$$\pi\Phi(e,t) = \phi(\pi e, t).$$

It is clear that if $\pi : E \to M$ is a fiber bundle, then π is a submersion. The converse implication is valid if the fibers are compact.

4.2.2 Lemma. *Let $\pi : E \to M$ be a submersion.*

(1) *Let g_M be a Riemannian metric on M. There exists a Riemannian metric g_E on E so that $\pi : (E, g_E) \to (M, g_M)$ is a Riemannian submersion.*

(2) *If all the fibers F_P are compact and if M is connected, then all the fibers are diffeomorphic and π defines a fiber bundle.*

Proof. We use a partition of unity to put a Riemannian metric on E. We may then define the horizontal distribution $\mathcal{H} := \mathcal{V}^\perp$ and split $TM = \mathcal{V} \oplus \mathcal{H}$. We then have $\pi_* : \mathcal{H}_e \to TM_{\pi e}$ is an isomorphism for any $e \in E$. We define a new Riemannian metric on E such that π will be a Riemannian submersion by setting:

$$ds^2_\mathcal{H} = \pi_*^{-1} ds^2_M \text{ and } ds^2_E := ds^2_\mathcal{V} \oplus ds^2_\mathcal{H}.$$

Let $m := \dim(M)$ and let $P_0 \in M$. Let $\mathcal{U} \subset \mathbb{R}^m$ be a coordinate neighborhood so that P_0 is the origin in $\mathcal{U}$. Let $F_0 := \pi^{-1}P_0$ be the fiber over P_0. If $y \in \mathbb{R}^m$, then y determines a constant vector field Y on $\mathcal{U}$ and the horizontal lift $\mathcal{H}(Y)$ determines a smooth vector field on $\pi^{-1}(\mathcal{U})$. Let $\phi(e,t,y)$ be the flow of this vector field; ϕ is defined on a neighborhood of $F_0 \times \{0\} \times \{0\}$ in $F_0 \times \mathbb{R} \times \mathbb{R}^m$ and is characterized by the properties:

$$\phi(e,0,y) = e \text{ and } \phi_*(\partial_t) = \mathcal{H}Y(\phi(e,t,y)).$$

We use rescaling to see that $\phi(e,ts,y) = \phi(e,t,sy)$. Since the fiber F_0 is compact, there exists $\varepsilon > 0$ so that $\phi(e,t,y)$ is well defined for

$$e \in F_0,\ t \in [-1,1] \text{ and } |y| \in B_\varepsilon := \{y \in \mathbb{R}^m : |y| < \varepsilon\}.$$

Let $T(y,e) := \phi(e,1,y)$ for $y \in B_\varepsilon$ and $e \in F_0$. If ε is sufficiently small, then T provides a diffeomorphism from $B_\varepsilon \times F_0$ to $\pi^{-1}(B_\varepsilon)$. Since $\pi_*\mathcal{H}Y$ is the constant vector field Y,

$$\pi T(ty,e) = \pi\phi(e,1,ty) = \pi\phi(e,t,y) = ty.$$

This defines the flow for Y which starts at the origin. Therefore, T intertwines π and projection on the first factor and provides the desired local trivialization. This shows all the fibers F_P are diffeomorphic for $P \in B_\varepsilon$. Since M is assumed to be connected, all the fibers are diffeomorphic. □

Let $\pi : E \to M$ be a fiber bundle with fiber F and let $\psi : N \to M$ be a smooth map. Define the *pull-back fiber bundle*:

$$\psi^*E := \{(P,e) \in N \times E : \psi(P) = \pi e\}.$$

Let $\pi_1(P,e) := P$ and $\pi_2(P,e) = e$ be projection on the first and second factors. We have the *pushout diagram*:

$$\begin{array}{ccc} \psi^*E & \xrightarrow{\pi_2} & E \\ \downarrow \pi_1 & \circ & \downarrow \pi \\ N & \xrightarrow{\psi} & M \end{array}.$$

Let $(\mathcal{O}_i, \psi_i)$ be trivializations of E over M. Then

$$\psi^*(E)|_{\psi^{-1}(\mathcal{O})_i} = \psi^{-1}(\mathcal{O}_i) \times F$$

and, consequently, ψ^*E is a fiber bundle over N with the same fiber F. If Φ_{ij} are the transition cocycles of E, then $\psi^*\Phi_{ij} := \Phi_{ij} \circ \psi$ are the transition cocycles of ψ^*E. We note:

$$\mathrm{Id}^*(E) = E \text{ and } (\psi \circ \tilde{\psi})^*(E) = \tilde{\psi}^*(\psi^*(E)).$$

Two maps ψ_0 and ψ_1 from N to M are said to be *homotopic* if there is a smooth map

$$\Psi : N \times [0,1] \to M \text{ so } \psi(\cdot,0) = \psi_0(\cdot) \text{ and } \psi(\cdot,1) = \psi_1(\cdot).$$

We say that $\psi : N \to M$ is a *homotopy equivalence* if there exists a smooth map $\phi : M \to N$ so that $\phi \circ \psi$ is homotopic to the identity map on N and so that $\psi \circ \phi$ is homotopic to the identity map on M. The following is often called the *homotopy property for fiber bundles*:

4.2.3 Lemma. *Let $\pi : E \to M$ be a fiber bundle over M with fiber F. Let ψ_0 and ψ_1 be homotopic maps from N to M. Then $\psi_0^* E$ is isomorphic to $\psi_1^* E$. If E is a vector or principle bundle, then this isomorphism can be chosen to be a vector or principle bundle isomorphism.*

Proof. As this result is well known, in the interests of brevity we will only sketch the proof. Suppose that the fiber F is compact. Let $\tilde{N} := N \times [0,1]$. Let

$$\pi_1 : \tilde{N} \to N \text{ and } \pi_2 : \tilde{N} \to [0,1]$$

be the projections on the first and second factors. Let $\Psi : \tilde{N} \to M$ be a homotopy from ψ_0 to ψ_1, and let $\tilde{E} := \Psi^* E$. Let $\tilde{\pi} : \tilde{E} \to \tilde{N}$ be the natural projection. Let

$$ds^2_{\tilde{N}} := ds^2_N + dt^2$$

be a product Riemannian metric on $\tilde{N}$. We use Lemma 4.2.2 to put a Riemannian metric $g_{\tilde{E}}$ on $\tilde{E}$ so that $\tilde{\pi} : (\tilde{E}, g_{\tilde{E}}) \to (\tilde{N}, g_{\tilde{N}})$ is a Riemannian submersion. Let ∂_t be the coordinate vector field on $N \times [0,1]$. Let $\tilde{\phi}(\tilde{e}, t)$ be the flow for the horizontal lift $\mathcal{H}(\partial_t)$ on $\tilde{E}$. Then

$$\begin{aligned} &\tilde{\pi}\tilde{\phi}(\tilde{e}, t) = (\pi_1(\tilde{\pi}\tilde{e}), \pi_2(\tilde{\pi}\tilde{e}) + t) \text{ so} \\ &\tilde{\phi}(\tilde{e}, t) \in \tilde{\pi}^{-1}\{(\pi_1(\tilde{\pi}\tilde{e})) \times [0,1]\}. \end{aligned}$$

For fixed $\tilde{e}$, the flow takes values in a compact set and hence exists for all $0 \le t \le \pi_2(\tilde{\pi}(\tilde{e}))$. Thus, the map $\tilde{e} \to \tilde{\phi}(\tilde{e}, 1)$ defines the desired fiber preserving diffeomorphism from

$$\tilde{\pi}^{-1}(N \times \{0\}) = \psi_0^* E \text{ to } \tilde{\pi}^{-1}(N \times \{1\}) = \psi_1^* E.$$

The argument for general (F, N) involves showing there exists an open cover $\mathcal{O}_i$ of N so that $\tilde{E}$ is trivial over $\mathcal{O}_i \times [0,1]$. A partition of unity argument is then used to construct the desired diffeomorphism.

If $\tilde{E}$ is a vector bundle, then we use a partition of unity to put a positive definite inner product on the fibers of $\tilde{E}$ and to define a Riemannian connection ∇ on $\tilde{E}$. Parallel translation with respect to this connection along the lines $t \to (\cdot, t)$ in $\tilde{N}$ then defines the desired vector bundle isomorphism between $\tilde{E}|_{N\times 0}$ and $\tilde{E}|_{N\times 1}$. The construction is similar in the context of principle bundles. □

We say that M is *contractible* if the identity map on M is homotopic to the constant map or equivalently if there exists a map $\Psi : M \times [0,1] \to M$ so $\Psi(P,0) = P$ and $\Psi(P,1) = P_0$.

4.2.4 Lemma. *Let M be a contractible smooth manifold.*

(1) *If E is a fiber bundle over M, then E is a trivial fiber bundle.*

(2) *If V is a vector bundle over M, then V is a trivial vector bundle.*

(3) *If E is a principle bundle over M, then E is a trivial principle bundle.*

(4) *Suppose that E is a fiber bundle over M. Let $\tilde{E}$ be a fiber bundle over E. Suppose that $\tilde{E}$ is trivial over one fiber of E. Then $\tilde{E}$ is trivial over E.*

(5) *Let g be pseudo-Riemannian metric of signature (p,q) on M. There exists a normalized orthonormal frame $\mathcal{B} := \{e_1^-, ..., e_p^-, e_1^+, ..., e_q^+\}$ for TM.*

Proof. Let Id be the identity map on M and c be the constant map which sends every point of M to the basepoint, i.e. $c(P) = P_0$ for all $P \in M$. By assumption Id and c are homotopic. Let E be a fiber bundle over M with fiber $F = \pi^{-1}(P_0)$. Since Id and c are homotopic, $E = \mathrm{Id}^* E$ and $M \times F = c^* E$ are isomorphic. Assertions (1), (2), and (3) now follow.

Let E be a fiber bundle over M. By Assertion (1), $E = M \times F$ is trivial. Let π_2 be projection on the second factor. Since M is contractible, π_2 is a homotopy equivalence. Let $\tilde{E}$ be a fiber bundle over E. Then $\tilde{E}$ is isomorphic to $\pi_2^*(\tilde{E}|_F)$. If $\tilde{E}$ is trivial over the fiber, then $\tilde{E}$ is trivial over E; Assertion (4) now follows.

Let $O(TM)$ be the principle $O(p,q)$ bundle of normalized orthonormal frames of TM. We use Assertion (1) to see $O(TM) = M \times O(p,q)$ is trivial and thereby compete the proof of Assertion (4). □

Let V_1 and V_2 be two vector bundles over M. We say that $\psi : V_1 \to V_2$ is a vector bundle morphism if ψ is a smooth fiber preserving map which is linear on the fibers. We say that a vector bundle V_1 is a subbundle of a vector bundle V if the fibers of V_1 are linear subspaces of V and if the inclusion map from V_1 to V is a vector bundle morphism. The following observations concerning vector bundles are well known; we refer to [7, 33, 109] for details:

4.2.5 Lemma. *Let V_i be smooth vector bundles of dimensions k_i over a manifold M.*

(1) *The fiberwise direct sum and tensor product define smooth vector bundles $V_1 \oplus V_2$ and $V_1 \otimes V_2$ over M.*

(2) *Let $\psi : V_1 \to V_2$ be a vector bundle morphism. Suppose* $\operatorname{rank}\psi$ *is constant. Then* $\ker\psi$, $\operatorname{range}\psi$, $\frac{V_1}{\ker\psi}$, *and* $\frac{V_2}{\operatorname{range}\psi}$ *are smooth vector bundles over M.*

(3) *Suppose that $V_1 \subset \ell \cdot 1$ be a sub-bundle of the trivial bundle. Let $\phi(x) := \pi_1^{-1}(x) \in \mathrm{Gr}_{k_1}(\ell)$. Then the map $x \to \phi(x)$ is smooth.*

We remark that Lemma 4.2.5 continues to hold in the continuous (as opposed to the smooth) category.

A *section* s to a fiber bundle $\pi : E \to M$ is simply a map $s : M \to E$ so that $\pi \circ s = \mathrm{Id}_M$; all maps are assumed smooth unless specifically noted otherwise. (Using a partition of unity and the Stone–Weierstrass theorem one can show that every continuous map or section can be uniformly approximated by a smooth map or section).

Let V be a vector bundle over M of fiber dimension k. A local *frame* $\mathcal{B} := \{s_1, ..., s_k\}$ for V over an open set $\mathcal{O} \subset M$ is a collection of smooth sections defined over $\mathcal{O}$ so that $\{s_1(P), ..., s_k(P)\}$ are linearly independent elements of V_P for every $P \in \mathcal{O}$. Such local frames define local trivializations $\Phi_{\mathcal{B}} : \mathcal{O} \times \mathbb{R}^k \to \pi^{-1}(\mathcal{O})$:

$$\Phi_{\mathcal{B}}(P, \lambda) := \lambda_1 s_1(P) + ... + \lambda_k s_k(P).$$

On the other hand, if given a local trivialization $\Phi : \mathcal{O} \times \mathbb{R}^k \to \pi^{-1}(\mathcal{O})$, then we can define a local frame $\mathcal{B}_\Phi$ by defining:

$$s_i(P) := \Phi(P, (0, ..., 0, 1, 0, ..., 0)).$$

If we put a smooth positive definite inner product on the fibers of V, then there are some additional constructions which are useful; such a bundle is said to be a *Riemannian vector bundle* in the real setting or a *unitary vector bundle* in the complex setting; by an abuse of notation, we will often say that V is unitary even in the real setting to have a consistent notation. Given a local frame for V, we can use the *Gram–Schmidt process* to construct an orthonormal frame.

4.2.6 Lemma. *Let V be a smooth unitary vector bundle over M.*

(1) *Let V_1 be a smooth subbundle of V. Then $V_1^\perp$ is a smooth complementary subbundle of V.*

(2) *Let $\psi : V \to V$ be a vector bundle morphism which is self-adjoint on the fibers. Assume that $\dim\ker\psi$ is constant. For each point $P \in M$, we can diagonalize $\psi(P)$; let V_0, V_+, and V_- be the span of the eigenvectors corresponding to the zero, positive, and negative eigenvalues. Then V_0, V_+, and V_- are smooth orthogonal subbundles of V.*

Proof. Let $\dim(V_1) = k$. Let $\{e_1, ..., e_k\}$ be a local orthonormal frame for V_1 over an open set $\mathcal{O} \subset M$. Let

$$\pi(v) := (v, e_1)e_1 + ... + (v, e_k)e_k.$$

Then $v \to \pi(v)$ is a smooth vector bundle morphism from $V \to V$ over $\mathcal{O}$. Since π defines orthogonal projection on V_1, π is self-adjoint. Furthermore, $\dim\ker(\pi) = \dim V - k$ is constant. Thus, by Lemma 4.2.5, $V_1^\perp = \ker\pi$ is a smooth subbundle of V; Assertion (1) now follows.

We use the *Cauchy integral formula* to prove Assertion (2). To prove V_0, V_+, and V_- are smooth subbundles, we may work locally. Fix a point $P \in M$. We let $\mathcal{K}$ be a contractible compact subset of M with P in the interior of $\mathcal{K}$. We choose an orthonormal frame for V over $\mathcal{K}$ which trivializes V. Thus, we may assume without loss of generality $V = \mathcal{K} \times \mathbb{R}^k$; the complex case is similar. We may then regard $\psi(P) = \psi_{ij}$ as a smooth map from $\mathcal{K}$ to the space $\mathbb{M}_k(\mathbb{R})$ of real $k \times k$ matrices. By Lemma 1.3.4, the eigenvalues of $\psi(P)$ vary continuously with P. Since the rank of ψ is constant, eigenvalues do not cross the origin. Thus, we can choose $\varepsilon > 0$ so that

$$\operatorname{spec}(\psi)(P) \subset [-\varepsilon^{-1}, -\varepsilon] \,\dot\sqcup\, \{0\} \sqcup [\varepsilon, \varepsilon^{-1}].$$

We put curves γ_-, γ_0, and γ_+ in the complex plane $\mathbb{C}$ around these disjoint closed subsets of $\mathbb{R}$ which are oriented counterclockwise.

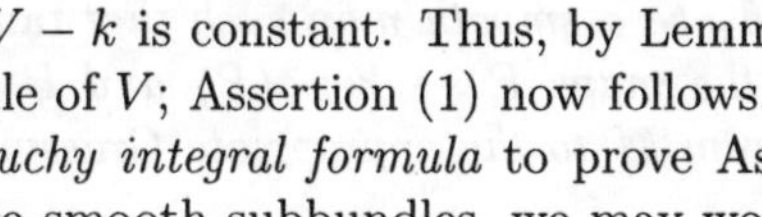

γ_- γ_0 γ_+

Let π^-, π^0, and π^+ be orthogonal projection on V_-, V_0, and V_+. We express these spectral projections in the form:

$$\begin{aligned}
\pi^- &:= \tfrac{1}{2\pi\sqrt{-1}} \textstyle\int_{\gamma_-} (\lambda - \psi)^{-1} d\lambda, \\
\pi^0 &:= \tfrac{1}{2\pi\sqrt{-1}} \textstyle\int_{\gamma_0} (\lambda - \psi)^{-1} d\lambda, \text{ and} \\
\pi^+ &:= \tfrac{1}{2\pi\sqrt{-1}} \textstyle\int_{\gamma_+} (\lambda - \psi)^{-1} d\lambda.
\end{aligned}$$

This shows the spectral projections are smooth. Since they have constant rank, Assertion (2) now follows from Lemma 4.2.5. □

Let $\mathbb{M}_k$ be the set of $m \times m$ matrices and let $\mathbb{S}_k$ be the set of $m \times m$ self-adjoint matrices. We omit the proof of the following result concerning the continuity of maps into Grassmannians in the interests of brevity.

4.2.7 Lemma. *Let M be a smooth connected manifold.*

(1) *Let $\phi : M \to \mathbb{S}_k$ be a smooth map so that the eigenvalues $\{\lambda_1, ..., \lambda_\ell\}$ of ϕ are constant. Then:*
   1a) *The multiplicities μ_i of each eigenvalue λ_i are constant.*
   1b) *The eigenspaces $\ker\{\phi(P) - \lambda_i\}$ form vector bundles over M.*
   1c) *The maps $\phi_i(P) := \ker\{\phi(P) - \lambda_i\} \in \mathrm{Gr}_{\mu_i}(\mathbb{R}^k)$ are smooth.*

(2) *Let ϕ_i be smooth maps from M to Grassmannians $Gr_{k_i}(\mathbb{R}^\ell)$. Suppose that $\dim\{\phi_1(P) \cap \phi_2(P)\} = r$ is constant on M. Then the map $\phi_1 \cap \phi_2 : M \to \mathrm{Gr}_r(M)$ is smooth.*

(3) *Let $\phi : M \to \mathbb{M}_k$ be a smooth map such that $\operatorname{rank} \phi(P)$ is constant on M. Then the maps $P \to \ker \phi(P)$ and $P \to \operatorname{range} \phi(P)$ are smooth maps from M to the appropriate Grassmannians.*

Again, we note that Lemma 4.2.7 holds in the continuous (as opposed to the smooth) category as well.

The homotopy groups $\pi_i(\cdot)$ of the fiber, total space, and base space are related by the *long exact sequence of a fibration*; we suppress the role of the basepoints in the interests of notational simplicity. We refer to Spanier [139, Theorem 7.2.10 page 377] for the proof of the following result:

4.2.8 Lemma. *Let $\pi : E \to M$ be a fiber bundle with fiber F. Assume F, E, and M are connected manifolds. Let $i : F \to E$ be the natural inclusion map. Let δ be the so-called connecting homomorphism. Then there is a natural long exact sequence of homotopy groups:*

$$...\pi_i(F) \xrightarrow{i_*} \pi_i(E) \xrightarrow{\pi_*} \pi_i(M) \xrightarrow{\delta} \pi_{i-1}(F)...$$

Morse theory shows that any smooth manifold is a CW-complex. The following result then follows from a more general result due to Whitehead - see Spanier [139, Corollary 24 page 405] for details.

4.2.9 Lemma. *Let $\psi : N \to M$ be a smooth map. Suppose that N and M are connected and that ψ_* is an isomorphism from $\pi_i(N)$ to $\pi_i(M)$ for all i. Then ψ is a homotopy equivalence.*

We follow the treatment in [24] and use Lemma 4.2.8 and Lemma 4.2.9 to study the homotopy type of the space forms $S^\pm(\mathbb{R}^{(p,q)})$, the homotopy type of the complex space forms $\mathbb{CP}^{(p,q)}$, and the homotopy type of the para-complex space forms $\widetilde{\mathbb{CP}}^p$ which were defined in Section 3.6:

Proof of Theorem 4.1.1. We must show that:

(1) $S^+(\mathbb{R}^{(p,q)})$ is homotopy equivalent to S^{q-1}.
(2) $\mathbb{CP}^{(p,q)}$ is homotopy equivalent to $\mathbb{CP}^{(0,q)}$.
(3) $\widetilde{\mathbb{CP}}^p$ is homotopy equivalent to S^p.

If v is spacelike (i.e. $(v,v) > 0$), then we may define

$$|v| := \sqrt{(v,v)} \in (0,\infty).$$

We then have $\frac{v}{|v|} \in S^+(V)$. Choose a normalized orthonormal basis

$$\begin{aligned} &\{e_1^-,...,e_p^-,e_1^+,...,e_q^+\} \text{ for } \mathbb{R}^{(p,q)}; \text{ set} \\ &V^- := \operatorname{span}\{e_1^-,...,e_p^-\} \text{ and } V^+ := \operatorname{span}\{e_1^+,...,e_q^+\}. \end{aligned} \tag{4.2.9.a}$$

We identify S^{q-1} with $S^+(\mathbb{R}^{(0,q)})$. The inclusion $i : V^+ \subset V^+ \oplus V^-$ induces a natural inclusion

$$i : S^{q-1} \subset S^+(\mathbb{R}^{(p,q)}). \tag{4.2.9.b}$$

Let $\pi^\pm$ be orthogonal projection on the subspaces $V^\pm$. If $v \in S^+(V)$, then

$$(\pi^+v,\pi^+v) \geq (\pi^+v,\pi^+v) + (\pi^-v,\pi^-v) = (v,v) = 1;$$

thus π^+v is spacelike. Define a retract r from $S^+(\mathbb{R}^{(p,q)})$ to S^{q-1} by:

$$r(v) := \frac{\pi^+v}{|\pi^+v|}. \tag{4.2.9.c}$$

Then $r \circ i$ is the identity on S^{q-1}. Since

$$\begin{aligned} &(tv + (1-t)\pi^+v, tv + (1-t)\pi^+v) = (\pi^+v,\pi^+v) + t^2(\pi^-v,\pi^-v) \\ \geq &(\pi^+v,\pi^+v) + (\pi^-v,\pi^-v) = (v,v) = 1, \end{aligned}$$

we have $tv + (1-t)\pi^+v$ is spacelike for $t \in [0,1]$. We show $i \circ r$ is homotopic to the identity on $S^+(\mathbb{R}^{(p,q)})$ and thereby complete the proof of Assertion (1) by defining the homotopy:

$$F(t,v) := \frac{tv+(1-t)\pi^+v}{|tv+(1-t)\pi^+v|}. \tag{4.2.9.d}$$

We may identify $\mathbb{R}^{(2r,2s+2)}$ with $\mathbb{C}^{(r,s+1)}$ and thereby choose the subspaces $V^{\pm}$ described in Display (4.2.9.a) to be complex. Then the maps

$$i : V^+ \to V^+ \oplus V^- \text{ and } \pi^+ : V^+ \oplus V^- \to V^+$$

are complex linear. Thus, the maps i and r of Equations (4.2.9.b) and (4.2.9.c) are equivariant with respect to the circle action and descend to maps

$$[i] : \mathbb{CP}^{(0,q)} \to \mathbb{CP}^{(p,q)} \text{ and } [r] : \mathbb{CP}^{(p,q)} \to \mathbb{CP}^{(0,q)}.$$

We have that $[r] \circ [i] = [r \circ i]$ is the identity on $\mathbb{CP}^{(0,q)}$ while the homotopy F of Equation (4.2.9.d) descends to a homotopy $[F]$ between the identity on $\mathbb{CP}^{(p,q)}$ and $[i] \circ [r]$. Assertion (2) now follows.

We use Lemma 4.2.8 and Lemma 4.2.9 to study the homotopy type of the para-complex projective spaces $\widetilde{\mathbb{CP}}^p$. We use the long exact sequence of the fibration

$$\tilde{S}^1 \to S^+(\tilde{\mathbb{C}}^{p+1}) \to \widetilde{\mathbb{CP}}^p$$

defined in Equation (3.6.5.b) to see we have a long exact sequence of homotopy groups

$$\ldots \pi_i(\tilde{S}^1) \to \pi_i(S^+(\mathbb{R}^{(p+1,p+1)})) \to \pi_i(\widetilde{\mathbb{CP}}^p) \to \pi_{i-1}(\tilde{S}^1) \ldots$$

The map $t \to \cosh(t) + i\sinh(t)$ identifies $\tilde{S}^1$ with $\mathbb{R}$. Thus, in particular $\pi_i(\tilde{S}^1) = 0$ for all i and thus π_* is an isomorphism from $\pi_i(S^+(\mathbb{R}^{(p+1,p+1)}))$ to $\pi_i(\widetilde{\mathbb{CP}}^p)$. Thus, by Lemma 4.2.9, π is a homotopy equivalence. Assertion (3) now follows from Assertion (1). □

We used the following result concerning equivariant vector fields on the sphere in our discussion of the Szabó operator in Section 3.8.

Proof of Theorem 4.1.2. We follow the argument of Szabó [150] to establish the following two facts:

(1) Let $\vec{s}$ be a continuous tangent vector field on S^{m-1}. Then there exists $x \in S^{m-1}$ so that $\vec{s}(x) = -\vec{s}(-x)$.

(2) Let $A(x)$ be a continuous map from S^{m-1} to the space of self-adjoint linear maps of $\mathbb{R}^m$. Assume that $A(-x) = -A(x)$, that $A(x)x = 0$, and that $\dim\ker(A)$ is constant on S^{m-1}. Then $A(x) = 0$ for all $x \in S^{m-1}$.

Suppose that Assertion (1) fails, i.e. that the vector field $\vec{s}(x) + \vec{s}(-x)$ never vanishes. Let

$$f(x) := \tfrac{\vec{s}(x)+\vec{s}(-x)}{|\vec{s}(x)+\vec{s}(-x)|}$$

be the corresponding unit tangent vector field on S^{m-1}. We note $f(x) \perp x$ for all $x \in S^{m-1}$. We show f is a degree 1 map by constructing the following homotopy f_ϵ connecting f to the identity map:

$$f_\epsilon(x) = \cos(\epsilon)f(x) + \sin(\epsilon)x.$$

On the other hand, as $f(x) = f(-x)$, f descends to induce a map $[f]$ from $\mathbb{RP}^{m-1}$ to S^{m-1}. This shows that the degree of f is even. This contradiction establishes Assertion (1).

Since $A(x)x = 0$, $A(x)$ preserves $x^\perp$. We let $\tilde{A}(x)$ denote the restriction of $A(x)$ to $x^\perp$. Since $\tilde{A}(x)$ is self-adjoint, $\tilde{A}$ is diagonalizable. We let $E_-(x)$, $E_0(x)$, and $E_+(x)$ denote the span of the eigenvectors with negative, zero, and positive eigenvalues. Since $\dim\ker A(x)$ is constant, we may use Lemma 4.2.5 to see that E_-, E_0, and E_+ are vector bundles over S^{m-1}. This gives a decomposition of the tangent bundle of the sphere in the form:

$$T(S^{m-1}) = E_- \oplus E_0 \oplus E_+.$$

Suppose that $\dim E_+ > 0$. Fix a point $x_0 \in S^{m-1}$. We apply Lemma 4.2.4. Since $S^{m-1} - \{x_0\}$ is contractible, the vector bundle E_+ is trivial over $S^{m-1} - \{x_0\}$ and we can choose a continuous unit section s_+ to E_+ on $S^{m-1} - \{x_0\}$. Let ψ_+ be a continuous function on S^{m-1} which vanishes only at x_0. Then $s_1 := \psi_+ s_+$ is a continuous section to E_+ which vanishes only at x_0. Since $A(-x) = -A(x)$, the section $s_2(x) := s_1(-x)$ is a continuous section to E_- which vanishes only at $-x_0$. As $A(x)$ is self-adjoint, $E_+ \perp E_-$. Let

$$s(x) := s_1(x) + s_2(x).$$

Since s_1 vanishes only at x_0 and s_2 vanishes only at $-x_0$, s is a nowhere vanishing vector field on S^{m-1}. Since $s(-x) = s(x)$, this contradicts Assertion (1). Thus, we conclude $\dim E_+ = 0$. Since $E_-(x) = E_+(-x)$, we also have $\dim E_- = 0$. Consequently, $A \equiv 0$. □

Under certain circumstances, the cohomology of the total space of a fiber bundle can be expressed in terms of the cohomology of the base and of the fiber. Let $\mathcal{R}$ be a principal ideal domain which we use define the cohomology groups $H^*(\cdot;\mathcal{R})$; in practice $\mathcal{R}$ will be $\mathbb{Z}_2$, $\mathbb{Z}_3$, or $\mathbb{Z}$. We refer to Spanier [139, Theorem 5.7.9, page 259] for the proof of the following result:

4.2.10 Theorem (Lerray–Hirsch). *Let $\pi : E \to M$ be a fiber bundle. Let i_P be the natural inclusion of the fiber $F_P := \pi^{-1}(P)$ in E. Assume that $H^*(F;\mathcal{R})$ is a free finitely generated $\mathcal{R}$ module. We suppose given classes*

$\{U_1, ..., U_\nu\}$ in $H^*(E;\mathcal{R})$ so that $\{i_P^*U_1, ..., i_P^*U_\nu\}$ is a free $\mathcal{R}$ module basis for the cohomology $H^*(F_P;\mathcal{R})$ of the fiber over every point P of M. Then:

(1) *π^* is an injective map from $H^*(M;\mathcal{R})$ to $H^*(E;\mathcal{R})$.*

(2) *$H^*(E;\mathcal{R})$ is a free $\pi^*H^*(M;\mathcal{R})$ module on the generators $\{U_i\}$.*

4.3 Characteristic classes and K-theory

In this section, we shall review some of the basic material concerning point set topology, characteristic classes, and K-theory which we shall need. As our purpose is primarily expository in this section, we shall refer to [7, 33, 109, 120] for further details and omit proofs of much of the standard material that we shall present.

We used several results from point set topology in Section 2.3; we state these results for completeness as follows. We refer to [12, Theorem 3.21] for the proof of Theorem 4.3.1 and to [139, Theorem 4.8.16] for the proof of Theorem 4.3.2.

4.3.1 Theorem. *Let f be a continuous bijective map from a compact space X to a Hausdorff space Y. Then f is a homeomorphism.*

4.3.2 Theorem [Invariance of domain]. *Let f be a continuous injective map from a manifold M of dimension m to a manifold N of dimension n. Then $m \le n$. If $m = n$, then f is an open map.*

Let $\mathbb{F} = \mathbb{R}$ or $\mathbb{F} = \mathbb{C}$ be either the real or the complex field. Let $\text{Vect}_{\mathbb{F}}^k(M)$ be the set of equivalence classes of $\mathbb{F}$ vector bundles over a smooth manifold M. Let $\text{Gr}_k(\mathbb{F}^n)$ be the Grassmannian manifold of all k planes in $\mathbb{F}^n$. The *classifying bundle* $\gamma_k \in \text{Vect}_{\mathbb{F}}^k(\text{Gr}_k(\mathbb{F}^n))$ and the orthogonal complement $\gamma_k^\perp \in \text{Vect}_{\mathbb{F}}^{n-k}(\text{Gr}_k(\mathbb{F}^n))$ are defined by:

$$\begin{aligned} &\gamma_k := \{(\pi, v) \in \text{Gr}_k(\mathbb{F}^n) \times \mathbb{F}^n : v \in \pi\} \text{ and} \\ &\gamma_k^\perp := \{(\pi, v) \in \text{Gr}_k(\mathbb{F}^n) \times \mathbb{F}^n : v \perp \pi\}. \end{aligned} \tag{4.3.3.a}$$

4.3.4 Lemma. *Let M be a smooth manifold of dimension m. There exists an integer $n = n(m, k, \mathbb{F})$ so that if $n \ge n(m, k, \mathbb{F})$, then:*

(1) *If $V \in \text{Vect}_{\mathbb{F}}^k(M)$, then there exists a smooth map $f : M \to Gr_k(\mathbb{F}^n)$ so that V is isomorphic to $f^*(\gamma_k)$.*

(2) *If $f_i : M \to Gr_k(\mathbb{F}^n)$ are two maps so that $f_1^*(\gamma_k)$ and $f_2^*(\gamma_k)$ are isomorphic vector bundles, then f_1 is homotopic to f_2.*

Let $[M, Gr_k(\mathbb{F}^n)]$ be the set of homotopy classes of maps from M to $Gr_k(\mathbb{F}^n)$. Theorem 4.3.4 shows that $\text{Vect}_{\mathbb{F}}^k(*)$ is a *representable functor*, i.e. the map $f \to f^*(\gamma_k)$ induces a natural equivalence of functors:

$$\text{Vect}_{\mathbb{F}}^k(*) = [*, Gr_k(\mathbb{F}^n)] \text{ for } n \geq n(m, k, \mathbb{F}) \text{ and } \dim(*) \leq m.$$

This focuses attention on the topology of the Grassmannians. We postpone until Lemma 4.3.12 the description of the cohomology rings of these spaces for general k. However if $k = 1$, then we may identify $Gr_1(\mathbb{F}^n)$ with the projective space $\mathbb{FP}^{n-1}$. We extend the natural inclusion

$$i(x_1, ..., x_n) := (x_1, ..., x_n, 0)$$

of $\mathbb{F}^n$ in $\mathbb{F}^{n+1}$ to an inclusion $[i]$ of $\mathbb{FP}^{n-1}$ in $\mathbb{FP}^n$. We describe the cohomology ring structure of the projective spaces as follows:

4.3.5 Lemma.

(1) $[i]^* : H^j(\mathbb{RP}^n; \mathbb{Z}_2) \to H^j(\mathbb{RP}^{n-1}; \mathbb{Z}_2)$ *is an isomorphism for* $j < n$.

(2) $[i]^* : H^j(\mathbb{CP}^n; \mathbb{Z}) \to H^j(\mathbb{CP}^{n-1}; \mathbb{Z})$ *is an isomorphism for* $j < 2n$.

(3) *There exist classes* $x_1 \in H^1(\mathbb{RP}^n; \mathbb{Z}_2)$ *and* $x_2 \in H^2(\mathbb{CP}^n; \mathbb{Z})$ *so that*

3a) $H^*(\mathbb{RP}^n; \mathbb{Z}_2) = \mathbb{Z}_2[x_1]/\{x_1^{n+1} = 0\}$.

3b) $H^*(\mathbb{CP}^n; \mathbb{Z}) = \mathbb{Z}[x_2]/\{x_2^{n+1} = 0\}$.

The *characteristic classes* are cohomological invariants of $\text{Vect}_{\mathbb{R}}^k(M)$ and $\text{Vect}_{\mathbb{C}}^k(M)$. We summarize their basic properties as follows:

4.3.6 Lemma (Stiefel-Whitney classes). *Let* M *be a smooth manifold. To every* $V \in \text{Vect}_{\mathbb{R}}^k(M)$, *we can associate a cohomology class* $w(V)$ *so that:*

(1) *We have* $w(V) = 1 + w_1(V) + ... + w_k(V)$, *where* $w_i(V) \in H^i(M; \mathbb{Z}_2)$.

(2) *If* $V_i \in \text{Vect}_{\mathbb{R}}^{k_i}(M)$, *then* $w(V_1 \oplus V_2) = w(V_1)w(V_2)$, *i.e.*
$w_i(V_1 \oplus V_2) = \sum_{i=j+k} w_j(V_1)w_k(V_2)$.

(3) *If* $f : N \to M$ *and if* $V \in \text{Vect}_{\mathbb{R}}^k(M)$, *then* $f^*(w(V)) = w(f^*(V))$.

(4) *The class* $w_1(\gamma_1)$ *generates* $H^1(\mathbb{RP}^n; \mathbb{Z}_2) = \mathbb{Z}_2$.

4.3.7 Lemma (Chern classes). *Let* M *be a smooth manifold. To every* $V \in \text{Vect}_{\mathbb{C}}^k(M)$, *we can associate a cohomology class* $c(V)$ *so that:*

(1) *We have* $c(V) = 1 + c_1(V) + ... + c_k(V)$, *where* $c_i(V) \in H^{2i}(M; \mathbb{Z})$.

(2) *If* $V_i \in \text{Vect}_{\mathbb{C}}^{k_i}(M)$, *then* $c(V_1 \oplus V_2) = c(V_1)c(V_2)$, *i.e.*
$c_i(V_1 \oplus V_2) = \sum_{i=j+k} c_j(V_1)c_k(V_2)$.

(3) *If* $f : N \to M$ *and if* $V \in \text{Vect}_{\mathbb{C}}^k(M)$, *then* $f^*(c(V)) = c(f^*(V))$.

(4) *The class* $c_1(\gamma_1)$ *generates* $H^2(\mathbb{CP}^n; \mathbb{Z}) = \mathbb{Z}$.

We have the following relationships:

4.3.8 Lemma. *Let $V \in \text{Vect}^k_{\mathbb{C}}(M)$, let $V^* \in \text{Vect}^k_{\mathbb{C}}(M)$ be the dual bundle, and let $V_{\mathbb{R}} \in \text{Vect}^{2k}_{\mathbb{R}}(M)$ be the underlying real vector bundle. Then*

(1) $c_i(V^*) = (-1)^i c_i(V)$.

(2) *$w_{2i}(V_{\mathbb{R}})$ is the mod 2 reduction of $c_i(V)$ and $w_{2i+1}(V_{\mathbb{R}}) = 0$.*

Let $V \in \text{Vect}^k_{\mathbb{F}}(M)$. The *associated projective bundle* $\mathbb{P}V$ is the space of lines in V; the fiber $\mathbb{P}V_x$ being just projective space on V_x. Let $\pi_P : \mathbb{P}V \to M$ be the natural projection. Let $\mathbb{L}_V$ be the associated line bundle over $\mathbb{P}V$;

$$\mathbb{L}_V = \{(\langle v\rangle, w) \in \mathbb{P}V \times V : \pi_P(\langle v\rangle) = \pi_V w \text{ and } w \in \langle v\rangle\}.$$

To have a consistent notation, we make the following definitions:

(1) Suppose $\mathbb{F} = \mathbb{R}$. Let $\mathcal{R} = \mathbb{Z}_2$ and let $cw_i(V) := w_i(V)$.

(2) Suppose $\mathbb{F} = \mathbb{C}$. Let $\mathcal{R} = \mathbb{Z}$ and let $cw_i(V) := c_i(V)$.

The following result follows from Theorem 4.2.10; see, for example, Equation (20.7) in [34]. It can be regarded as a generalization of Theorem 4.3.5.

4.3.9 Theorem. *Let $V \in \text{Vect}^k_{\mathbb{F}}(M)$.*

(1) *The map $\pi_P^* : H^*(M;\mathcal{R}) \to H^*(\mathbb{P}V;\mathcal{R})$ is injective.*

(2) *Let $x := cw_1(\mathbb{L}_V)$. Then $H^*(\mathbb{P}V;\mathcal{R})$ is a free $\pi_P^*\{H^*(M;\mathcal{R})\}$ module on the elements $\{1, x, ..., x^{k-1}\}$ with the algebra structure given by $x^k - x^{k-1}\pi_P^*\{cw_1(V)\} + ... + (-1)^k \pi_P^*\{cw_k(V)\} = 0$.*

Not every vector bundle splits as the direct sum of line bundles; see, for example, Theorem 4.3.13. However, the *splitting principle* allows us to reduce calculations in characteristic classes for general vector bundles to the case of direct sums of line bundles. Let $V \in \text{Vect}^k_{\mathbb{F}}(M)$. The *Flag bundle* $\text{Flag}(V)$ can be defined inductively:

(1) Suppose that $k = 1$. We set $\text{Flag}(V) = M$.

(2) Suppose that $k = 2$. We set $\text{Flag}(V) := \mathbb{P}V$.

(3) Suppose that $k \geq 3$. Let $M_1 := \mathbb{P}V$. Then $\mathbb{L}_V^\perp$ is a $k-1$ dimensional subbundle of $\pi_P^*(V)$ and we define $\text{Flag}(V) := \text{Flag}(\mathbb{L}_V^\perp)$.

Alternatively, we can define:

$$\text{Flag}(V)_x := \{0 \subset \sigma_1 \subset \sigma_2 ... \subset \sigma_{k-1} \subset \sigma_k = V_x\}$$

to be the set of complete k-flags where each σ_i is an i dimensional subspace of V_x. Let $\pi_F : \text{Flag}(V) \to M$ be the natural projection.

4.3.10 Theorem (Splitting principle). *Let* $V \in \mathrm{Vect}_{\mathbb{F}}^k(M)$. *Then:*

(1) $\pi_F^* : H^*(M;\mathcal{R}) \to H^*(\mathrm{Flag}(V);\mathcal{R})$ *is injective.*

(2) $\pi_F^*(V) = L_1 \oplus ... \oplus L_k$ *is the direct sum of line bundles over* $\mathrm{Flag}(V)$.

(3) $cw(\pi_F^*(V)) = \prod_i (1 + cw_1(L_i))$.

One can use Lemmas 4.3.4 and 4.3.10 to see that the Stiefel-Whitney classes and the Chern classes are uniquely determined by the properties of Lemmas 4.3.6 and 4.3.7.

We use tensor product to define a group structure on $\mathrm{Vect}_{\mathbb{F}}^1(M)$. The first Stiefel-Whitney class completely detects real line bundles and the first Chern class completely detects complex line bundles:

4.3.11 Lemma. *If* $\mathbb{F} = \mathbb{R}$, *then let* $\varepsilon = 1$. *If* $\mathbb{F} = \mathbb{C}$, *then let* $\varepsilon = 2$.

(1) *If* $V_i \in \mathrm{Vect}_{\mathbb{F}}^1(M)$, *then* $cw_1(V_1 \otimes V_2) = cw_1(V_1) + cw_1(V_2)$.

(2) *The map* $V \to cw_1(V)$ *defines a group isomorphism from* $\mathrm{Vect}_{\mathbb{F}}^1(M)$ *to* $H^\varepsilon(M;\mathcal{R})$ *which is a natural equivalence of functors.*

Since $\gamma_k \oplus \gamma_k^\perp$ is a trivial bundle over $Gr_k(\mathbb{R}^n)$, we have $w(\gamma_k)w(\gamma_k^\perp) = 1$. Thus, $w(\gamma_k^\perp) = w(\gamma_k)^{-1}$. We let $w(\gamma_k)^{-1} = \sum_i w_i^\perp(\gamma_k)$ be this formal element. For example, if $k = 2$, then we have:

$$\begin{aligned}
&w_1^\perp(\gamma_2) = w_1(\gamma_2),\\
&w_2^\perp(\gamma_2) = w_1(\gamma_2)^2 + w_2(\gamma_2),\\
&w_3^\perp(\gamma_2) = w_1(\gamma_2)^3, \text{ and}\\
&w_4^\perp(\gamma_2) = w_1(\gamma_2)^4 + w_1(\gamma_2)^2 w_2(\gamma_2) + w_2(\gamma_2)^2.
\end{aligned}$$

Since $\dim(\gamma_k^\perp) = n - k$, we have for dimensional reasons that

$$w_i^\perp(\gamma_k) = 0 \text{ for } i > n - k. \tag{4.3.11.a}$$

Borel [31] showed that the cohomology algebra of the Grassmannian is generated by the characteristic classes of the classifying bundle and that there are no other relations others than that imposed by Equation (4.3.11.a):

4.3.12 Theorem. *Let* $x_i := w_i(\gamma_k) \in H^i(\mathrm{Gr}_k(\mathbb{R}^n);\mathbb{Z}_2)$ *and let* $x_j^\perp$ *be formal classes defined by the relation* $(1 + x_1 + ... + x_k)(1 + x_1^\perp + ...) = 1$ *in the formal power series ring* $\mathbb{Z}_2[[x_1, ..., x_k]]$. *Let* $\mathcal{I}_{k,n-k}$ *be the ideal of* $\mathbb{Z}_2[[x_1, ..., x_k]]$ *which is generated by the classes* $x_j^\perp$ *for* $j > n - k$. *Then*

$$H^*(\mathrm{Gr}_k(\mathbb{R}^k);\mathbb{Z}_2) = \mathbb{Z}_2[[x_1, ..., x_k]]/\mathcal{I}_{k,n-k}.$$

The following result of Stong [149] concerning splittings of the canonical bundle over Grassmannians played a central role in the discussion in Section 3.7 of the higher order Jacobi operator and is a very useful geometrical fact.

4.3.13 Theorem. *Let $2 \leq k \leq n-k$.*

(1) *The classifying bundle γ_k over $Gr_k(\mathbb{F}^n)$ does not contain a proper subbundle.*

(2) *The orthogonal complement $\gamma_k^\perp$ over $Gr_k(\mathbb{F}^n)$ contains a proper subbundle if and only if $(k,n,\mathbb{F})$ is $(2,7,\mathbb{R})$ or $(3,8,\mathbb{R})$; in these special cases $\gamma_k^\perp$ only splits off a line bundle.*

We assumed that $k \leq n-k$ in Theorem 4.3.13 to ensure $\dim \gamma_k \leq \dim \gamma_k^\perp$. If $k=1$, then we are dealing with projective spaces. If $\mathbb{F}=\mathbb{R}$, then the dimensions of the maximal splittings is closely related to the result of Adams cited in Theorem 1.4.2. We recall the Adams number $\nu(m)$ which was defined in Display (1.4.1.a):

$$\nu(a2^s) = \nu(2^s) \text{ for } a \equiv 1 \bmod 2,$$
$$\nu(1)=0,\ \nu(2)=1,\ \nu(4)=3,\ \nu(8)=7, \text{ and } \nu(16\tilde{b}) = \nu(\tilde{b})+8.$$

4.3.14 Lemma.

(1) *Let $\gamma_1^\perp = V_1 \oplus V_2$ be a non-trivial decomposition of $\gamma_1^\perp$ over $\mathbb{RP}^{m-1}$. If $\dim(V_1) \leq \dim(V_2)$, then $\dim(V_1) \leq \nu(m)$.*

(2) *If $\nu(m) \geq 1$, we can decompose $\gamma_1^\perp = L_1 \oplus ... \oplus L_{\nu(m)} \oplus V_2$, where the L_i are suitably chosen line bundles.*

Proof. Let $\pi : S^{m-1} \to \mathbb{RP}^{m-1}$. Since $\pi^*(\gamma_1^\perp) = TS^{m-1}$, Assertion (1) follows from Theorem 1.4.2. To prove Assertion (2), we use Lemma 3.2.4 to construct an algebraic curvature tensor R on V so that $\tilde{\mathcal{J}}_R(x)$ has eigenvalues of multiplicity $\mu_0 \geq \mu_1 = ... = \mu_{\nu(m)} = 1$. Since $\tilde{\mathcal{J}}_R(-x) = \tilde{\mathcal{J}}_R(x)$, this descends to define the desired decomposition of $\gamma_1^\perp$ over $\mathbb{RP}^{m-1}$. □

There is an analogous theorem in the complex setting that is due to Glover, Homer, and Stong [89]; Theorem 4.1.3 is a generalization of this result.

4.3.15 Theorem. *Let γ_1 be the classifying line bundle over $\mathbb{CP}^{n-1}$.*

(1) *If n is odd, then $\gamma_1^\perp$ is indecomposable.*

(2) *If n is even, then $\gamma_1^\perp$ at most splits off a line bundle and such a decomposition is possible.*

Steenrod squares will play a crucial role in what follows. They have the following properties; see Steenrod [148, page 1] for details.

4.3.16 Theorem.

(1) *For all integers* $i \geq 0$ *and* $n \geq 0$, Sq^i *is a natural transformation of functors from* $H^n(M;\mathbb{Z}_2)$ *to* $H^{n+i}(M;\mathbb{Z}_2)$ *which is a homomorphism.*

(2) *Let* $x \in H^n(M;\mathbb{Z}_2)$. *Then we have* $Sq^0(x) = x$, $Sq^n(x) = x^2$, *and* $Sq^i(x) = 0$ *for* $i > n$.

(3) *The total Steenrod square* $Sq = \sum_i Sq^i$ *is a unital ring homomorphism of* $H^*(M;\mathbb{Z}_2)$.

There are similar power operations on $H^*(M;\mathbb{Z}_p)$ when p is an odd prime. Again, we refer to Steenrod [148, page 76] for details.

4.3.17 Theorem. *Let* p *be an odd prime.*

(1) *For all integers* $i \geq 0$ *and* $n \geq 0$, $\mathcal{P}^i$ *is a natural transformation of functors from* $H^n(M;\mathbb{Z}_p)$ *to* $H^{n+2i(p-1)}(M;\mathbb{Z}_p)$ *which is a homomorphism.*

(2) *Let* $x \in H^n(M;\mathbb{Z}_p)$. *Then* $\mathcal{P}^0(x) = x$. *If* $n = 2i$, *then* $\mathcal{P}^i x = x^p$. *If* $2i > n$, *then* $\mathcal{P}^i(x) = 0$.

(3) *The total power operator* $\mathcal{P} := \sum_i \mathcal{P}^i$ *is a unital ring homomorphism of* $H^*(M;\mathbb{Z}_p)$.

If V is a complex vector bundle, then let V^* be the dual vector bundle. Let $\tilde{\mathcal{P}} := \sum_i (-1)^i \mathcal{P}^i$ be the alternating sum of the reduced $\mathbb{Z}_3$ power operations; the change of sign is crucial here but does not affect the fact that $\tilde{\mathcal{P}}$ is a unital ring homomorphism.

4.3.18 Lemma.

(1) *If* $V \in \mathrm{Vect}^k_{\mathbb{F}}(M)$, *then* $Sq(cw_k(V)) = cw(V)cw_k(V)$ *in* $H^*(M;\mathbb{Z}_2)$.

(2) *If* $V \in \mathrm{Vect}^k_{\mathbb{C}}(M)$, *then* $\tilde{\mathcal{P}}(c_k(V)) = c(V)c(V^*)c_k(V)$ *in* $H^*(M;\mathbb{Z}_3)$.

Proof. Since the secondary operations are natural with respect to pullback, we may use the splitting principle described in Theorem 4.3.10 to assume without loss of generality that V decomposes as a sum of line bundles. Since the Stiefel-Whitney and Chern classes are multiplicative with respect to products and since Sq and $\tilde{\mathcal{P}}$ are unital ring homomorphisms, we may assume without loss of generality that $k = 1$ so $V \in \mathrm{Vect}^1_{\mathbb{F}}(M)$. We use naturality and Lemma 4.3.4 to see that we may assume that $V = \gamma_1$ is the classifying line bundle over $\mathbb{FP}^n$ for some large n. Since $H^3(\mathbb{CP}^n;\mathbb{Z}_2) = 0$, we have that $sq_1(c_1(\gamma_1)) = 0$ in $H^3(\mathbb{CP}^n;\mathbb{Z}_2)$. We use Theorems 4.3.16 and

4.3.17 to compute:

$$\begin{aligned}
Sq(w_1(\gamma_1)) &= (\mathrm{Id} + sq_1)\{w_1(\gamma_1)\} = w_1(\gamma_1) + w_1(\gamma_1)^2 \\
&= (1 + w_1(\gamma_1))w_1(\gamma_1) \text{ if } \mathbb{F} = \mathbb{R}, \\
Sq(c_1(\gamma_1)) &= (\mathrm{Id} + sq_1 + sq_2)\{c_1(\gamma_1)\} = c_1(\gamma_1) + c_1(\gamma_1)^2 \\
&= (1 + c_1(\gamma_1))c_1(\gamma_1) \text{ if } \mathbb{F} = \mathbb{C}, \text{ and} \\
\tilde{\mathcal{P}}c_1(\gamma_1) &= (\mathrm{Id} - \mathcal{P}_1)\{c_1(\gamma_1)\} = c_1(\gamma_1) - c_1(\gamma_1)^3 \\
&= (1 + c_1(\gamma_1))(1 - c_1(\gamma_1))c_1(\gamma_1) \text{ if } \mathbb{F} = \mathbb{C}. \quad \square
\end{aligned}$$

Stavrov [146] has extended Stong's result as follows:

4.3.19 Lemma. *Let $V \in \mathrm{Vect}_{\mathbb{R}}^k(M)$. Suppose that $w_j(V) = 0$ for $j > j_0$. Then $Sq(w_{j_0}(V)) = w(V)w_{j_0}(V)$.*

Proof. Again, by using the splitting principle, we may suppose without loss of generality that we may decompose V as a direct sum of line bundles in the form: $V = L_1 \oplus ... \oplus L_k$. For any j and arbitrary V, $Sq_j(V)$ is a symmetric function of the variables $\{w_1(L_1), ..., w_1(L_k)\}$ and which can therefore be expressed in terms of the elementary symmetric functions. Thus, there exists a universal polynomial $F_{j,k} \in \mathbb{Z}_2[w_1, ..., w_k]$ so

$$Sq_j(V) = F_{j,k}(w_1(V), ..., w_k(V)).$$

We have assumed that $w_j(V) = 0$ for $j > j_0$. Thus,

$$Sq_j(V) = F_{j,k}(w_1(V), ..., w_{j_0}(V), 0, ..., 0).$$

We can determine this new polynomial by taking L_i to the classifying bundle over the ith factor of $\mathbb{RP}^\infty \times ... \times \mathbb{RP}^\infty$ for $i \leq j_0$ and by taking $L_i = \mathbf{1}$ for $i > j_0$. Lemma 4.3.19 then follows from Lemma 4.3.18. $\square$

Let $\mathrm{Vect}_{\mathbb{F}}(\cdot)$ be the set of isomorphism classes of $\mathbb{F}$ vector bundles of arbitrary dimensions:

$$\mathrm{Vect}_{\mathbb{F}}(\cdot) = \cup_{k=1}^{\infty} \mathrm{Vect}_{\mathbb{F}}^k(\cdot).$$

This inherits the natural binary operations $\oplus$ and $\otimes$. Cancellation fails in general; an isomorphism between $V \oplus V_1$ and $V \oplus \tilde{V}_1$ does not in general imply that V_1 is isomorphic to $\tilde{V}_1$. For example,

$$\mathbf{1} \oplus TS^m = T(\mathbb{R}^{m+1})|_{S^m} = \mathbf{1} \oplus m \cdot \mathbf{1}.$$

By Theorem 1.4.2, except for $m = 1, 3, 7$, TS^m is not isomorphic to $m \cdot \mathbf{1}$. To force $\text{Vect}_{\mathbb{F}}(\cdot)$ to become a ring, we must allow cancellation and formal differences. More precisely, let $K\mathbb{F}(\cdot)$ be the Grotendieck ring generated by $\text{Vect}_{\mathbb{F}}(\cdot)$. If $V \in \text{Vect}_{\mathbb{F}}(\cdot)$, then we let $[V]$ be the corresponding element of $K\mathbb{F}(\cdot)$. We have $[V_1] = [V_2]$ in $K\mathbb{F}(\cdot)$ if and only if there exists a vector bundle V so that $V \oplus V_1 = V \oplus V_2$. Thus, for example, $[TS^m] = m[\mathbf{1}]$ in $K\mathbb{R}(S^m)$. The map $V \to \dim(V)$ extends to a group homomorphism from $K\mathbb{F}(\cdot) \to \mathbb{Z}$ and the reduced K theory $\widetilde{K\mathbb{F}}(\cdot)$ is the kernel of this map.

There is a useful geometric characterization of $\widetilde{K\mathbb{F}}(\cdot)$. If $V \in \text{Vect}_{\mathbb{F}}^k(\cdot)$, then $V \oplus \mathbf{1} \in \text{Vect}_{\mathbb{F}}^{k+1}(\cdot)$ is called the *stabilization* of V. Let M be a smooth manifold of dimension m. We say that k is in the *stable range* if $k > m$ and $\mathbb{F} = \mathbb{R}$ or if $2k > m$ and $\mathbb{F} = \mathbb{C}$.

4.3.20 Lemma. *Let M be a smooth manifold of dimension m and let k be in the stable range.*

(1) *The map $V \to V \oplus \mathbf{1}$ is bijective from $\text{Vect}_{\mathbb{F}}^k(M)$ to $\text{Vect}_{\mathbb{F}}^{k+1}(M)$.*

(2) *The map $V \to [V] - k[\mathbf{1}]$ is bijective from $\text{Vect}_{\mathbb{F}}^k(M)$ to $\widetilde{K\mathbb{F}}(M)$.*

We can use Theorem 4.3.20 to describe the additive group structure on $\widetilde{K\mathbb{F}}(M)$ a bit more geometrically. Let k be in the stable range and let $V_i \in \text{Vect}_{\mathbb{F}}^k(M)$. Find $V_3 \in \text{Vect}_{\mathbb{F}}^k(M)$ so that $V_1 \oplus V_2 = V_3 \oplus k \cdot \mathbf{1}$. Then the induced addition on $\text{Vect}_{\mathbb{F}}^k(M)$ is defined by $V_1 + V_2 = V_3$, i.e.:

$$\{[V_1] - k[\mathbf{1}]\} + \{[V_2] - k[\mathbf{1}]\} = \{[V_3] - k[\mathbf{1}]\} \text{ in } \widetilde{K\mathbb{F}}(M).$$

Similarly, given $V_1 \in \text{Vect}_{\mathbb{F}}^k(M)$, we may find $V_2 \in \text{Vect}_{\mathbb{F}}^k(M)$ so that we have $V_1 \oplus V_2 = 2k \cdot \mathbf{1}$. The inverse is then given by $-V_1 = V_2$, i.e.

$$-\{[V_1] - k[\mathbf{1}]\} = \{[V_2] - k[\mathbf{1}]\} \text{ in } \widetilde{K\mathbb{F}(M)}.$$

Following Adams [1], we let $\phi(n)$ to be the number of integers s such that $1 \le s \le n$ and $s \equiv 0, 1, 2, 4 \bmod 8$. Then $\phi(n + 8) = 4 + \phi(n)$. We refer to [1, Theorem 7.4] for the proof of the following result:

4.3.21 Theorem (Adams). *The element $[\gamma_1] - [\mathbf{1}]$ has order $2^{\phi(n)}$ and generates $\widetilde{K\mathbb{R}}(\mathbb{RP}^n)$.*

Let $\mathfrak{so}(m)$ be the vector space of skew-symmetric linear maps of $\mathbb{R}^m$. We use Theorem 4.3.21 to complete the proof of Theorem 2.1.4 by proving the following slightly more general result.

4.3.22 Lemma. *Let $R : \mathbb{R}^m \times \mathbb{R}^m \to \mathfrak{so}(m)$ be a bilinear map where $m \geq 3$. Suppose there exists a vector $0 \neq v \in \mathbb{R}^m$ so that $R(x,y)v \neq 0$ for every orthonormal set $\{x,y\}$. Then $m = 4$ or $m = 8$.*

Proof. Define the map ψ from $S^{m-1} \times \mathbb{R}^m$ to $\mathbb{R}^m$ by:

$$\psi(x,y) := \{R(x,y) + (x,y)\,\mathrm{Id}\}v.$$

If $\psi(x,y) = 0$, then $R(x,y)v + (x,y)v = 0$. Consequently,

$$(R(x,y)v, v) + (x,y)(v,v) = 0.$$

As $R(x,y)$ takes values in $\mathfrak{so}(m)$, $(R(x,y)v,v) = 0$ and thus $(x,y)(v,v) = 0$ so $(x,y) = 0$. Thus, we have that $R(x,y)v = 0$ so $y = 0$. Therefore, the map $(x,y) \to (x, \psi(x,y))$ defines an injective linear map:

$$\Psi : m \cdot \mathbf{1} \to m \cdot \mathbf{1} \text{ over } S^{m-1}.$$

As $\Psi(-x,y) = -\Psi(x,y)$ is an odd map, Ψ descends to define a isomorphism

$$\psi : m \cdot \gamma_1 \to m \cdot \mathbf{1} \text{ over } \mathbb{RP}^{m-1}. \tag{4.3.22.a}$$

We set $n = m - 1$. Equation (4.3.22.a) and Theorem 4.3.21 then shows that $2^{\phi(n)}$ divides $n+1$. We have the following values of ϕ and n:

Table 4.3.23

n	1	2	3	4	5	6	7	8	9	10	11	12	13	14	15	16
$\phi(n)$	1	2	2	3	3	3	3	4	5	6	6	7	7	7	7	8

Since $2^{\phi(n)}$ is growing exponentially, $2^{\phi(n)} > n+1$ for $n \geq 8$. Thus, $2^{\phi(n)}$ divides $n+1$ only in the following two cases:

(1) If $n = 3$ and $m = 4$, then $2^{\phi(n)} = 4$ divides $n + 1 = 4$.

(2) If $n = 7$ and $m = 8$, then $2^{\phi(n)} = 8$ divides $n + 1 = 8$. □

The following technical observation will be useful in doing computations in $H^*(\mathbb{RP}^n; \mathbb{Z}_2)$. Let $c_\nu(r) \in \{0,1\}$ be the coefficient of 2^ν in the 2-adic expansion of r, i.e.

$$r = \textstyle\sum_\nu c_\nu(r) 2^\nu.$$

4.3.24 Lemma. *The coefficient of x^i is non-zero in $(1+x)^r$ in $\mathbb{Z}_2[x]$ if and only if all 2-adic coefficients of i which are non zero are also non-zero in r, i.e. if $c_\nu(i) \neq 0$, then $c_\nu(r) \neq 0$.*

Proof. Since $(1+x)^{2^s} = 1 + x^{2^s}$, we have that:

$$(1+x)^r = \textstyle\prod_{\nu : c_\nu(r) \neq 0} (1 + x^{2^\nu}).$$

The Lemma now follows. □

4.4 Symmetric vector bundles

In this section, we shall complete the proof of Theorem 4.1.3. We begin by establishing some notational conventions. Let $\mathbb{F}$ denote either the field of real numbers ($\mathbb{F} = \mathbb{R}$) or the field of complex numbers ($\mathbb{F} = \mathbb{C}$). Let

$$\mathbf{1}^n := \mathbb{FP}^{n-1} \times \mathbb{F}^n$$

be the trivial n plane bundle over the projective space $\mathbb{FP}^{n-1}$ of lines in $\mathbb{F}^n$. Let $V \in \text{Vect}_{\mathbb{F}}^k(\mathbb{FP}^{n-1})$ be a subbundle of $\mathbf{1}^n$ of dimension k, where $0 < k < n$. Let $V^\perp \subset \mathbf{1}^n$ be the complementary bundle. We say that V is *geometrically symmetric* if $\sigma \subset V(\tau)$ implies $\tau \subset V(\sigma)$ for any $\sigma, \tau \in \mathbb{FP}^{n-1}$. Let

$$\mathcal{R} = \mathbb{Z}_2 \text{ and } cw(V) = w(V) \in H^*(\mathbb{RP}^{n-1}; \mathcal{R}) \text{ if } \mathbb{F} = \mathbb{R},$$
$$\mathcal{R} = \mathbb{Z} \text{ and } cw(V) = c(V) \in H^*(\mathbb{CP}^{n-1}; \mathcal{R}) \text{ if } \mathbb{F} = \mathbb{C}.$$

Let $x = cw_1(\gamma_1) \in H^*(\mathbb{FP}^{n-1}; \mathcal{R})$, where $\gamma_1 \in \text{Vect}_{\mathbb{F}}^1(\mathbb{FP}^{n-1})$ is the classifying line bundle. Then

$$H^*(\mathbb{FP}^{n-1}; \mathcal{R}) = \mathcal{R}[x]/x^n = 0.$$

We can regard $cw(V)$ as an element of $\mathcal{R}[x]$; as $0 < k < n$, the truncating relation $x^n = 0$ plays no role.

Let $P \in \mathcal{R}[x]$ be a polynomial of degree k with $P(0) = 1$. We decompose

$$P(x) = a_0 + a_1 x + ... + a_k x^k \text{ where } a_0 = 1.$$

We say that P is *algebraically symmetric* if

$$(a_0, ..., a_k) = \pm(a_k, ..., a_0) \text{ i.e. } x^k P(\tfrac{1}{x}) = \pm P(x).$$

4.4.1 Lemma. *If $V \subset \mathbf{1}^n$ is a geometrically symmetric k plane bundle over $\mathbb{FP}^{n-1}$ for $0 < k < n$, then $cw(V)$ is algebraically symmetric and in particular $cw_k(V) = \pm x^k$ in $H^k(\mathbb{FP}^{n-1}; \mathcal{R})$.*

Proof. Suppose first that $k = 1$. Let $\phi(\sigma) = V(\sigma)$ define a continuous map from $\mathbb{FP}^{n-1}$ to itself. As V is symmetric, $\phi^2 = \text{Id}$ and $V = \phi^*(\gamma_1)$. Thus:

$$cw(V) = 1 + \phi^* x = 1 \pm x.$$

We therefore suppose that $2 \le k \le n-1$. Let $\mathbb{P}(V)$ and $\mathbb{L}_V$ be the associated projective and line bundles defined in Section 4.3:

$$\mathbb{P}(V) := \{(\sigma, \tau) \in \mathbb{FP}^{n-1} \times \mathbb{FP}^{n-1} : \tau \subset V(\sigma)\} \text{ and}$$
$$\mathbb{L}(V) := \{(\sigma, \tau, z) \in \mathbb{FP}^{n-1} \times \mathbb{FP}^{n-1} \times \mathbb{F}^n : z \in \tau \subset V(\sigma)\}.$$

Denote the two canonical projections from $\mathbb{P}(V)$ to $\mathbb{FP}^{n-1}$ by

$$\pi_1(\sigma,\tau)=\sigma \text{ and } \pi_2(\sigma,\tau)=\tau.$$

Let $x_i := \pi_i^* x \in H^*(\mathbb{P}(V);\mathcal{R})$. Since $\mathbb{L}_V = \pi_2^*(\gamma_1)$, $x_2 = cw_1(\mathbb{L}_V)$. Thus, by Theorem 4.3.9 we have that:

$$\text{(4.4.1.a)}\qquad \begin{array}{c}\pi_1^* \text{ is injective from } H^*(\mathbb{FP}^{n-1};\mathcal{R}) \text{ to } H^*(\mathbb{P}(V);\mathcal{R}),\\ H^*(\mathbb{P}(V);\mathcal{R}) \text{ is a free } \pi_1^*\{H^*(\mathbb{FP}^{n-1};\mathcal{R})\}\\ \text{module on the set } \{1,x_2,...,x_2^{k-1}\}, \text{ and}\\ 0 = x_2^k - x_2^{k-1}\pi_1^*\{cw_1(V)\} + ... + (-1)^k\pi_1^*\{cw_k(V)\}.\end{array}$$

We expand $cw(V)=\sum_i a_i x^i$, where $a_0 = 1$ and rewrite the final identity of Display (4.4.1.a) in the form:

$$\text{(4.4.1.b)}\qquad a_0x_2^k - a_1x_2^{k-1}x_1 + ... + (-1)^k a_k x_1^k = 0.$$

As V is a geometrically symmetric vector bundle, the map $\phi(\sigma,\tau) := (\tau,\sigma)$ which interchanges the factors is an involution of $\mathbb{P}(V)$ which intertwines the projections π_1 and π_2. We apply ϕ^* to Equation (4.4.1.b) to see:

$$\text{(4.4.1.c)}\qquad a_0x_1^k - a_1x_1^{k-1}x_2 + ... + (-1)^k a_k x_2^k = 0.$$

We use Display (4.4.1.a) to see that Equations (4.4.1.b) and (4.4.1.c) are linear multiples of each other so

$$(a_0,-a_1,a_2,...,(-1)^k a_k) = c(a_k,-a_{k-1},a_{k-2},...,(-1)^k a_0).$$

Since $a_0 = 1$, c is a unit in $\mathcal{R}=\mathbb{Z}$ or $\mathcal{R}=\mathbb{Z}_2$ so $c=\pm 1$. □

Lemma 4.4.1 shows that $cw_k(V)=\pm x^k$ if V is geometrically symmetric. Assertion (1) of Theorem 4.1.3 will therefore follow from:

4.4.2 Lemma. *Let $V \subset \mathbf{1}^n$ be a k plane bundle over $\mathbb{FP}^{n-1}$ for $0<k<n$. Suppose that $cw_k(V)=\pm x^k$ in $H^*(\mathbb{FP}^{n-1};\mathcal{R})$.*

(1) *We have $cw(V)cw(V^\perp)=1\pm x^n$ in $\mathcal{R}[x]$.*

(2) *If $2k\le n$, then $cw(V)=(1+x)^k$ in $\mathbb{Z}_2[x]$.*

Proof. We may expand:

$$cw(V)=\textstyle\sum_{i\le k} a_i x^i \text{ and } cw(V^\perp)=\sum_{i\le n-k} a_i^\perp x^i.$$

Assume that $a_k = \pm 1$. If $\mathbb{F} = \mathbb{R}$, take the prime $p = 2$; otherwise, let the prime p be arbitrary and reduce mod p. We have

$$cw(V)cw(V^{\perp}) = 1 \text{ in } \mathbb{Z}_p[x]/(x^n = 0).$$

If $a_{n-k}^{\perp} = 0$ in $\mathbb{Z}_p$, then the truncating relation $x^n = 0$ plays no role so:

$$cw(V)cw(V^{\perp}) = 1 \text{ in } \mathbb{Z}_p[x]. \tag{4.4.2.a}$$

Since $a_k = \pm 1$ and $a_0^{\perp} = 1$, we have that:

$$\text{degree}_{\mathbb{Z}_p[x]}\{c(V)c(V^{\perp})\} \geq \text{degree}_{\mathbb{Z}_p[x]}\{c(V)\} \geq k.$$

Thus, Equation (4.4.2.a) is false and we may conclude that $a_{n-k}^{\perp}$ is a unit in $\mathbb{Z}_p$. Suppose that $\mathbb{F} = \mathbb{R}$. Since $p = 2$ and $a_{n-k}^{\perp}$ is a unit in $\mathcal{R} = \mathbb{Z}_2$, $a_{n-k}^{\perp} = 1$. Suppose that $\mathbb{F} = \mathbb{C}$. Since the reduction of $a_{n-k}^{\perp}$ is a unit in $\mathbb{Z}_p$ for any p, $a_{n-k}^{\perp} = \pm 1$ in $\mathcal{R} = \mathbb{Z}$. Assertion (1) follows since

$$a_k a_{n-k}^{\perp} = \pm 1 \text{ and}$$
$$cw(V)cw(V^{\perp}) = 1 \text{ in } H^*(\mathbb{FP}^{n-1}; \mathcal{R}) = \mathcal{R}[x]/(x^n = 0).$$

Since $cw_k(V) = x^k$ in $H^*(\mathbb{FP}^{n-1}; \mathbb{Z}_2)$, by Lemma 4.3.18 we have:

$$x^k \cdot cw(V) = Sq(x^k) = x^k(1+x)^k \text{ in } \mathbb{Z}_2[x]/(x^n = 0).$$

Since $2k \leq n$, we can divide by x^k to see that

$$cw(V) = (1+x)^k \text{ in } \mathbb{Z}_2[x]/(x^k = 0).$$

Since $cw_k(V) = x^k$ in $H^*(\mathbb{FP}; \mathbb{Z}_2)$, the truncation plays no role and

$$cw(V) = (1+x)^k \text{ in } \mathbb{Z}_2[x]. \quad \square$$

The following is a useful technical observation:

4.4.3 Lemma. *Let $n = a2^s$ for a odd.*

(1) *If $(1+x)^k$ divides $1+x^n$ in $\mathbb{Z}_2[x]$, then $k \leq 2^s$.*
(2) *The polynomial $1 + x^{2^s}$ is irreducible in $\mathbb{Z}[x]$.*

Proof. To prove Assertion (1), we factor

$$1 + x^n = (1 + x^{2^s})q(x) \text{ in } \mathbb{Z}_2[x] \text{ where}$$
$$q(x) := (1 + x^{2^s} + x^{2\cdot 2^s} + x^{3\cdot 2^s} + ... + x^{(a-1)\cdot 2^s}).$$

We have $q(0) = 1$ and $q(1) = a = 1$ since a is odd. Thus, q has no roots in $\mathbb{Z}_2$. Since $(1+x)^k$ divides $1 + x^n$, we see that $(1+x)^k$ divides $(1 + x^{2^s}) = (1+x)^{2^s}$ in $\mathbb{Z}_2[x]$ and hence $k \leq 2^s$.

Suppose we have non-trivial factorization

$$1 + x^{2^s} = p_1(x)p_2(x) \text{ in } \mathbb{Z}[x]$$

as the product of two polynomials of positive degrees a_i with $a_1 + a_2 = 2^s$. Let $\tilde{p}_i$ be the mod 2 reduction of p_i. Since

$$(1+x)^{2^s} = \tilde{p}_1(x)\tilde{p}_2(x) \text{ in } \mathbb{Z}_2[x]$$

we must have degree$(\tilde{p}_1)$ + degree$(\tilde{p}_2) = 2^s$ and thus $\tilde{p}_i(x) = (1+x)^{a_i}$ in $\mathbb{Z}_2[x]$. Since $a_i > 0$, $\tilde{p}_i(1) = 0$ in $\mathbb{Z}_2$ for $i = 1, 2$. Thus, $p_i(1)$ is divisible by 2 in $\mathbb{Z}$ for $i = 1, 2$. This implies that $2 = p_1(1)p_2(1)$ is divisible by 4 in $\mathbb{Z}$ which is false. □

Assertion (2a) of Theorem 4.1.3 will follow from Lemma 4.4.1 and from the following Lemma.

4.4.4 Lemma. *Suppose* $\mathbf{1}^n = \oplus_{0 \leq i \leq \ell} V_i$ *for* $V_i \in \text{Vect}_{\mathbb{F}}^{k_i}(\mathbb{FP}^{(n-1)})$, *where* $\ell \geq 1$ *and where* $k_0 \geq ... \geq k_\ell \geq 1$. *Suppose that* $cw_{k_j}(V_j) = \pm x^j$ *for some* $0 \leq j \leq \ell$. *Let* $n = a2^s$, *where* a *is odd. Then* $k_1 + ... + k_\ell \leq 2^s$.

Proof. We use Lemma 4.4.2 to see that

$$1 + x^n = cw(V_j)cw(V_j^\perp) = cw(V_j)\textstyle\prod_{i \neq j} cw(V_i) \text{ in } \mathbb{Z}_2[x].$$

This implies $cw_{k_i}(V_i) = x^{k_i}$ for $0 \leq i \leq \ell$. If $i \geq 1$, then $2k_i \leq k_i + k_0 \leq n$. Thus, by Lemma 4.4.2, $cw(V_i) = (1+x)^{k_i}$ for $1 \leq i \leq \ell$ so:

$$cw(V_1)...cw(V_\ell) = (1+x)^{k_1+...+k_\ell} \text{ divides } 1 + x^n \text{ in } \mathbb{Z}_2[x].$$

We use Lemma 4.4.3 to see $k_1 + ... + k_\ell \leq 2^s$. □

The remainder of this section is devoted to the proof of Assertion (2b) in Theorem 4.1.3. We therefore set $\mathbb{F} = \mathbb{C}$, $\mathcal{R} = \mathbb{Z}$, and $cw = c$.

4.4.5 Lemma. *Let $V \subset \mathbf{1}^n$ be a k plane bundle over $\mathbb{CP}^{n-1}$, for $0 < k < n$. Let $n = a2^s$ for a odd.*

(1) *Suppose that $c_k(V) = \pm x^k$ in $H^*(\mathbb{CP}^{n-1}; \mathbb{Z})$.*
   1a) *Let $\bar{k} = n - k$. Then $c_{\bar{k}}(V^\perp) = \pm x^{\bar{k}}$.*
   1b) *If $0 < 2k \leq n$, then $c(V)$ divides $1 \pm x^{2^s}$ in $\mathbb{Z}[x]$.*
   1c) *The polynomial $c(V)$ is algebraically symmetric in $\mathbb{Z}[x]$.*

(2) *$c(V)$ is algebraically symmetric if and only if $c_k(V) = \pm x^k$.*

(3) *If $V = V_1 \oplus V_2$, then $c(V)$ is algebraically symmetric if and only if both $c(V_1)$ and $c(V_2)$ are algebraically symmetric.*

(4) *Assume that $0 < 2k \leq n$ and that $c_k(V) = \pm x^k$.*
   4a) *If n is even and if $n \geq 6$, then $c(V)c(V^*) = (1 - x^2)^k$ in $\mathbb{Z}_3[x]$.*
   4b) *If n is odd, then $k = 1$.*
   4c) *If n is even, then $k \leq 2$.*

Proof. By Lemma 4.4.2, $c(V)c(V^\perp) = 1 \pm x^n$ in $\mathbb{Z}[x]$. Assertion (1a) now follows. We factor

$$1 \pm x^n = (1 \pm x^{2^s})q_\pm \text{ in } \mathbb{Z}[x] \text{ for}$$
$$q_\pm = 1 \mp x^{2^s} + \cdots + x^{(a-1)2^s}.$$

Since $\mathbb{Z}[x]$ is a unique factorization domain and since $c(V)$ divides $1 \pm x^n$, we have a corresponding factorization $c(V) = p_1 p_2$, where p_1 divides $(1 \pm x^{2^s})$ and p_2 divides $q_\pm$. Since a is odd, the reduction of $q_\pm$ mod 2 has no roots in $\mathbb{Z}_2$. Suppose $2k \leq n$. By Lemma 4.4.2, $c(V) \equiv (1 + x)^k$ mod 2. Thus, $p_2 = 1$ which proves Assertion (1b). We factor

$$1 - x^{2^s} = (1 - x)(1 + x)(1 + x^2)...(1 + x^{2^{s-1}})$$

into irreducible polynomials in $\mathbb{Z}[x]$. Thus, $c(V)$ is the product of algebraically symmetric factors and hence is algebraically symmetric and Assertion (1c) follows in this special case.

Suppose that $2k \geq n$. Let $\bar{k} = n - k = \dim(V^\perp)$. We then have

$$2\bar{k} = 2n - 2k \leq 2n - n \leq n.$$

By Assertion (1a), $c_{\bar{k}}(V^\perp) = \pm x^{\bar{k}}$. Thus, the argument given above shows $c(V^\perp)$ is symmetric. We complete the proof of Assertion (1c) by computing:

$$x^k c(V)\{\tfrac{1}{x}\} = \frac{x^n(1 \pm x^{-n})}{x^{\bar{k}} c(V^\perp)\{\frac{1}{x}\}} = \frac{\pm(1 \pm x^n)}{\pm c(V^\perp)\{x\}} = \pm c(V)\{x\}.$$

Assertion (2) is immediate from Assertion (1).

Suppose that $V = V_1 \oplus V_2$. Since $c(V_1)c(V_2) = c(V)$,

$$c_{k_1}(V_1)c_{k_2}(V_2) = c_k(V).$$

Thus, $c_k(V) = \pm x^k$ if and only if $c_{k_i}(V_i) = \pm x^{k_i}$ for $i = 1, 2$. Assertion (3) now follows from Assertion (2).

Suppose $c_k(V) = \pm x^k$. We adapt an argument of [89] to establish Assertion (4a). By Assertion (2), $c(V)$ is algebraically symmetric. Let p be the mod 3 reduction of $c(V)$. Then $p(-x)$ is the mod 3 reduction of $c(V^*)$. We apply the mod 3 operations and use Lemma 4.3.18 to compute:

$$\pm(x - x^3)^k = \mathcal{P}(\pm x^k) = \pm x^k p(x)p(-x) \text{ in } \mathbb{Z}_3[x]/(x^n = 0).$$

Since the constant terms of $(1 - x^2)^k$ and $p(x)p(-x)$ are $+1$, we can divide by x^k and ignore the sign ambiguity to see that:

$$(1 - x^2)^k = p(x)p(-x) \text{ in } \mathbb{Z}_3[x]/(x^{n-k} = 0). \tag{4.4.5.a}$$

We express

$$(1 - x^2)^k = 1 + a_1 x + ... + (-1)^k x^{2k}, \text{ and}$$
$$p(x)p(-x) = 1 + b_1 x + ... + (-1)^k x^{2k}.$$

Since the polynomials $(1 - x^2)^k$ and $p(x)p(-x)$ are algebraically symmetric and since the leading term is $(-1)^k x^{2k}$, we have:

$$\begin{aligned}(a_0, ..., a_{2k}) &= (-1)^k(a_{2k}, ..., a_0) \text{ and}\\ (b_0, ..., b_{2k}) &= (-1)^k(b_{2k}, ..., b_0).\end{aligned} \tag{4.4.5.b}$$

Since $k \le n - k$, we use Equation (4.4.5.a) to see that

$$a_0 = b_0, ..., a_{k-1} = b_{k-1}.$$

We may then use the duality result of Display (4.4.5.b) to see

$$a_{k+1} = b_{k+1}, ..., a_{2k} = b_{2k}.$$

If $2k < n$, then $k < n - k$ and hence $a_k = b_k$ as well and Assertion (4a) follows. We therefore suppose that $2k = n$ and that Assertion (4a) fails. Then:

$$(1 - x^2)^k - p(x)p(-x) = \pm x^k \text{ in } \mathbb{Z}_3[x]. \tag{4.4.5.c}$$

Decompose $n = a2^s$. By Lemma 4.4.4, $k \leq 2^s$. Since $2k = n$, we have that $a = 1$, that $n = 2^s$, and that $k = 2^{s-1}$. Let $p^\perp$ be the mod 3 reduction of $c(V^\perp)$. By Lemma 4.4.2, $c(V)c(V^\perp) = 1 \pm x^{2^s}$ in $\mathbb{Z}[x]$. Since the polynomial $1 + x^{2^s}$ is irreducible in $\mathbb{Z}[x]$ and since we have a non-trivial factorization, we must have $c(V)c(V^\perp) = 1 - x^{2^s}$. The only factorization as the product of two polynomials of order 2^{s-1} has the factors $1 \pm x^{2^{s-1}} = 1 \pm x^k$. Thus, $p(V) = 1 + x^k$ or $p(V) = 1 - x^k$ mod 3. Since $n \geq 6$ and $n = 2^s$, we have $n \geq 8$ so $k \geq 4$. Thus,

$$(1 - x^2)^k - p(x)p(-x) = (1 - x^2)^k - (1 \pm x^k)^2 = -kx^2 + O(x^4)$$

which contradicts equation (4.4.5.c) since k is not divisible by 3.

If n is odd, then $k \leq 1$ by Lemma 4.4.4. This proves Assertion (4b). We now suppose that n is even. If $n \leq 4$, then we are done so we assume $n \geq 6$. Let p be the mod 3 reduction of $c(V)$. By Assertion (1b), p divides $1 \pm x^{2^s}$. The polynomial $1 + x^{2^s}$ has no roots in $\mathbb{Z}_3$ for $s \geq 1$. Since

$$1 - x^{2^s} = (1 - x^2)(1 + x^2)(1 + x^4)...(1 + x^{2^{s-1}})$$

and since $1 + x^{2c}$ has no roots in $\mathbb{Z}_3$, we see that $1 - x^{2^s}$ has two roots in $\mathbb{Z}_3$. Thus, p has at most 2 roots in $\mathbb{Z}_3$. Consequently, $p(x)p(-x) = (1 - x^2)^k$ has at most 4 roots in $\mathbb{Z}_3$ (counted with multiplicity). It now follows that $k \leq 2$. □

Assertion (2b) of Theorem 4.1.3 will follow from Lemma 4.4.1 and from the following result:

4.4.6 Lemma. *Suppose* $\mathbf{1}^n = \oplus_i V_i$ *for* $V_i \in \operatorname{Vect}_{\mathbb{C}}^{k_i}(\mathbb{CP}^{(n-1)})$, *where* $\ell \geq 1$ *and where* $k_0 \geq ... \geq k_\ell \geq 1$. *Suppose that* $c(V_j) = x^{k_j}$ *for some* j.

(1) *If* n *is odd, then* $\ell = 1$ *and* $k_1 = 1$.

(2) *If* n *is even, then* $\ell = 1$ *and* $k_1 \leq 2$ *or* $\ell = 2$ *and* $k_1 = k_2 = 1$.

Proof. We apply Lemma 4.4.5 throughout. We apply Lemma 4.4.2 to see

$$\textstyle\prod_i c(V_i) = 1 \pm x^n \text{ in } \mathbb{Z}[x].$$

Thus, $c(V_i) = \pm x^{k_i}$ for all values of i and more generally

$$c_{k_{i_1}+...+k_{i_\nu}}(V_{i_1} \oplus ... \oplus V_{i_\nu}) = \pm x^{k_{i_1}+...+k_{i_\nu}}$$

for any proper subcollection of indices $\{i_1, ..., i_\nu\}$.

Assume that Lemma 4.4.6 fails. Thus, $\ell \geq 1$. If $i \geq 1$, then

$$2k_i \leq k_i + k_0 \leq n$$

so $k_i \leq 1$ if n is odd and $k_i \leq 2$ if n is even. We consider various cases:

(1) Suppose $n = 3$. Then $\ell = 2$ and $k_0 = k_1 = k_2 = 1$. We then have $c(V_i) = 1 \pm x$ for $0 \leq i \leq 2$. This is false as there is no factorization of the form:

$$1 = c(V_0 \oplus V_1 \oplus V_2) = (1+x)^3 \text{ in } \mathbb{Z}_2[x]/x^3.$$

(2) Suppose $n = 4$. Then $\ell = 3$ and $k_0 = k_1 = k_2 = k_3 = 1$. Thus, $c(V_i) = 1 \pm x_i$; this is false as there is no factorization of the form:

$$1 \pm x^4 = (1 \pm x)(1 \pm x)(1 \pm x)(1 \pm x) \text{ in } \mathbb{Z}[x].$$

(3) Suppose $n \geq 5$ is odd. Then $k_i = 1$ for $1 \leq i \leq \ell$. Suppose $\ell \geq 2$. We apply Assertion (4b) of Lemma 4.4.4 to $V = V_1 \oplus V_2$ and $k = k_1 + k_2$ to show this case is impossible:

$$2(k_1 + k_2) = 4 \leq n \Rightarrow 2 = k_1 + k_2 = \dim(V_1 \oplus V_2) \leq 1.$$

(4) Suppose $n \geq 8$ is even. Then $k_i \leq 2$ for $1 \leq i \leq \ell$. Since Lemma 4.4.6 fails, $\ell > 1$. Let $0 < i < j \leq \ell$. We apply Assertion (4c) of Lemma 4.4.4 to $V = V_i \oplus V_j$ and $k = k_i + k_j$ for $0 < i < j \leq \ell$ to see:

$$2(k_i + k_j) \leq 8 \leq n \Rightarrow k_i + k_j \leq 2.$$

Thus, $k_i = 1$ for $1 \leq i \leq \ell$. As Lemma 4.4.6 fails, $\ell \geq 3$. We apply Assertion (4c) of Lemma 4.4.4 to $V = V_1 \oplus V_2 \oplus V_3$ and $k = k_1 + k_2 + k_3$ to show this case is impossible:

$$2(k_1 + k_2 + k_3) = 6 \leq n \Rightarrow 3 = k_1 + k_2 + k_3 \leq 2.$$

(5) Suppose $n = 6$. Since Lemma 4.4.6 fails, $k_0 \leq 3$. Thus, $2k_i \leq 6$ for all i and we apply Assertion (4c) of Lemma 4.4.4 to see $k_i \leq 2$ for $0 \leq i \leq \ell$. Thus, by Lemma 4.4.2, $cw(V_i) = (1+x)^{k_i}$ in $\mathbb{Z}_2[x]$ for all i. This leads to the following false factorization:

$$1 + x^6 = \textstyle\prod_i c(V_i) = \prod_i (1+x)^{k_i} = (1+x)^6 \text{ in } \mathbb{Z}_2[x]. \quad \square$$

4.5 Odd maps of constant rank

This section is devoted to the proof of Theorem 4.1.4. We first establish some notational conventions. Let X be a smooth manifold on which $\mathbb{Z}_2$ acts smoothly and without fixed points. Let $\tilde{X} := X/\mathbb{Z}_2$ be the quotient manifold and let π_X be the associated double covering projection from X to $\tilde{X}$. Let γ_X be the associated real line bundle over $\tilde{X}$ so that $S(\gamma_X) = X$;

$$\gamma_X = \{(x, z) \in X \times \mathbb{R} \text{ where we identify } (x, \lambda) \sim (-x, -\lambda)\}.$$

For example if $\mathbb{Z}_2 = \{\pm 1\}$ acts on S^n by scalar multiplication, then we may identify the quotient $\tilde{S}^n = S^n/\mathbb{Z}_2$ with the real projective space $\mathbb{RP}^n$ and we may identify γ_S with the classifying line bundle γ_1 over $\mathbb{RP}^n$.

Let $T : X \to \mathfrak{so}(m)$. If $T(-x) = -T(x)$, then T is said to be an *odd map.* If $\operatorname{rank} T(x) = r$ for all $x \in X$, then T is said to have *constant rank* r. Let $T : X \to \mathfrak{so}(m)$ be an odd map of constant rank r. The $\mathbb{Z}_2$ ambiguity plays no role so $\ker(T)$ and $\operatorname{range}(T)$ descend to define vector bundles U_0 and U_1 over $\tilde{X}$:

$$\begin{aligned} U_0 &:= \{\langle x\rangle \times v \in \tilde{X} \times \mathbb{R}^m : T(x)v = 0\}, \text{ and} \\ U_1 &:= \{\langle x\rangle \times v \in \tilde{X} \times \mathbb{R}^m : v \in \operatorname{range}(T(x))\}. \end{aligned} \tag{4.5.1.a}$$

Since $T(x)$ is skew-adjoint, $\ker\{T(x)\} \perp \operatorname{range}\{T(x)\}$. Since we are in the Riemannian setting, we have an orthogonal direct sum decomposition

$$U_0 \oplus U_1 = \mathbf{1}^n. \tag{4.5.1.b}$$

Since $\ker(T(x)) \cap \operatorname{range}(T(x)) = \{0\}$, $T(x)$ defines an isomorphism from $\operatorname{range}\{T(x)\}$ to $\operatorname{range}\{T(x)\}$ for any $x \in X$. Since $T(-x) = -T(x)$, T does not descend to define an isomorphism from U_1 to U_1. Instead, after taking into account the $\mathbb{Z}_2$ equivariance, we see that T descends to define an isomorphism

$$U_1 \xrightarrow{\approx} U_1 \otimes \gamma_X. \tag{4.5.1.c}$$

Let $\phi(n)$ be the number of integers s such that $1 \le s \le n$ and $s \equiv 0, 1, 2, 4$ mod 8. By Theorem 4.3.21,

$$\widetilde{K\mathbb{R}}(\mathbb{RP}^n) = \{[\gamma_1] - [\mathbf{1}]\} \cdot \mathbb{Z}_{2^{\phi(n)}}.$$

4.5.2 Lemma. *Let* $\mathbf{1}^n = U_0 \oplus U_1$ *over* $\mathbb{RP}^n$. *Assume* $U_1 \otimes \gamma_1 = U_1$. *Decompose* $[U_i] = a_i\{[\gamma_1] - [\mathbf{1}]\} + \dim(U_i)[\mathbf{1}]$ *in* $K\mathbb{R}(\mathbb{RP}^n)$, *where* $a_i \geq 0$ *is well defined mod* $2^{\phi(n)}$. *Choose* $j = j(n)$ *so* $2^j \leq n < 2^{j+1}$.

(1) *We have* $2a_1 \equiv \dim(U_1)$ *mod* $2^{\phi(n)}$.
(2) *We have* $w(U_i) = (1+x)^{a_i}$.
(3) *We have* $a_0 + a_1 \equiv 0$ *mod* 2^{j+1}.
(4) *If* $n \geq 9$, *then* $j + 2 \leq \phi(n)$.
(5) *If* $n \geq 9$, *then* $a_1 \equiv \frac{1}{2}\dim(U_1)$ *mod* 2^{j+1}.

Proof. By assumption, U_1 is isomorphic to $U_1 \otimes \gamma_1$. To prove Assertion (1), we apply Theorem 4.3.21 to equate the coefficients of $([\gamma_1] - [\mathbf{1}])$ mod $2^{\phi(n)}$ in the following expansions:

$$\begin{aligned}[U_1] &= a_1([\gamma_1] - [\mathbf{1}]) + \dim(U_1)[\mathbf{1}] \\ [U_1 \otimes \gamma_1] &= (\dim(U_1) - a_1)([\gamma_1] - [\mathbf{1}]) + \dim(U_1)[\mathbf{1}].\end{aligned}$$

Let $x := w_1(\gamma_1) \in H^1(\mathbb{RP}^n; \mathbb{Z}_2)$. By Lemma 4.3.5, the cohomology of projective space is a truncated polynomial ring:

$$H^*(\mathbb{RP}^n; \mathbb{Z}_2) = \mathbb{Z}_2[x]/x^{n+1}.$$

We may use Lemma 4.3.20 to see that if ν is sufficiently large, then

$$U_i \oplus (\nu + a_i) \cdot \mathbf{1} = a_i\gamma_1 \oplus (\nu + \dim(U_i)) \cdot \mathbf{1} \text{ in } \mathrm{Vect}_{\mathbb{R}}(\mathbb{RP}^n).$$

We establish Assertion (2) by checking:

$$\begin{aligned}w(U_i) &= w\{U_i \oplus (\nu + a_i) \cdot \mathbf{1}\} \\ &= w\{a_i\gamma_1 \oplus (\nu + \dim(U_i)) \cdot \mathbf{1}\} = w(a_i\gamma_1) \\ &= (1+x)^{a_i} \text{ for } i = 0, 1.\end{aligned}$$

By assumption, $U_0 \oplus U_1 = \mathbf{1}^n$. Thus:

$$1 = w(U_0)w(U_1) = (1+x)^{a_0}(1+x)^{a_1} = (1+x)^{a_0+a_1}.$$

Express $a_0 + a_1 = 2^s a$, where a is odd. Since

$$1 = (1+x)^{a_1+a_2} = 1 + x^{2^s} + \ldots \text{ in } \mathbb{Z}_2[x]/x^{n+1},$$

we must have $n < 2^s$. Since $2^j \leq n$, 2^{j+1} divides 2^s and Assertion (3) follows. The function ϕ is growing roughly linearly while the function j is growing logarithmically; the estimate of Assertion (4) follows from the following table:

Table 4.5.2.a

n	1	2	3	4	5	6	7	8	9	10	11	12	13	14	15	16
$\phi(n)$	1	2	2	3	3	3	3	4	5	6	6	7	7	7	7	8
$j(n)$	0	1	1	2	2	2	2	3	3	3	3	3	3	3	3	4

Assertion (5) now follows from Assertions (1) and (4). □

We can now obtain our first result bounding the rank by establishing Assertion (1) of Theorem 4.1.4. This result, due to Stavrov [146], generalizes earlier work of Gilkey, Leahy and Sadofsky [83] and of Zhang [163, 164].

4.5.3 Lemma. *Let $T : S^n \to \mathfrak{so}(n+1+\alpha)$ be an odd map of constant rank $r > 0$, where $n \geq 9$ and $0 \leq \alpha \leq n$. Choose k so $2^k \leq \frac{r}{2} < 2^{k+1}$. Then:*

(1) $r \leq 2\alpha$.

(2) 2^k *divides at least one element of the set* $\{n+1, ..., n+1+\alpha-\frac{r}{2}\}$.

(3) *If* $\alpha = 2$ *and* $r = 4$, *then* n *is odd.*

Proof. By Lemma 1.2.14, r is even. We adopt the notation of Display (4.5.1.a) to define bundles U_i over $\mathbb{RP}^n$. We use Theorem 4.3.21 to find a non-negative integer $\tilde{a}_1$ so that:

$$[U_1] - r[\mathbf{1}] = \hat{a}_1\{[\gamma_1] - [\mathbf{1}]\} \text{ in } \widetilde{K\mathbb{R}}(\mathbb{RP}^n).$$

Choose j so $2^j \leq n < 2^{j+1}$. Choose a_1 such that $0 \leq a_1 < 2^{j+1}$ and such that $a_1 \equiv \hat{a}_1 \bmod 2^{j+1}$. By Lemma 4.5.2, $a_1 \equiv \frac{r}{2} \bmod 2^{j+1}$. As

$$r \leq n+1+\alpha \leq 2n+1 < 2^{j+2},$$

we have $\frac{r}{2} < 2^{j+1}$ so

$$a_1 = \tfrac{r}{2} > 0.$$

Thus, by Lemma 4.5.2,

$$w(U_1) = (1+x)^{\hat{a}_1} = (1+x)^{\frac{r}{2}} \text{ in } H^*(\mathbb{RP}^n; \mathbb{Z}_2).$$

We apply Equation (4.5.1.b) to see that $1 = w(U_0)w(U_1)$. Thus, we see that:

$$w(U_0) = \frac{1}{(1+x)^{\frac{r}{2}}} \text{ in } \mathbb{Z}_2[x]/(x^{n+1} = 0).$$

We have $\dim U_0 = n+1+\alpha-r$. Therefore, for dimensional reasons, we have

$$w(U_0)_q = 0 \text{ in } H^q(\mathbb{RP}^n; \mathbb{Z}_2) \text{ if } q > n+1+\alpha-r. \tag{4.5.3.a}$$

Consequently, we may multiply Equation (4.5.3.a) by $(1+x)^{\frac{r}{2}-1}$ to see:

(4.5.3.b) $$\left\{\frac{1}{1+x}\right\}_q = 0 \text{ in } H^q(\mathbb{RP}^n;\mathbb{Z}_2) \text{ if } q > n+\alpha-\tfrac{r}{2}.$$

We take $q_0 := n+1+\alpha-\frac{r}{2}$. Since x^{q_0} is a monomial of $\frac{1}{1+x}$, we use Equation (4.5.3.b) and Lemma 4.3.5 to prove Assertion (1) by noting we have:

$$n+1+\alpha-\tfrac{r}{2} \geq n+1.$$

Choose k so $2^k \leq \frac{r}{2} < 2^{k+1}$. We may then express

$$\tfrac{r}{2} = 2^k + \varepsilon \text{ for } 0 \leq \varepsilon < 2^k,$$
$$n = n_1 2^k + n_2 \text{ for } 0 \leq n_1 \text{ and } 0 \leq n_2 < 2^k.$$

Since $\frac{r}{2} \leq \alpha \leq n$, $k \leq j$ so $n_1 \geq 1$. We multiply Equation (4.5.3.a) by $(1+x)^\varepsilon$ to see

$$\left\{\frac{1}{1+x^{2^k}}\right\}_q = 0 \text{ in } H^q(\mathbb{RP}^n;\mathbb{Z}_2) \text{ if } q > n+1+\alpha-r+\varepsilon.$$

As $n_1 \geq 1$, $x^{n_1 2^k}$ appears in $\frac{1}{1+x^{2^k}}$. As $n_1 2^k \leq n$, there is no truncation. This shows that:

$$n_1 2^k \leq n+1+\alpha-r+\varepsilon.$$

Since $n = n_1 2^k + n_2$, we may estimate that:

$$0 \leq n_2+1+\alpha-r+\varepsilon = n_2+1+\alpha-\tfrac{r}{2}-2^k$$

so $n_2+1+\alpha-\frac{r}{2} \geq 2^k$ and hence $n+1+\alpha-\frac{r}{2} \geq (n_1+1)2^k$. Since $n+1 > n_1 2^k$, we have some integer in the collection $\{n+1, ..., n+1+\alpha-\frac{r}{2}\}$ must equal $(n_1+1)2^k$; this proves Assertion (2).

If $\alpha = 2$ and $r = 4$, then $2^k = 2$ and (2) implies 2 divides $n+1$ so n is odd as claimed in Assertion (3). □

The estimates of Lemma 4.5.3 are sharp if $\alpha = 1, 2$.

4.5.4 Lemma.

(1) *There exists an odd map of constant rank* 2 *from* S^n *to* $\mathfrak{so}(n+2)$.

(2) *If n is odd, then there exists an odd map of constant rank* 4 *from* S^n *to* $\mathfrak{so}(n+3)$.

Proof. Let $\{e_1, ..., e_\nu\}$ be the usual orthonormal basis for $\mathbb{R}^\nu$. Suppose that $x = x_1e_1 + ... + x_{n+1}e_{n+1} \in S^n$. To prove Assertion (1), we define:

$$T(x)y := (x, y)e_{n+2} - (e_{n+2}, y)x \text{ for } y \in \mathbb{R}^{n+2} \text{ and } x \in \mathbb{R}^{n+1}.$$

Let $x \in S^n$. Since $\{x, e_{n+2}\}$ is a linearly independent set, we use Lemma 1.2.2 to see that $T(x)$ has constant rank 2. It is clear that $T(-x) = -T(x)$. We show that T is skew-adjoint by computing:

$$(T(x)y, z) = (x, y)(e_{n+2}, z) - (e_{n+2}, y)(x, z) = -(y, T(x)z).$$

Suppose that n is odd. Let J be the usual almost complex structure on $\mathbb{R}^{n+1} = \mathbb{C}^{\frac{1}{2}(n+1)}$. For $y \in \mathbb{R}^{n+3}$ and $x \in \mathbb{R}^{n+1}$, we define:

$$T(x)y := (x, y)e_{n+2} - (e_{n+2}, y)x + (Jx, y)e_{n+3} - (e_{n+3}, y)Jx.$$

Let $x \in S^n$. Since $\{x, Jx, e_{n+2}, e_{n+3}\}$ is a linearly independent set, we use Lemma 1.2.2 to see $\text{rank}(T) = 4$. It is clear that $T(-x) = -T(x)$. We show that T is skew-adjoint by computing:

$$\begin{aligned}(T(x)y, z) =&(x, y)(e_{n+2}, z) - (e_{n+2}, y)(x, z)\\ &+(Jx, y)(e_{n+3}, z) - (e_{n+3}, y)(Jx, z)\\ =&-(y, T(x)z). \quad \square\end{aligned}$$

We conclude our study of odd maps of constant rank from projective spaces by establishing Assertion (2) of Theorem 4.1.4:

4.5.5 Lemma. *If $T : S^8 \to \mathfrak{so}(10)$ is an odd map of constant rank r, then $r \leq 2$.*

Proof. We adopt the notation of Lemma 4.5.2. Let $u_i := \dim(U_i)$; $u_1 = r$. We have $u_0 + u_1 = 10$ and $a_0 + a_1 \equiv 0 \bmod 16$. We use Table 4.5.2.a to see that $\phi(8) = 4$. Thus, $2a_1 \equiv \dim(U_1) = r \bmod 16$. There are four cases:

(1) Suppose $u_1 = 10$. Then $a_1 \equiv 5 \bmod 8$, $a_0 \equiv 3 \bmod 8$, and $u_0 = 0$. Thus, $w(U_0) = (1 + x)^{a_0} = 1 + x + ...$ so $u_0 \geq 1$ which is false.
(2) Suppose $u_1 = 8$. Then $a_1 \equiv 4 \bmod 8$, $a_0 \equiv 4 \bmod 8$, and $u_0 = 2$. Thus, $w(U_0) = 1 + x^4 + ...$ so $u_0 \geq 4$ which false.
(3) Suppose $u_1 = 6$. Then $a_1 \equiv 3 \bmod 8$, $a_0 \equiv 5 \bmod 8$, and $u_0 = 4$. Thus, $w(U_0) = (1 + x)^{a_0} = 1 + x + x^4 + x^5 + ...$ so $u_0 \geq 5$ which is false.

(4) Suppose $u_1 = 4$. Then $a_1 \equiv 2 \bmod 8$, $a_0 \equiv 6 \bmod 8$, and $u_0 = 6$. If $a_1 \equiv 2 \bmod 16$, then $a_0 \equiv 14 \bmod 16$ so

$$w(U_0) = (1+x)^{2+4+8} = 1 + x^2 + x^4 + x^6 + x^8$$

which implies $u_0 \geq 8$ which is false. On the other hand, if $a_1 \equiv 10 \bmod 16$, then $w(U_1) = (1+x)^{2+8} = 1 + x^2 + x^8$ so $u_1 \geq 8$ which is false. □

Let $\mathrm{Gr}_2^+(\mathbb{R}^m)$ be the Grassmannian of oriented 2 planes in $\mathbb{R}^m$; we assume $m \geq 5$ henceforth. We reverse the orientation on $\pi \in \mathrm{Gr}_2^+(\mathbb{R}^m)$ to define $-\pi \in \mathrm{Gr}_2^+(\mathbb{R}^m)$. The map $\pi \to -\pi$ defines a fixed point free action by $\mathbb{Z}_2$; the quotient

$$\mathrm{Gr}_2(\mathbb{R}^m) := \mathrm{Gr}_2^+(\mathbb{R}^m)/\mathbb{Z}_2$$

is the unoriented Grassmannian of 2 planes in $\mathbb{R}^m$. We let

$$\bar{\gamma}_1 := \{\mathrm{Gr}_2^+(\mathbb{R}^m) \times \mathbb{R}\}/\mathbb{Z}_2$$

be the associated real line bundle. Let $n = m - 2$ and let

$$i(e) := \mathrm{span}\{e, e_m\} \text{ map } S^n \to \mathrm{Gr}_2^+(\mathbb{R}^m).$$

The map i descends to define a corresponding inclusion $i : \mathbb{RP}^n \subset \mathrm{Gr}_2(\mathbb{R}^m)$. The cohomology rings $H^*(\mathbb{RP}^n; \mathbb{Z}_2)$ and $H^*(\mathrm{Gr}_2(\mathbb{R}^n); \mathbb{Z}_2)$ are given in Lemmas 4.3.5 and 4.3.12. We let

$$\begin{aligned} x &:= w_1(\gamma_1) \in H^1(\mathbb{RP}^n; \mathbb{Z}_2), \\ \bar{x} &:= w_1(\gamma_2) \in H^1(\mathrm{Gr}_2(\mathbb{R}^m); \mathbb{Z}_2), \text{ and} \\ \bar{y} &:= w_2(\gamma_2) \in H^2(\mathrm{Gr}_2(\mathbb{R}^m); \mathbb{Z}_2). \end{aligned}$$

4.5.6 Lemma. *Let $m = n + 2$ and let $i : \mathbb{RP}^n \to \mathrm{Gr}_2(\mathbb{R}^m)$. Then*

(1) *We have $i^*\bar{\gamma}_1 = \gamma_1$ and $i^*\bar{x} = x$.*

(2) *We have $i^*\gamma_2 = \gamma_1 \oplus \mathbf{1}$ and $i^*\bar{y} = 0$.*

(3) *We have $Sq(\bar{x}) = \bar{x} + \bar{x}^2$ and $Sq(\bar{y}) = \bar{y} + \bar{x}\bar{y} + \bar{y}^2$.*

Proof. Since γ_1 is the real line bundle associated to the covering projection $S^n \to \mathbb{RP}^n$, since $\bar{\gamma}_1$ is the real line bundle associated to the covering projection $\mathrm{Gr}_2^+(\mathbb{R}^m) \to \mathrm{Gr}_2(\mathbb{R}^m)$, and since $i : S^n \to \mathrm{Gr}_2^+(\mathbb{R}^m)$ is equivariant with respect to the $\mathbb{Z}_2$ action, we have $i^*\bar{\gamma}_1 = \gamma_1$ so:

$$i^* w_1(\bar{\gamma}_1) = w_1(i^*\bar{\gamma}_1) = w_1(\gamma_1) = x.$$

This shows that $w_1(\bar\gamma_1) \neq 0$. Since $\bar x$ generates $H^1(\mathrm{Gr}_2(\mathbb{R}^m); \mathbb{Z}_2) = \mathbb{Z}_2$, we have $w_1(\bar\gamma_1) = \bar x$ and the Assertion (1) follows.

Since e_m is a global section to $i^*\gamma_2$ over $\mathbb{RP}^n$, $i^*\gamma_2$ decomposes as a direct sum of two line bundles $i^*\gamma_2 = L \oplus \mathbf{1}$. Since we have

$$w_1(\gamma_1) = x = i^*\bar x = i^* w_1(\gamma_2) = w_1(i^*\gamma_2) = w_1(L),$$

we use Lemma 4.3.11 to see $\gamma_1 = L$ and thus $i^*\gamma_2 = \gamma_1 \oplus \mathbf{1}$. We may then complete the proof by computing

$$i^* y = i^* w_2(\gamma_2) = w_2(i^*\gamma_2) = w_2(\gamma_1 \oplus 1) = w_2(\gamma_1) = 0.$$

We use Lemma 4.3.18 to prove Assertion (3) by computing:

$$\begin{aligned} Sq(w_1(\bar\gamma_1)) &= Sq(\bar x) = \bar x w(\gamma_1) = \bar x + \bar x^2 \\ Sq(w_2(\gamma_2)) &= Sq(\bar y) = \bar y w(\gamma_2) = \bar y(1 + \bar x + \bar y). \qquad \square \end{aligned}$$

We define the formal class $\bar z \in H^*(\mathrm{Gr}_2(\mathbb{R}^m); \mathbb{Z}_2)$ so that

$$1 = (1 + \bar x + \bar y)(1 + \bar z_1 + \bar z_2 + \ldots).$$

We then have recursively that $\bar z_k = \bar x \bar z_{k-1} + \bar y \bar z_{k-2}$. We compute:

Table 4.5.6.a

$\bar z_1 = \bar x$	$\bar z_2 = \bar x^2 + \bar y$
$\bar z_3 = \bar x^3$	$\bar z_4 = \bar x^4 + \bar x^2 \bar y + \bar y^2$
$\bar z_5 = \bar x^5 + \bar x \bar y^2$	$\bar z_6 = \bar x^6 + \bar x^4 \bar y + \bar y^3$
$\bar z_7 = \bar x^7$	$\bar z_8 = \bar x^8 + \bar x^6 \bar y + \bar x^4 \bar y^2 + \bar y^4$

By Lemma 4.3.12, $H^*(\mathrm{Gr}_2(\mathbb{R}^m); \mathbb{Z}_2)$ is $\mathbb{Z}_2[\bar x, \bar y]$ modulo the relations:

$$\bar z_i = 0 \text{ for } i \geq m - 1.$$

We begin our study of odd maps of constant rank r over $Gr_2^+(\mathbb{R}^m)$ by establishing Assertion (3) of Theorem 4.1.4:

4.5.7 Lemma. *Let $T : \mathrm{Gr}_2^+(\mathbb{R}^m) \to \mathfrak{so}(m)$ be an odd map of constant rank r. If $m = 5$, or $m = 6$, or $m = 9$, then $r \leq 2$.*

Proof. The map i^*T is an odd map of constant rank r which defines bundles i^*U_i over $\mathbb{RP}^n$. We adopt the notation of Lemma 4.5.2 and complete the proof by considering the following cases:

Case 1: Suppose $m = 5$, $n = 3$, $j = 1$, $\phi(n) = 2$, $r = 4$, and $u_0 = 1$. Then:

$$a_1 \equiv 2 \bmod 2^{\phi(n)-1} = 2 \text{ so } a_0 \equiv 0 \bmod 2.$$

Therefore, $w(i^*U_0) = (1+x)^{a_0} = 1$ in $H^1(\mathbb{RP}^3; \mathbb{Z}_2)$. Since i^* is injective on H^1, $w_1(U_0) = 0$ and hence by Lemma 4.3.11, $U_0 = \mathbf{1}$. Consequently,

$$5 \cdot \mathbf{1} = U_0 \oplus U_1 = \mathbf{1} \oplus U_1 \text{ so } w(U_1) = 1.$$

On the other hand, by Equation (4.5.1.c) $U_1 \otimes \bar{\gamma}_1 = U_1$ so:

$$5 \cdot \bar{\gamma}_1 = \bar{\gamma}_1 \oplus U_1 \text{ so } (1+\bar{x})^5 = 1 + \bar{x}.$$

This shows that $\bar{x}^4 = 0$ in $H^4(\mathrm{Gr}_2(\mathbb{R}^5); \mathbb{Z}_2)$. We now use Lemma 4.3.12 and Table 4.5.6.a to obtain the desired contradiction:

$$\bar{x}^4 \in \mathrm{span}\{\bar{z}_4\} \text{ so } \bar{x}^4 = c(\bar{x}^4 + \bar{x}^2\bar{y} + \bar{y}^2) \text{ in } \mathbb{Z}_2[\bar{x}, \bar{y}].$$

Case 2: Suppose $m = 6$, $n = 4$, $j = 2$, $\phi(n) = 3$, $r = 6$, and $u_0 = 0$. Then

$$a_1 \equiv 3 \bmod 2^{\phi(n)-1} = 4 \text{ so } a_0 \equiv 1 \bmod 4.$$

Consequently, $w(i^*U_0) = (1+x)^{a_0} = 1 + x + \ldots$ so $u_0 \geq 1$ which is false.

Case 3: Suppose $m = 6$, $n = 4$, $j = 2$, $\phi(n) = 3$, $r = 4$, and $u_0 = 2$. Then

$$a_1 \equiv 2 \bmod 2^{\phi(n)-1} = 4.$$

If $a_1 \equiv 2 \bmod 8$, then $a_0 \equiv 6 \bmod 8$. We derive a contradiction by checking:

$$w(i^*U_0) = (1+x)^6 = 1 + x^2 + x^4 \text{ so } u_0 \geq 4.$$

Thus, we must have that $a_1 \equiv 6 \bmod 8$ and $a_0 \equiv 2 \bmod 8$. Thus, we have $w(i^*U_0) = (1+x)^2$. Since $w_1(i^*U_0) = 0$, $w_1(U_0) = 0$. Since $w_2(i^*U_0) = x^2$,

$$\begin{aligned} &w(U_0) = (1 + \bar{x}^2 + \alpha\bar{y}) \text{ and} \\ &w(U_1) = w(U_0)^{-1} = 1 + (\bar{x}^2 + \alpha\bar{y}) + (\bar{x}^2 + \alpha\bar{y})^2 + (\bar{x}^2 + \alpha\bar{y})^3 + \ldots \end{aligned}$$

Since $u_1 = 4$, we have $(\bar{x}^2 + \alpha\bar{y})^3 = 0$ for dimensional reasons. Thus, $(\bar{x}^2 + \alpha\bar{y})^3 \in \mathrm{span}\{\bar{x}\bar{z}_5, \bar{z}_6\}$ which means that we can express

$$\begin{aligned} &\bar{x}^6 + \alpha(\bar{x}^4\bar{y} + \bar{x}^2\bar{y}^2 + \bar{y}^3) \\ =&c_0(\bar{x}^6 + \bar{x}^2\bar{y}^2) + c_1(\bar{x}^6 + \bar{x}^4\bar{y} + \bar{y}^3) \text{ in } \mathbb{Z}_2[\bar{x}, \bar{y}]. \end{aligned}$$

This leads to the following inconsistent set of equations in $\mathbb{Z}_2$:

$$c_0 + c_1 = 1, \ c_0 = \alpha, \ c_1 = \alpha.$$

Case 4: Suppose $m = 9$, $n = 7$, $j = 2$, $\phi(n) = 3$, $r = 8$, and $u_0 = 1$. Then

$$a_1 \equiv 4 \bmod 2^{\phi(n)-1} = 4 \text{ so } a_0 \equiv 0 \bmod 4.$$

We argue as in Case 1 above to see $w(i^*U_0) = (1+x)^{a_0} = 1 + O(x^4)$ so $w_1(i^*U_0) = 0$. Thus, $U_0 = \mathbf{1}$ so

$$9 \cdot \mathbf{1} = U_0 \oplus U_1 = \mathbf{1} \oplus U_1 \text{ so } w(U_1) = 1.$$

Consequently, we have $9 \cdot \bar{\gamma}_1 = \bar{\gamma}_1 \oplus U_1$ so $(1+\bar{x})^9 = 1 + \bar{x}$ and $\bar{x}^8 = 0$ in $H^8(\mathrm{Gr}_2(\mathbb{R}^9); \mathbb{Z}_2)$. Thus, by Lemma 4.3.12, we obtain the desired contradiction:

$$\begin{aligned} &\bar{x}^8 \in \operatorname{span}\{\bar{z}_8\} \text{ so} \\ &\bar{x}^8 = c(\bar{x}^8 + \bar{x}^6\bar{y} + \bar{x}^4\bar{y}^2 + \bar{y}^4\} \text{ in } \mathbb{Z}_2[\bar{x}, \bar{y}]. \end{aligned}$$

Case 5: Suppose $m = 9$, $n = 7$, $j = 2$, $\phi(n) = 3$, $r = 6$, and $u_0 = 3$. Then

$$a_1 \equiv 3 \bmod 2^{\phi(n)-1} = 4 \text{ so } a_0 \equiv 1 \bmod 4.$$

If $a_1 \equiv 7 \bmod 8$, then $w(i^*(U_1)) = (1+x)^7 = 1 + x + ... + x^7$ and thus $u_1 \geq 7$ for dimensional reasons which is false. If $a_1 \equiv 3 \bmod 8$, then $a_0 \equiv 5 \bmod 8$ so $w(i^*(U_0)) = (1+x)^5 = 1 + x + x^4 + x^5$ which implies $u_0 \geq 5$ which is false.

Case 6: Suppose $m = 9$, $n = 7$, $j = 2$, $\phi(n) = 3$, $r = 4$, and $u_0 = 5$. Then

$$a_1 \equiv 2 \bmod 2^{\phi(n)-1} = 4 \text{ so } a_0 \equiv 2 \bmod 4.$$

If $a_1 \equiv 6 \bmod 8$, then $w(i^*(U_1)) = (1+x)^6 = 1 + x^2 + x^4 + x^6$ so $u_1 \geq 6$ which is false. If $a_1 \equiv 2 \bmod 8$, then $a_0 \equiv 6 \bmod 8$ so $w(i^*(U_0)) = (1+x)^6$ and $u_0 \geq 6$ which is false. □

We now turn our attention briefly to odd maps $T : \mathrm{Gr}_2^+(\mathbb{R}^m) \to \mathfrak{so}(m)$ which have constant (complex) eigenvalues; such maps necessarily have constant rank. Let $\{\lambda_i, \mu_i\}$ be the eigenvalues and associated multiplicities of the self-adjoint operator $T(x)^2$; since $T(x)^2$ is non positive, the eigenvalues λ_i are real and non-positive. The eigenvalue 0 plays a distinguished role. Let $\lambda_0 = 0$; T is invertible if and only if $\mu_0 = 0$. The remaining multiplicities are all even as T itself is skew-adjoint. We order the μ_i so

$$\mu_1 \geq ... \geq \mu_\ell \geq 2 \text{ and set } \vec{\mu} = (\mu_0, ..., \mu_\ell).$$

We denote the associated eigenbundles over $Gr_2(\mathbb{R}^m)$ by

$$E_i(\pi) := \{y \in \mathbb{R}^m : T(\pi)^2 y = \lambda_i y\}; \ m \cdot \mathbf{1} = \oplus_i E_i.$$

As T preserves the eigenbundles E_i of T^2 and as T is an odd map, we have

$$E_i = E_i \otimes \bar{\gamma}_1 \text{ for } i \geq 1.$$

We establish Assertion (4) of Theorem 4.1.4 with the following result:

4.5.8 Lemma. *Let $T : Gr_2^+(\mathbb{R}^m) \to \mathfrak{so}(m)$ be an odd map so T has constant eigenvalues and constant rank $r > 2$. Let $\{\lambda_i, \mu_i\}$ be the associated multiplicities of the self-adjoint operator T^2, where $\lambda_0 = 0$ and $\mu_1 \geq ... \geq \mu_l \geq 1$.*

(1) *If $m = 7$, then $\vec{\mu} = (1,6)$, $\vec{\mu} = (1,4,2)$, or $\vec{\mu} = (3,4)$.*
(2) *If $m = 8$, then $\vec{\mu} = (0,8)$, $\vec{\mu} = (0,6,2)$, or $\vec{\mu} = (2,6)$.*

Proof. To prove Assertion (1), we must show the eigenvalue multiplicities

$$\vec{\mu} = (3,2,2) \text{ or } \vec{\mu} = (1,2,2,2)$$

are impossible if $m = 7$.

Suppose that $\vec{\mu} = (3,2,2)$. Let $i : \mathbb{RP}^5 \to \mathrm{Gr}_2(\mathbb{R}^7)$, let $V_i := i^*(E_i)$, and let $v_i := \dim(V_i)$. We then have a decomposition

$$\begin{aligned} &7 \cdot \mathbf{1} = V_0 \oplus V_1 \oplus V_2 \text{ over } \mathbb{RP}^5 \text{ where} \\ &v_0 = 3,\ v_1 = 2,\ v_2 = 2, \\ &V_1 \otimes \gamma_1 = V_1, \text{ and } V_2 \otimes \gamma_2 = V_2. \end{aligned} \tag{4.5.8.a}$$

We show this is false. We have $n = 5$, $\phi(5) = 3$, and $j(5) = 2$. We express

$$[V_i] = a_i\{[\gamma_1] - \mathbf{1}\} + v_i[\mathbf{1}] \text{ in } K\mathbb{R}(\mathbb{RP}^5).$$

Since $V_i = V_i \otimes \gamma_1$ for $i = 1, 2$, we use Lemma 4.5.2 to see:

$$a_1 \equiv 1 \bmod 2^{\phi(5)-1} = 4 \text{ and } a_2 \equiv 1 \bmod 2^{\phi(5)-1} = 4.$$

If $a_1 \equiv 5 \bmod 8$, then $w(V_1) = (1+x)^5 = 1 + x + x^4 + x^5$ which implies $v_1 \geq 5$ which is false. Similarly, $a_2 \equiv 5 \bmod 8$ is impossible. Thus, $a_1 \equiv 1 \bmod 8$ and $a_2 \equiv 1 \bmod 8$. Thus, $a_0 \equiv 6 \bmod 8$ so

$$w(V_0) = (1+x)^6 = 1 + x^2 + x^4 \text{ and } v_0 \geq 4$$

which is false.

Suppose that $\vec{\mu} = (1,2,2,2)$. We set $V_0 := i^*(E_0 \oplus E_3)$, $V_1 := i^*(E_1)$, and $V_2 := i^*(E_2)$ to construct a decomposition of the form given in Display (4.5.8.a); we have already shown this is impossible. This completes the proof of Assertion (1).

To prove Assertion (2) of Lemma 4.5.8, we must show that the eigenvalue decompositions:

$$\begin{aligned} &\mu = (0,4,4),\ \mu = (0,4,2,2),\ \mu = (0,2,2,2,2) \\ &\mu = (2,4,2),\ \mu = (2,2,2,2),\ \mu = (4,4),\ \mu = (4,2,2) \end{aligned}$$

are impossible if $m = 8$. In these 7 cases, we can combine eigenspaces to construct bundles $V_i \in \mathrm{Vect}^4_{\mathbb{R}}(\mathbb{RP}^6)$ so that

$$8 \cdot \mathbf{1} = V_0 \oplus V_1 \text{ over } \mathbb{RP}^6 \text{ where}$$
$$v_0 = v_1 = 4 \text{ and } V_1 \otimes \gamma_1 = V_1.$$

We have $n = 6$, $\phi(6) = 3$, and $j(6) = 2$. We decompose

$$[V_i] = a_i\{[\gamma_i] - \mathbf{1}\} + v_i[\mathbf{1}] \text{ in } K\mathbb{R}(\mathbb{RP}^6).$$

Since $V_1 \otimes \gamma_1 = V_1$, we use Lemma 4.5.2 to see

$$a_1 \equiv 2 \mod 2^{\phi(n)-1} = 4.$$

If $a_1 \equiv 6 \bmod 8$, then

$$w(V_1) = (1+x)^6 = 1 + x^2 + x^4 + x^6 \text{ so } v_1 \geq 6$$

which is false. If $a_1 \equiv 2 \bmod 8$, then $a_0 \equiv 6 \bmod 8$ so we may derive the final contradiction by arguing that

$$w(V_0) = (1+x)^6 = 1 + x^2 + x^4 + x^6 \text{ so } v_0 \geq 6. \quad \square$$

We now show that the results of Lemmas 4.5.7 and 4.5.8 are sharp:

4.5.9 Lemma.

(1) *Let $m \geq 2$. There exists an odd map of constant rank 2 from $\mathrm{Gr}_2^+(\mathbb{R}^m)$ to $\mathfrak{so}(m)$.*

(2) *Let $m = 8$. There exist odd maps with constant eigenvalues from $\mathrm{Gr}_2^+(\mathbb{R}^8)$ to $\mathfrak{so}(8)$ with eigenvalue structures given by $\vec{\mu} = (0,8)$, $\vec{\mu} = (0,6,2)$, or $\vec{\mu} = (2,6)$.*

(3) *Let $m = 7$. There exist odd maps with constant eigenvalues from $\mathrm{Gr}_2^+(\mathbb{R}^7)$ to $\mathfrak{so}(7)$ with eigenvalue structures given by $\vec{\mu} = (1,6)$, by $\vec{\mu} = (1,4,2)$, and by $\vec{\mu} = (3,4)$.*

4.5.10 Remark. Assertion (3) also follows from Theorem 2.9.1.

Proof. Let $\{u, v\}$ be an oriented orthonormal basis for $\pi \in \mathrm{Gr}_2^+(\mathbb{R}^m)$. We prove Assertion (1) by defining:

$$T(\pi)y := (v, y)u - (u, y)v.$$

Let the Clifford algebra Cliff(8) be the universal unital algebra generated by $\mathbb{R}^8$ subject to the Clifford commutation rules

$$v_1 * v_2 + v_2 * v_1 = -2(v_1, v_2)\,\mathrm{Id}; \tag{4.5.10.a}$$

we refer to Section 1.4 for further details. We use Theorem 1.4.4 to see that:

$$\mathrm{Cliff}(8) = \mathbb{M}_{16}(\mathbb{R}). \tag{4.5.10.b}$$

Let c be the representation of Cliff(8) on $\mathbb{R}^{16}$ which is defined by Equation (4.5.10.b). Let $\{e_i\}$ be the standard orthonormal basis for $\mathbb{R}^8$. Let G be the finite group generated by elements $\{c_1, ..., c_8, \iota\}$ subject to the relations:

$$c_i^2 = \iota,\ \iota c_i = c_i \iota,\ \iota^2 = \mathrm{Id}, \text{ and } c_i c_j = \iota c_j c_i \text{ if } i \neq j.$$

Then $c_i \rightarrow c(e_i)$ and $\iota \rightarrow -\mathrm{Id}$ defines an action of G on $\mathbb{R}^{16}$. By averaging over this action of G, we can choose a metric on $\mathbb{R}^{16}$ so that this action by G is an isometric action. Since $c(e_i)^2 = -\mathrm{Id}$, we then have that $c(e_i)$ is skew-adjoint for $1 \leq i \leq 8$. Since $c(a_1e_1 + ... + a_8e_8) = a_1c(e_1) + ... + a_8c(e_8)$, $c(v)$ is skew-adjoint for any $v \in \mathbb{R}^8$.

Let $orn := e_1 * ... * e_8 \in \mathrm{Cliff}(8)$ be the orientation class. We use the Clifford commutation relations to see that $c(orn)c(e_i) = -c(e_i)c(orn)$ and that $c(orn)^2 = \mathrm{Id}$. Thus, we may decompose

$$\begin{aligned} \mathbb{R}^{16} &= \mathbb{R}^8_+ \oplus \mathbb{R}^8_- \text{ where} \\ c(orn) &= \pm\,\mathrm{Id} \text{ on } \mathbb{R}^8_\pm \text{ and} \\ c(v) &: \mathbb{R}^8_\pm \rightarrow \mathbb{R}^8_\mp \text{ for any } v \in \mathbb{R}^8. \end{aligned}$$

These are the half spin representations. Fix an element $v_- \in S(\mathbb{R}^8_-)$. For $x, y \in \mathbb{R}^8$ and $v \in \mathbb{R}^8_+$, we define:

$$\begin{aligned} T_0(x,y)v &= \tfrac{1}{2}\{c(x)c(y) - c(y)c(x)\}v \\ T_1(x,y)v &= (c(x)v_-, v)c(y)v_- - (c(y)v_-, v)c(x)v_-. \end{aligned}$$

If $\{x, y\}$ is an orthonormal set, then $T_0(x,y) = c(x)c(y)$.

Both $T_0(x,y)$ and $T_1(x,y)$ preserve $\mathbb{R}^8_+$. We show that $T_0(x,y)$ and $T_1(x,y)$ belong to $\mathfrak{so}(\mathbb{R}^8_+)$ by computing:

$$\begin{aligned} (T_0(x,y)v_1, v_2) &= -\tfrac{1}{2}(c(y)v_1, c(x)v_2) + \tfrac{1}{2}(c(x)v_1, c(y)v_2) \\ &= -(v_1, T_0(x,y)v_2) \\ (T_1(x,y)v_1, v_2) &= (c(x)v_-, v_1)(c(y)v_-, v_2) \\ &\quad - (c(y)v_-, v_1)(c(x)v_-, v_2) \\ &= -(v_1, T_1(x,y)v_2). \end{aligned}$$

We have that $T_i(x,y) = -T_i(y,x)$ and that T_i is bilinear in (x,y) for $i=0,1$. Thus, T_0 and T_1 induce odd maps from $\mathrm{Gr}_2^+(\mathbb{R}^8)$ to $\mathfrak{so}(\mathbb{R}^8_+)$. Let $\{x,y\}$ be an oriented orthonormal basis for $\pi \in \mathrm{Gr}_2^+(\mathbb{R}^8)$. Let

$$\pi_1(x,y) := \mathrm{span}\{c(y)v_-, c(x)v_-\}.$$

We check that $T_0 = T_1$ are rotations through an angle of 90 degrees on $\pi_1(x,y)$ by computing:

$$\begin{aligned} T_0(x,y)c(y)v_- &= -c(x)v_-, \ T_1(x,y)c(y)v_- = -c(x)v_-,\\ T_0(x,y)c(x)v_- &= c(y)v_-, \ T_1(x,y)c(x)v_- = c(y)v_-. \end{aligned}$$

Furthermore, $T_1 = 0$ on $\pi_1^\perp$ while T_0 is a rotation through an angle of 90 degrees on $\pi_1^\perp$. Let $T := \alpha_0 T_0 + (\alpha_1 - \alpha_0)T_1$, where α_0 and α_1 are real constants. Then T is an odd map from $\mathrm{Gr}_2^+(\mathbb{R}^8)$ to $\mathfrak{so}(\mathbb{R}^8_+)$ such that:

$$T(\pi)^2 = -\alpha_0^2 \,\mathrm{Id} \text{ on } \pi_1^\perp \text{ and } T(\pi)^2 = -\alpha_1^2 \,\mathrm{Id} \text{ on } \pi_1.$$

Thus, T^2 has constant eigenvalues. We have $\dim \pi_1 = 2$ and $\dim \pi_1^\perp = 6$. Assertion (2) now follows by choosing the α_0 and α_1 suitably to ensure we have multiplicities

$$\vec{\mu} = (0,8), \ \vec{\mu} = (0,6), \text{ or } \vec{\mu} = (0,6,2).$$

Suppose $m = 7$. Let $\mathbb{R}^7 := \mathrm{span}\{e_1, ..., e_7\} = e_8^\perp$. Choose an element $e_+ \in S(\mathbb{R}^8_+)$ to define $\mathbb{R}^7_+ := e_+^\perp$. Let $\{x,y\}$ be an oriented orthonormal basis for a 2 plane π in $\mathbb{R}^7$. Then $\{x, y, e_8\}$ is an orthonormal subset of $\mathbb{R}^8_+$. We use the Clifford commutation relations given in Equation (4.5.10.a) and fact that $c(v)$ is skew-adjoint for $v \in \mathbb{R}^8$ to see that

$$\{e_+, c(x)c(y)e_+, c(x)c(e_8)e_+, c(y)c(e_8)e_+\}$$

is an orthonormal set. We let

$$\begin{aligned} T_0(\pi) &:= \tfrac{1}{2}\{c(x)c(y) - c(y)c(x)\} = c(x)c(y),\\ E_0(\pi) &:= \mathrm{span}\{e_+, c(x)c(y)e_+\},\\ E_1(\pi) &:= \mathrm{span}\{c(x)c(e_8)e_+, c(y)c(e_8)e_+\} \subset \mathbb{R}^7_+, \text{ and}\\ E_2(\pi) &:= (E_0(\pi) \oplus E_1)^\perp \subset \mathbb{R}^7_+. \end{aligned}$$

The subspaces $E_i(\pi)$ are invariant under the action of $T_0(\pi)$. We have $E_1(\pi) \subset \mathbb{R}^7_+$ and $E_2(\pi) \subset \mathbb{R}^7_+$. We let

$$R_i(\pi) := T_0(\pi)\rho_i$$

be the restriction of $T_i(\pi)$ to $E_i(\pi)$, where $\rho_i(\pi)$ is orthonormal projection on $E_i(\pi)$. We note that $E_i(-\pi) = E_i(\pi)$ so $R_i(-\pi) = -R_i(\pi)$. We let

$$T := c_1 R_1 + c_2 R_2.$$

Then T is an odd map from $\mathrm{Gr}_2^+(\mathbb{R}^7)$ to $\mathfrak{so}(\mathbb{R}^7_+)$ with constant eigenvalues. We may now choose constants (c_1, c_2) to ensure T has the appropriate eigenvalue multiplicities. □

Before beginning the proof of Assertion (5) of Theorem 4.1.4, we must establish some technical results.

4.5.11 Lemma.

(1) *Let $m \geq 3$. Suppose that $\binom{m-1-b}{b} = 0$ in the field $\mathbb{Z}_2$ for all b such that $1 \leq b \leq \frac{m}{2} - 1$. Then m is a power of 2.*

(2) *Suppose that $\sum_{a+b+c=q}(u_1 + u_2)^a {u_1}^b {u_2}^c = 0$ in the polynomial ring $\mathbb{Z}_2[u_1, u_2]$. Then $q + 3$ is a power of 2.*

(3) *Suppose that $\sum_{b+c=q} u_1^b u_2^c = \sum_{a+b+c=q}(u_1 + u_2)^a u_1^b u_2^c$ in the polynomial ring $\mathbb{Z}_2[u_1, u_2]$. Then $q + 2$ is a power of 2.*

Proof. Suppose that $\binom{m-1-b}{b} = 0$ in $\mathbb{Z}_2$ for all b so that $1 \leq b \leq \frac{m}{2} - 1$. We suppose that m is not a power of 2 and argue for a contradiction. As m is not a power of 2, m has a non-trivial 2-adic expansion of the form:

$$m = 2^j + c_{j-1}2^{j-1} + \ldots + c_{n+1}2^{n+1} + 2^n \text{ where } j > n.$$

Let $b = 2^n$. If $n = 0$, then $b = 1$ and $b \leq \frac{m}{2} - 1$. If $n \geq 1$, then

$$2b \leq 2^j \leq m - 2^n \leq m - 2.$$

Thus, b is an admissible value. We compute:

$$\begin{aligned} m - 1 - b =& 2^j + c_{j-1}2^{j-1} + \ldots + c_{n+1}2^{n+1} - 1 \\ =& \ldots + 2^n + 2^{n-1} + \ldots + 1. \end{aligned}$$

Thus, by Lemma 4.3.24, $(1 + x)^{m-1-b} = 1 + \ldots + x^{2^n} + \ldots$ so

$$\binom{m-1-b}{b} = 1 \text{ in } \mathbb{Z}_2.$$

This contradiction establishes Assertion (1).

Assertions (2) and (3) follow from work of Stong [149]; for the convenience of the reader, we reproduce his argument. We set $u_1 = u$, $u_2 = 1$, and define:

(4.5.11.a) $$\Phi_q := \textstyle\sum_{0\le a+b\le q}(u+1)^a u^b \text{ in } \mathbb{Z}_2[u].$$

We have the identity

(4.5.11.b) $$u^{n+1}+1 = (u+1)(u^n + u^{n-1} + \ldots + u + 1) \text{ in } \mathbb{Z}_2[u].$$

We multiply Equation (4.5.11.a) by $(u+1)$ and use Equation (4.5.11.b) with $n = q - a$ to compute:

$$\begin{aligned}(u+1)\Phi_q =& \textstyle\sum_{0\le a\le q}(u+1)^a(u+1)(u^{q-a}+u^{q-a-1}+\ldots+u+1)\\ =& \textstyle\sum_{0\le a\le q}(u+1)^a(u^{q+1-a}+1)\\ =& u^{q+1}\textstyle\sum_{0\le a\le q}(\frac{u+1}{u})^a + \sum_{0\le a\le q}(u+1)^a.\end{aligned}$$

We sum the geometric series which are involved to see

$$\begin{aligned}(u+1)\Phi_q =& u^{q+1}\{1+\tfrac{u+1}{u}\}^{-1}\{1+(\tfrac{u+1}{u})^{q+1}\}\\ &+\{1+(u+1)\}^{-1}\{1+(u+1)^{q+1}\}.\end{aligned}$$

Setting $1+\frac{u+1}{u} = \frac{1}{u}$ then yields

(4.5.11.c) $$\begin{aligned}(u+1)\Phi_q =& u^{q+2}\{1+(\tfrac{u+1}{u})^{q+1}\} + \tfrac{1}{u}\{1+(u+1)^{q+1}\}\\ =& \tfrac{1}{u}\{u^{q+3}+u^2(u+1)^{q+1}+1+(u+1)^{q+1}\}\\ =& \tfrac{1}{u}\{u^{q+3}+1+(u^2+1)(u+1)^{q+1}\}\\ =& \tfrac{1}{u}\{u^{q+3}+1+(u+1)^{q+3}\}.\end{aligned}$$

We multiply Equation (4.5.11.c) by u to see:

(4.5.11.d) $$u(u+1)\Phi_q = 1+u^{q+3}+(1+u)^{q+3} \text{ in } \mathbb{Z}_2[u].$$

If the assumptions of Assertion (2) hold, then $\Phi_q = 0$. We use Equation (4.5.11.d) and Lemma 4.3.24 to complete the proof of Assertion (2) since

$$(1+u)^{q+3} = 1+u^{q+3}.$$

If the assumptions of Assertion (3) hold, then $\Phi_q = \sum_{0\le b\le q} u^b$. Thus:

(4.5.11.e) $$u(u+1)\Phi_q = u(u+1)(u^q+\ldots+1) = u(u^{q+1}+1).$$

We use Equations (4.5.11.d) and (4.5.11.e) to see that $q+2$ is a power of 2 by computing:

$$\begin{aligned}&u(u^{q+1}+1) = 1+u^{q+3}+(1+u)^{q+3},\\ &(1+u)^{q+3} = 1+u+u^{q+2}+u^{q+3} = (1+u)(1+u^{q+2}),\\ &(1+u)^{q+2} = 1+u^{q+2}. \quad \square\end{aligned}$$

Theorem 4.1.4 (5) will follow from the following Lemma:

4.5.12 Lemma. *Let $T : \mathrm{Gr}_2^+(\mathbb{R}^m) \to \mathfrak{so}(m+2)$ be an odd map of constant rank $r = 4$. Let $m \geq 10$ be even. Let $U_1 := \mathrm{range}(T) \in \mathrm{Vect}^4_{\mathbb{R}}(\mathrm{Gr}_2(\mathbb{R}^m))$.*

(1) *$w_1(U_1) = 0$ and $\bar{x}^3 + \bar{x}w_2(U_1) + w_3(U_1) = 0$.*

(2) *There exist constants S,T, and U in $\mathbb{Z}_2$ so that we have $w(U_1) = 1 + \bar{x}^2 + S\bar{y} + S\bar{x}\bar{y} + T\bar{x}^2\bar{y} + U\bar{y}^2$.*

(3) *One of the following possibilities holds:*
   3a) *$w(U_1) = 1 + \bar{x}^2$.*
   3b) *$w(U_1) = 1 + \bar{x}^2 + \bar{y} + \bar{x}\bar{y}$.*
   3c) *$w(U_1) = 1 + \bar{x}^2 + \bar{y}^2$.*

(4) *If $w(U_1) = 1 + \bar{x}^2$, then m is a power of 2.*

(5) *If $w(U_1) = 1 + \bar{x}^2 + \bar{y} + \bar{x}\bar{y}$, then $m + 2$ is a power of 2.*

(6) *We have $w(U_1) \neq 1 + \bar{x}^2 + \bar{y}^2$.*

Proof. We follow the discussion in Zhang [163]. We begin with some preliminary remarks. Let

$$U_0 := \ker(T) \in \mathrm{Vect}^{m-2}_{\mathbb{R}}(\mathrm{Gr}_2(\mathbb{R}^m)).$$

Then:

$$U_0 \oplus U_1 = (m+2) \cdot \mathbf{1} \text{ and } U_1 \otimes \bar{\gamma}_1 = U_1.$$

There are no relations in the cohomology algebra of the Grassmannian in degrees less than $m - 1$. More precisely, we can use Theorem 4.3.12 to see:

$$H^\nu(\mathrm{Gr}_2(\mathbb{R}^m); \mathbb{Z}_2) = \mathrm{span}_{\mathbb{Z}_2}\{\bar{x}^a\bar{y}^b\}_{a+2b=\nu} \text{ for } \nu \leq m-2. \tag{4.5.12.a}$$

We use the splitting principle described in Lemma 4.3.10 to prove Assertion (1). Let π_F be the projection from the Flag bundle $\mathrm{Flag}(U_1)$ to $\mathrm{Gr}_2(\mathbb{R}^m)$. Then π_F^* is a monomorphism in $\mathbb{Z}_2$ cohomology and we may decompose

$$\pi^*U_1 = L_1 \oplus L_2 \oplus L_3 \oplus L_4$$

as the direct sum of line bundles. Let

$$x_i := w_1(L_i),\ \check{\gamma}_1 := \pi_F^*\bar{\gamma}_1, \text{ and}$$
$$\check{x} := w_1(\pi_F^*\bar{\gamma}_1) = \pi_F^*(\bar{x}).$$

We use the fact that $U_1 = U_1 \otimes \bar{\gamma}_1$ to compute:

$$\begin{aligned} 0 =& w(\pi_F^*(U_1 \otimes \bar{\gamma}_1)) - w(\pi_F^*U_1) = \textstyle\prod_i w(L_i \otimes \check{\gamma}_1) - \prod_i w(L_i) \\ =& \textstyle\sum_i\{(\check{x} + x_i) - x_i\} + \sum_{i<j}\{(\check{x} + x_i)(\check{x} + x_j) - x_ix_j\} \\ &+ \textstyle\sum_{i<j<k}\{(\check{x} + x_i)(\check{x} + x_j)(\check{x} + x_k) - x_ix_jx_k\} \\ &+ (\check{x} + x_1)(\check{x} + x_2)(\check{x} + x_3)(\check{x} + x_4) - x_1x_2x_3x_4. \end{aligned} \tag{4.5.12.b}$$

The linear terms in Equation (4.5.12.b) yield no information. We set the quadratic terms to zero to see:

$$0 = 6\breve{x}^2 + 3\breve{x}\textstyle\sum_i x_i = \pi_F^*(\bar{x}w_1(U_1)). \tag{4.5.12.c}$$

The cubic terms in Equation (4.5.12.b) yield no additional information. We set the quartic terms to zero and use Equation (4.5.12.c) to see:

$$\begin{aligned} 0 =& \breve{x}^4 + \breve{x}^3(\textstyle\sum_i x_i) + \breve{x}^2 \sum_{i<j} x_i x_j + \breve{x} \sum_{i<j<k} x_i x_j x_k \\ =& \pi_F^*\{\bar{x}^4 + \bar{x}^2 w_2(U_1) + \bar{x} w_3(U_1)\}. \end{aligned} \tag{4.5.12.d}$$

Since π_F^* is injective, we use Equations (4.5.12.c) and (4.5.12.d) to see:

$$\begin{aligned} 0 =& \bar{x}w_1(U_1), \text{ and} \\ 0 =& \bar{x}^4 + \bar{x}^2 w_2(U_1) + \bar{x} w_3(U_1). \end{aligned} \tag{4.5.12.e}$$

As $m \geq 8$, we use Equation (4.5.12.a) to see that we may divide the relations of Display (4.5.12.e) by $\bar{x}$ to complete the proof of Assertion (1) by showing

$$\begin{aligned} 0 &= w_1(U_1), \text{ and} \\ 0 &= \bar{x}^3 + \bar{x}w_2(U_1) + w_3(U_1). \end{aligned}$$

Let $i(x) := \operatorname{span}\{x, e_m\}$ define the natural inclusion i from $\mathbb{RP}^{m-2}$ to $\mathrm{Gr}_2(\mathbb{R}^m)$. We expand

$$[i^*(U_1)] = a_1\{[\gamma_1] - [\mathbf{1}]\} + 4 \cdot [\mathbf{1}] \text{ in } K\mathbb{R}(\mathbb{RP}^{m-2}).$$

Since $i^*(U_1) \otimes \gamma_1 = i^*(U_1)$, we may apply Lemma 4.5.2 to see that $2a_1 \equiv 4 \mod 2^{\phi(m-2)}$. Since $m \geq 10$,

$$\begin{aligned} &\phi(m-2) \geq \phi(8) = 4 \text{ so } a_1 \equiv 2 \bmod 8 \text{ and} \\ &w(i^*U_1) = (1+x)^2. \end{aligned}$$

By Lemma 4.5.6, $i^*\bar{x} = x$ and $i^*\bar{y} = 0$. Thus, we may expand:

$$w(U_1) = 1 + \bar{x}^2 + S\bar{y} + Q\bar{x}\bar{y} + T\bar{x}^2\bar{y} + U\bar{y}^2.$$

We apply Assertion (1) to see $Q = S$ and prove the Assertion (2).

We use Lemma 4.3.18 to see $w(U_1)w_4(U_1) = Sq\{w_4(U_1)\}$. We compute:

$$\begin{aligned}
&w(U_1)w_4(U_1)\\
&\quad=(1+\bar{x}^2+S\bar{y}+S\bar{x}\bar{y}+T\bar{x}^2\bar{y}+U\bar{y}^2)(T\bar{x}^2\bar{y}+U\bar{y}^2)\\
&\quad=T\bar{x}^2\bar{y}+T\bar{x}^4\bar{y}+ST\bar{x}^2\bar{y}^2+ST\bar{x}^3\bar{y}^2+T\bar{x}^4\bar{y}^2+TU\bar{x}^2\bar{y}^3\\
\text{(4.5.12.f)}\quad &\quad+U\bar{y}^2+U\bar{x}^2\bar{y}^2+SU\bar{y}^3+SU\bar{x}\bar{y}^3+TU\bar{x}^2\bar{y}^3+U\bar{y}^4,\\
&Sq\{w_4(U_1)\}=Sq\{T\bar{x}^2\bar{y}+U\bar{y}^2\}\\
&\quad=T(\bar{x}+\bar{x}^2)^2(\bar{y}+\bar{x}\bar{y}+\bar{y}^2)+U(\bar{y}+\bar{x}\bar{y}+\bar{y}^2)^2\\
&\quad=T\bar{x}^2\bar{y}+T\bar{x}^4\bar{y}+T\bar{x}^3\bar{y}+T\bar{x}^5\bar{y}+T\bar{x}^2\bar{y}^2+T\bar{x}^4\bar{y}^2\\
\text{(4.5.12.g)}\quad &\quad+U\bar{y}^2+U\bar{x}^2\bar{y}^2+U\bar{y}^4.
\end{aligned}$$

Since $w(U_1)w_4(U_1) = Sq\{w_4(U_1)\}$, we use Equation (4.5.12.a) to see that we may equate the coefficients of the monomials $\bar{x}^3\bar{y}$ and $\bar{y}^3$ in Equation (4.5.12.f) with the corresponding coefficients in Equation (4.5.12.g); this implies $0 = T$ and $SU = 0$. Assertion (3) now follows.

To prove Assertion (4), we suppose that $w(U_1) = 1 + \bar{x}^2$. Let $m = 2\ell$. Since $w(U_0)w(U_1) = 1$, we have

$$w(U_0) = w(U_1)^{-1} = 1 + \bar{x}^2 + ... + \bar{x}^{2\ell} + ...$$

Since $2\ell > m - 2$, $w_{2\ell}(U_0) = \bar{x}^{2\ell} = 0$ in $H^{2\ell}(\mathrm{Gr}_2(\mathbb{R}^m); \mathbb{Z}_2)$ for dimensional reasons. Thus, by Theorem 4.3.12, there exist constants c_1 and c_2 in $\mathbb{Z}_2$ so:

$$\bar{x}^{2\ell} = c_1\bar{x}w^{\perp}_{2\ell-1} + c_2 w^{\perp}_{2\ell} \text{ in } \mathbb{Z}_2[\bar{x}, \bar{y}].$$

Expand:

$$\text{(4.5.12.h)}\qquad w^{\perp} = (1+\bar{x}+\bar{y})^{-1} = \textstyle\sum_{\nu\geq 0}(\bar{x}+\bar{y})^{\nu} = \sum_{a,b}\binom{a+b}{b}\bar{x}^a\bar{y}^b.$$

Since $\bar{y}^{\ell}$ survives in $w^{\perp}_{2\ell}$, $c_2 = 0$. Consequently, $c_1 = 1$ and $\bar{x}^{m-1} = w^{\perp}_{m-1}$. We use Equation (4.5.12.h) to see

$$\text{(4.5.12.i)}\qquad \binom{m-1-b}{b} = 0 \text{ in } \mathbb{Z}_2 \text{ if } 1 \leq b \leq \tfrac{m}{2} - 1.$$

Thus, by Lemma 4.5.11, m is a power of 2. Assertion (4) now follows.

To prove Assertion (5), we suppose that $w(U_1) = 1 + \bar{x}^2 + \bar{y} + \bar{x}\bar{y}$. We apply the splitting principle described in Theorem 4.3.10 to the classifying bundle γ_2. Then $\pi_F^*(\gamma_2) = L_1 \oplus L_2$. Let $u_i := w_1(L_i)$. We compute:

$$\begin{aligned}
\pi_F^*(w(\gamma_2)) =&(1+u_1)(1+u_2),\\
\pi_F^*(\bar{x}) =&w_1(\pi_F^*(\gamma_2)) = u_1 + u_2,\\
\pi_F^*(\bar{y}) =&w_2(\pi_F^*(\gamma_2)) = u_1 u_2,\\
\pi_F^*(w^\perp) =&(1+u_1)^{-1}(1+u_2)^{-1} = \textstyle\sum_{b,c} u_1^b u_2^c,\\
\pi_F^*(w(U_1)) =&\pi_F^*\{(1+\bar{x})(1+\bar{x}+\bar{y})\}\\
=&(1+u_1+u_2)(1+u_1)(1+u_2) \text{ and}\\
\pi_F^*(w(U_0)) =&\textstyle\sum_{a,b,c}(u_1+u_2)^a u_1^b u_2^c.
\end{aligned}$$

For dimensional reasons, $\pi_F^*(w_{m-1}(U_0)) = 0$. Since π_F^* is injective and since range(π_F^*) generates the symmetric polynomials in $\mathbb{Z}_2[u_1, u_2]$, we can use Theorem 4.3.12 to see that one of the following two alternatives holds:

(1) We have $\pi_F^*(w_{m-1}^\perp) = \pi_F^*(w(U_0)_{m-1})$ in $\mathbb{Z}_2[u_1, u_2]$. This implies

$$\textstyle\sum_{b+c=m-1} u_1^b u_2^c = \sum_{a+b+c=m-1}(u_1+u_2)^a u_1^b u_2^c \text{ in } \mathbb{Z}_2[u_1, u_2].$$

We use Lemma 4.5.11 to see $m-1+2 = m+1$ is a power of 2. This is impossible as m is even.

(2) We have $\pi_F^*(w(U_0)_{m-1}) = 0$ in $\mathbb{Z}_2[u_1, u_2]$. This implies

$$0 = \textstyle\sum_{a+b+c=m-1}(u_1+u_2)^a u_1^b u_2^c \text{ in } \mathbb{Z}_2[u_1, u_2].$$

We use Lemma 4.5.11 to see $m-1+3 = m+2$ is a power of 2; this completes the proof of Assertion (5).

We complete the proof by showing that $w(U_1) \neq 1 + \bar{x}^2 + \bar{y}^2$. Since $\dim U_0 = m-2$, $w_m(U_0) = 0$ in $H^m(\mathrm{Gr}_2(\mathbb{R}^m); \mathbb{Z}_2)$ for dimensional reasons. Thus, $w_m(U_0) = c_1 w_m^\perp + c_2 \bar{x} w_{m-1}^\perp$. Since $\bar{x}^m$ survives in $w_m(U_0)$, in $w_m^\perp$, and in $\bar{x} w_{m-1}^\perp$, we see that $c_1 + c_2 = 1$. We compute:

$$\begin{aligned}
(\bar{x}+\bar{y})^{4j} &= \bar{x}^{4j} + 0\cdot\bar{x}^{4j-1}\bar{y} + 0\cdot\bar{x}^{4j-2}\bar{y}^2 + \ldots\\
(\bar{x}+\bar{y})^{4j+1} &= \bar{x}^{4j+1} + 1\cdot\bar{x}^{4j}\bar{y} + 0\cdot\bar{x}^{4j-1}\bar{y}^2 + \ldots\\
(\bar{x}+\bar{y})^{4j+2} &= \bar{x}^{4j+2} + 0\cdot\bar{x}^{4j+1}\bar{y} + 1\cdot\bar{x}^{4j}\bar{y}^2 + \ldots\\
(\bar{x}+\bar{y})^{4j+3} &= \bar{x}^{4j+3} + 1\cdot\bar{x}^{4j+2}\bar{y} + 1\cdot\bar{x}^{4j+1}\bar{y}^2 + \ldots
\end{aligned} \tag{4.5.12.j}$$

We sum the terms in Display (4.5.12.j) over j and recombine terms to see:

$$\begin{aligned}
w_{4k}^\perp &= \bar{x}^{4k} + 1\cdot\bar{x}^{4k-2}\bar{y} + 1\cdot\bar{x}^{4k-4}\bar{y}^2 + \ldots\\
w_{4k+1}^\perp &= \bar{x}^{4k+1} + 0\cdot\bar{x}^{4k-1}\bar{y} + 1\cdot\bar{x}^{4k-3}\bar{y}^2 + \ldots\\
w_{4k+2}^\perp &= \bar{x}^{4k+2} + 1\cdot\bar{x}^{4k}\bar{y} + 0\cdot\bar{x}^{4k-2}\bar{y}^2 + \ldots\\
w_{4k+3}^\perp &= \bar{x}^{4k+3} + 0\cdot\bar{x}^{4k+1}\bar{y} + 0\cdot\bar{x}^{4k-1}\bar{y}^2 + \ldots
\end{aligned} \tag{4.5.12.k}$$

Similarly, we compute:

$$
(4.5.12.l)\qquad
\begin{aligned}
(\bar{x}^2+\bar{y}^2)^{2j} &= \bar{x}^{4j} + 0\cdot\bar{x}^{4j-2}\bar{y}^2+\ldots\\
(\bar{x}^2+\bar{y}^2)^{2j+1} &= \bar{x}^{4j+2} + 1\cdot\bar{x}^{4j}\bar{y}^2+\ldots\\
w_{4k}(U_0) &= \bar{x}^{4k} + 1\cdot\bar{x}^{4k-4}\bar{y}^2+\ldots\\
w_{4k+2}(U_0) &= \bar{x}^{4k+2} + 0\cdot\bar{x}^{4k-2}\bar{y}^2+\ldots
\end{aligned}
$$

We use Equations (4.5.12.k) and (4.5.12.l) to see that

$$
\begin{aligned}
&w_{4k}(U_0) \neq w_{4k}^{\perp}, && w_{4k}(U_0) \neq \bar{x}w_{4k-1}^{\perp},\\
&w_{4k+2}(U_0) \neq w_{4k+2}^{\perp}, &\text{and}\ & w_{4k+2}(U_0) \neq \bar{x}w_{4k+1}^{\perp}. \quad \square
\end{aligned}
$$

Bibliography

[1] J. Adams, *Vector fields on spheres*, Annals of Math. **75** (1962), 603–632.

[2] W. Adkins and S. Weintraub, *Algebra - an approach via module theory*, Graduate texts in Mathematics #136, Springer-Verlag, New York (1992) ISBN 0-387-97839-9.

[3] D. V. Alekseevsky, *Riemannian manifolds with exceptional holonomy groups*, Funkts. Anal. Prilozh. **2** (1968), 1–10; English translation, Funct. Anal. Appl. **2** (1968), 97–105.

[4] D. V. Alekseevsky, N. Blažić, N. Bokan, and Z. Rakić, *Self-duality and pointwise Osserman manifolds*, Arch. Math. (Brno) **35** (1999), 193–201.

[5] W. Ambrose and I. M. Singer, *A theorem on holonomy*, Trans. Amer. Math. Soc. **75** (1953), 428-443.

[6] V. Apostolov and T. Draghici, *Almost Kaehler* 4 *manifolds with J invariant Ricci tensor and special Weyl tensor*, Q. J. Math. **51** (2000), 275–294.

[7] M. F. Atiyah, *K-Theory*, W. A. Benjamin, Inc., New York, 1967.

[8] M F Atiyah, R Bott, and V K Patodi, *On the heat equation and the index theorem*, Invent. Math. **19** (1973), 279–330.

[9] ———, Errata **28** (1975), 277–280.

[10] M. F. Atiyah, R. Bott, and A. Shapiro, *Clifford Modules*, Topology **3** **suppl. 1** (1964), 3–38.

[11] M. Baros and A. Romero, *Indefinite Kähler manifolds*, Math. Ann. **261** (1982), 55–62.

[12] J. Baum, *Pointset Topology*, Prentice-Hall (1964).

[13] M. Belger and G. Stanilov, *About the Riemannian geometry of some curvature operators*, Annuaire de l'Univ Sofia **1** (1995).

[14] J. Berndt and L. Vanhecke, *Two natural generalizations of locally symmetric spaces*, Differential Geom. Appl. **2** (1992), 57–80.

[15] ———, *Geodesic spheres and generalizations of symmetric spaces*, Boll. Un. Mat. Ital. A **7** (1993), 125–134.

[16] A. L. Besse, *Manifolds all of whose geodesics are closed*, Ergebnisse der Mathematik und ihrer Grenzgebiete, 93, Springer-Verlag, Berlin, 1978.

[17] ———, *Einstein manifolds*, Ergebnisse der Mathematik und ihrer Grenzgebiete, 3. Folge, 10, Springer-Verlag, Berlin, 1987.

[18] N. Blažić, *Paraquaternionic projective space and pseudo-Riemannian geometry*, Publ. Inst. Math. (Beograd) **60** (1996), 101–107.

[19] N. Blažić, N. Bokan and P. Gilkey, *A Note on Osserman Lorentzian manifolds*, Bull. London Math. Soc. **29** (1997), 227-230.

[20] N. Blažić, N. Bokan, P. Gilkey, and Z. Rakić, *Pseudo-Riemannian Osserman manifolds*, Balkan J. Geom. Appl **2** (1997), 1–12.

[21] N. Blažić, N. Bokan and Z. Rakić, *The first order PDE system for type III Osserman manifolds*, Publ. Inst. Math. (Beograd) (N.S.) **62** (1997), 113–119.

[22] ———, *Nondiagonalizable timelike (spacelike) Osserman* $(2,2)$ *manifolds.*, Saitama Math. J. **16** (1998), 15–22.

[23] ———, *A note on the Osserman conjecture and isotropic covariant derivative of curvature*, Proc. Amer. Math. Soc. **128** (2000), 245–253.

[24] N. Blažić and S. Vukmirović, *Solutions of Yang-Mills equations on generalized Hopf bundles*, Preprint 2001.

[25] E. Boeckx, *Einstein-like semi-symmetric spaces*, Arch. Math. (Brno) **29** (1993), 235–240.

[26] E. Boeckx, O. Kowalski and L. Vanhecke, *Riemannian manifolds of conullity two*, World Scientific Publishing Co., Inc., River Edge, NJ ISBN: 981-02-2768-X, 1996.

[27] A. Bonome, P. Castro, and E. García-Rio, *Generalized Osserman Four-dimensional manifolds*, preprint (2000).

[28] A. Bonome, P. Castro, E. García-Rio, L. Hervella and R. Vásquez-Lorenzo, *On Osserman semi-Riemannian manifolds*, preprint (1997).

[29] ———, *Nonsymmetric Osserman indefinite Kähler manifolds*, Proc. Amer. Math. Soc. **126** (1998), 2763–2769.

[30] ———, *Pseudo-Riemannian manifolds with simple Jacobi operators*, J. Math. Soc. Japan (to appear).

[31] A. Borel, *Sur la cohomologie des espaces fibrés principaux et des spaces homogènes de groupes de Lie compactes*, Ann. of Math. **57** (1953), 115–207.

[32] A. Borowiec, M. Ferraris, M. Francaviglia, I. Volovich, *Almost complex and almost product Einstein manifolds from a variational principle*, J. Math. Phys. **7** (1999), 3446–3464.

[33] R. Bott, *Lectures on $K(X)$*, Mathematics Lecture Note Series, W. A. Benjamin, New York (1969).

[34] R. Bott and T. Loring, *Differential Forms in Algebraic Topology*, Graduate Text in Mathematics, vol. 82, Springer-Verlag, Berlin, 1982.

[35] R. B. Brown and A. Gray, *Riemannian manifolds with holonomy group Spin(9)*, Differential geometry — in honor of Kentaro Yano (S. Kobayashi, M. Obata and T. Takahasi, eds.), Kinokuniya, Tokyo, 1972, pp. 41–59.

[36] R. Bryant and S. Salamon, *On the construction of some complete metrics with exceptional holonomy*, Duke Math. J. **58** (1989), 829–850.

[37] K. Burns and A. Katok, *Manifolds with non-positive curvature*, Ergodic Theory Dynam. Systems **5** (1985), 307–317.

[38] E. Calabi, *Métriques kählériennes et fibrés holomorphes*, Ann. Sci. Ecole Norm. Sup. Ann. Sci. Ecole Norm. Sup., 4e sér **12** (1979), 269–294.

[39] ______, *Isometric Families of Kähler Structures*, The Chern Symposium, 1979 (W.-Y. Hsiang, S. Kobayashi, I. M. Singer, A. Weinstein, J. Wolf and H.-H. Wu, eds.), Springer, New York-Berlin, 1980, pp. 23–39.

[40] P. Carpenter, A. Gray and T. J. Willmore, *The curvature of Einstein symmetric spaces*, Q. J. Math. Oxford **33** (1982), 45–64.

[41] B.-Y. Chen and L. Vanhecke, *Differential geometry of geodesic spheres*, J. Reine Angew. Math. **325** (1981), 29–67.

[42] Q.-S. Chi, *A curvature characterization of certain locally rank-one symmetric spaces*, J. Differential Geom **28** (1988), 187–202.

[43] ______, *Quaternionic Kähler manifolds and a characterization of two-point homogeneous spaces*, Illinois J. Math. **35** (1991), 408–418.

[44] ______, *Curvature characterization and classification of locally rank one symmetric spaces*, Pacific J. Math. **150** (1991), 31–42.

[45] V. Cruceanu, P. Fortuny, and P. M. Gadea, *A survey on paracomplex geometry*, Rocky Mountain Math. J. **26** (1996), 83–115.

[46] M. Dajczer and K. Nomizu, *On sectional curvature of indefinite metrics II*, Math. Ann. **247** (1980), 279–282.

[47] E. Damek and F. Ricci, *A class of nonsymmetric harmonic Riemannian spaces*, Bull. Amer. Math. Soc. **27** (1992), 139–142.

[48] M. Djorić and L. Vanhecke, *Almost Hermitian geometry, geodesic spheres and symmetries*, Math. J. Okayama Univ. **32** (1990), 187–206.

[49] H. Donnelly, *Symmetric Einstein spaces and spectral geometry*, Indiana Univ. Math. J. **24** (1974), 603–606.

[50] I. Dotti and M. Druetta, *Negatively curved homogeneous Osserman spaces*, Differential Geom. Appl. **11** (1999), 163–178.

[51] ______, *Osserman-p spaces of Iwasawa type*, Masaryk Univ., Brno, 1999, Differential geometry and applications (Brno, 1998), 61–72.

[52] ______, *Damek Ricci spaces satisfying the Osserman p condition*, preprint.

[53] S. Eilenberg and N. Steenrod, *Foundations of Algebraic Topology*, Princeton University Press, Princeton (1952).

[54] J.-H. Eschenburg, *A note on symmetric and harmonic spaces*, J. London Math. Soc. **21** (1980), 541–543.

[55] M. Falcitelli, A. Farinola, and S. Salamon, *Almost-Hermitian geometry*, Differential Geom. Appl. **4** (1994), 259–282.

[56] B. Fiedler, *About the structure of algebraic curvature tensors*, preprint.

[57] Th. Friedrich, I. Kath, A. Moroianu, and U. Semmelman, *On Nearly Parallel G_2 Structures*, J. Geom. Phys. **23** (1997), 259–286.

[58] P. M. Gadea, *The paracomplex projective spaces as symmetric and natural spaces*, Indian J. pure appl. Math. **23** (1992), 261–275.

[59] P. M. Gadea and J. M. Masqué, *Spaces of constant para-holomorphic sectional cuvature*, Pacific Journal of Math. **136** (1989), 85–101.

[60] ______, *Classification of non-flat para-Kählerian space forms*, Houston J. Math. **21** (1995), 89–94.

[61] K. Galicki and H. B. Lawson, *Quaternionic reduction and quaternionic orbifolds*, Math. Ann. **282** (1988), 1–21.

[62] K. Galicki and T. Nitta, *Non-zero scalar curvature generalizations of the ALE hyperkähler metrics*, J. Math. Phys. **33** (1992), 1765–1771.

[63] E. García-Rio and D. N. Kupeli, *Four-Dimensional Osserman Lorentzian Manifolds*, New developments in differential geometry. Proceedings of the colloquium on differential geometry, Debrecen, Hungary, July 26-30 **350** (1996), Math. Appl. 350 Kluwer Academic Publishers, 201–211.

[64] E. García-Rio, D. N. Kupeli, and M. Vázquez-Abal, *On a problem of Osserman in Lorentzian geometry*, Differential Geom. Appl. **7** (1997), 85–100.

[65] E. García-Rio, D. N. Kupeli, M. Vázquez-Abal, and R. Vázquez-Lorenzo, *Affine Osserman connections and their Riemann extensions*, Differential Geom. Appl. **11** (1999), 145–153.

[66] E. García-Rio, D. N. Kupeli, and R. Vázquez-Lorenzo, *Osserman manifolds in semi-Riemannian geometry*, Lecture notes in Mathematics, Springer Verlag (to appear).

[67] E. García-Rio, M. E. Vázquez-Abal and R. Vázquez-Lorenzo, *Non-symmetric Osserman pseudo-Riemannian manifolds*, Proc. Amer. Math. Soc. **126** (1998), 2771–2778.

[68] E. García-Rio and R. Vázquez-Lorenzo, *Four-Dimensional Osserman Symmetric Spaces*, Geometria Dedicata (to appear).

[69] G. W. Gibbons and C. N. Pope, *The positive action conjecture and asymptotically Euclidean metrics in quantum gravity*, Comm. Math. Phys. **66** (1979), 267–290.

[70] P. Gilkey, *Manifolds whose higher odd order curvature operators have constant eigenvalues at the basepoint*, J. Geom. Anal. **2** (1992), 151–156.

[71] ———, *Manifolds whose curvature operator has constant eigenvalues at the basepoint*, J. Geom. Anal. **4** (1994), 155–158.

[72] ———, *Invariance Theory, the heat equation, and the Atiyah-Singer index theorem*, CRC Press ISBN 0-8493-7874-4 (1994). Bibliography by H. Schroeder (Dortmund).

[73] ———, *Generalized Osserman Manifolds*, Abh. Math. Sem. Univ. Hamburg **68** (1998), 127–127.

[74] ———, *Riemannian manifolds whose skew-symmetric curvature operator has constant eigenvalues II*, in Differential geometry and applications (eds Kolar, Kowalski, Krupka, and Slovak) Publ Masaryk University Brno Czech Republic ISBN 80-210-2097-0 (1999), 73–87.

[75] ———, *Algebraic curvature tensors which are p Osserman*, Differential Geom. Appl. **14** (2001), 297–311.

[76] ———, *Bundles over projective spaces and algebraic curvature tensors*, J. Geom. (to appear).

[77] P. Gilkey and R. Ivanova, *Geometric consequences of some algebraic properties of the curvature tensor*, Bull. Mathematiques Soc. Math. Roumanie **45** (2000), 255–265.

[78] ———, *The geometry of the skew-symmetric curvature operator in the complex setting*, Contemp. Math., Proceedings of the Bilbao Conference (to appear).

[79] ———, *Complex IP pseudo-Riemannian algebraic curvature tensors*, Proceedings of the 2000 Warsaw Conference on Differential Geometry, Banach Center Publications (to appear).

[80] ———, *The Jordan normal form of Osserman algebraic curvature tensors.*, Results in Math. (to appear).

[81] ———, *The Jordan normal form of higher order Osserman algebraic curvature tensors*, preprint.

[82] P. Gilkey, J. Leahy, and JH. Park, *Spectral Geometry, Riemannian Submersions, and the Gromov-Lawson Conjecture*, Studies in Advanced Mathematics, Chapman & Hall Boca Raton, FL., ISBN 0-8493-8277-7, 1999.

[83] P. Gilkey, J. V. Leahy, and H. Sadofsky, *Riemannian manifolds whose skew-symmetric curvature operator has constant eigenvalues*, Indiana Univ. Math. J. **48** (1999), 615–634.

[84] P. Gilkey and U. Semmelman, *Proceedings of the Symposium in Contemporary Mathematics in honor of 125 years of faculty of Math at Univ Belgrade*, Neda Bokan Editor ISBN 86-7589-014-1, pp 1–12, 2000.

[85] P. Gilkey, G. Stanilov, and V. Videv, *Pseudo Riemannian Manifolds whose Generalized Jacobi Operator has Constant Characteristic Polynomial*, J. Geom. **62** (1998), 144–153.

[86] P. Gilkey and I. Stavrov, *Curvature tensors whose Jacobi or Szabó operator is nilpotent on null vectors*, preprint.

[87] P. Gilkey, A. Swann and L. Vanhecke, *Isoparametric geodesic spheres and a conjecture of Osserman regarding the Jacobi Operator*, Quart J. Math. Oxford **46** (1995), 299-320.

[88] P. Gilkey and T. Zhang, *Algebraic curvature tensors whose skew symmetric curvature operator has constant rank* 2, Periodica Mathematica Hungarica (to appear).

[89] H. Glover, W. Homer, and R. Stong, *Splitting the tangent bundle of projective space*, Indiana Univ. Math. J. **31** (1982), 161-166.

[90] A. Gray, *Classification des variétés approximativement kählériennes de courbure sectionelle holomorphe constante*, C. R. Acad. Sci. Paris **279** (1974), 797–800.

[91] ———, *Curvature identities for Hermitian and almost Hermitian manifolds*, Tohoku Math. J. **28** (1976), 601–612.

[92] A. Gray and T. J. Willmore, *Mean-value theorems for Riemannian manifolds*, Proc. Roy. Soc. Edinburgh **92A** (1982), 343–364.

[93] V. Guillemin and S. Sternberg, *An ultra-hyperbolic analogue of the Robertson-Kerr theorem*, Lett. Math. Phys. **12** (1986), 1–6.

[94] R. Harvey and H.B. Lawson, *Calibrated geometries*, Acta Math. **148** (1982), 47–157.

[95] S. Helgason, *Differential geometry, Lie groups and symmetric spaces*, Academic Press, New York, 1978.

[96] S. Ivanov and I. Petrova, *Conformally flat Einstein-like 4-manifolds and conformally flat Riemannian 4 manifolds all of whose Jacobi operators have parallel eigenspaces along every geodesic*, Annuare de l'Université de Sofia "St. Kliment Ohridski" **85 (1991)**, 55–64.

[97] ———, *Riemannian manifold in which the skew-symmetric curvature operator has pointwise constant eigenvalues*, Geom. Dedicata **70** (1998), 269–282.

[98] R. Ivanova, *Necessary and sufficient conditions for the point-wise constancy of a curvature operatos's characteristical coefficient*, Ann. de'l Universite de Sofia, Fac. de Math. et Inf. **88** (1994), 17–25.

[99] ———, *About a classification of some Einstein manifolds*, Mathematics and Education in Mathematics, Proc. XXIII Spring Conference of the Union of Bulgarian Mathematicians, Stara Zagora (1994), 175–182.

[100] ———, *On the geometry of some 4-dimensional Einstein manifolds*, Masaryk Univ., Brno, 1996, Differential geometry and applications (Brno, 1995), 35–43.

[101] ———, *On a property of the 4-dimensional Einstein manifolds*, Proceedings of the 4th International Congress of Geometry (Thessaloniki, 1996), 198–207.

[102] ———, *Point-wise constancy of the skew-symmetric curvature operator's characteristic coefficients*, Tensor (N.S.) **57** (1996), 97–104.

[103] ———, *Two systems of curvature conditions for some 4-dimensional Riemannian manifolds.*, God. Univ. Arkhit. Stroit. Geod. Sofiya Svitk II Mat. Mekh (1996/7), 33–44.

[104] ———, *An orthogonal tangential group of transformations of a 4 dimensional Riemannian manifold*, Math. Balkanica (N.S.) **11** (1997), 53–63.

[105] ———, *4-dimensional Riemannian manifolds characterized by a skew symmetric curvature operator*, Tensor (N.S.) **60** (1998), 293–302.

[106] ———, *Generalization of a property of some 4 dimensional Einstein manifolds*, Tensor (N.S.) **60** (1998), 303–308.

[107] R. Ivanova and G. Stanilov, *A skew-symmetric curvature operator in Riemannian geometry*, Conf. A: Mathematics and Theoretical Physics (Munich 1993), 391–395.

[108] G. Jensen and M. Rigoli, *Neutral surfaces in neutral four spaces*, Matematiche (Catania) **45** (1990), 407–443.

[109] M Karoubi, *K-theory. An introduction*, Grundlehren der Mathematischen Wissenschaften, Band 226 ISBN: 3-540-08090-2, Springer-Verlag, Berlin, 1978.

[110] I. Kath, *Almost complex algebraic curvature tensors in dimension 4*,

(Preprint).

[111] O. Kowalski, M. Sekizawa, and Z. Vlasek, *Can tangent sphere bundles over Riemannian manifolds have strictly positive curvature?*, preprint.

[112] O. Kowalski, F. Tricerri and L. Vanhecke, *Curvature homogeneous Riemannian manifolds*, J. Math. Pures et Appl. **71** (1992), 471–501.

[113] O. Kowalski and L. Vanhecke, *Ball-homogeneous and disk homogeneous Riemannian manifolds*, Math. Z. **180** (1982), 429–444.

[114] P. Kronheimer, *The construction of ALE spaces as hyper-Kähler quotients*, J. Differential Geom. **29** (1989), 665–683.

[115] K. Y. Lam and P. Yiu, *Linear spaces of real matrices of constant rank*, Linear Algebra Appl **195** (1993), 69–79.

[116] P. R. Law, *Neutral Einstein metrics in four dimensions*, J. Math. Phys. **32** (1991), 3039–3042.

[117] S. Lefschetz, *Introduction to topology, Princeton Mathematical Series*, Princeton University Press, vol. 11, 1949.

[118] S. Marchiafava, *Variétés riemanniennes dont le tenseur de courbure est celui d'un espace symétrique de rang un*, C. R. Acad. Sci. Paris **295** (1982), 463-466.

[119] R. Meshulam, *On K spaces of real matrices*, Linear and Multilinear Algebra **26** (1990), 39–41.

[120] J. Milnor and J. Stasheff, *Characteristic Classes*, Annals of Mathematics Studies, vol. 76, Princeton Univ. Press, 1974.

[121] K. Monks, *Groebner bases and the cohomology of Grassmann manifolds with application to immersion*, Bol. Soc. Mat. Mexicana **7** (2001), 123–136.

[122] A. Newlander and L. Nirenberg, *Complex analytic coordinates in almost complex manifolds*, Ann. Math. **65** (1957), 391–404.

[123] Z. Olszak, *On the existence of generalized complex space forms*, Israel J. Math. **65** (1989), 214–218.

[124] ———, *Four dimensional para-Hermitian manifolds*, Tensor (N.S.) **56** (1995), 215–226.

[125] B. O'Neill, *Semi-Riemanian geometry. With applications to relativity*, Pure and Applied Mathematics, Academic Press, New York, vol. 103 ISBN: 0-12-526740-1, 1983.

[126] R. Osserman, *Curvature in the eighties*, Amer. Math. Monthly **97** (1990), 731–756.

[127] R. Osserman and P. Sarnak, *A new curvature invariant and entropy of geodesic flows*, Invent. Math. **77** (1984), 455–462.

[128] JH Park, *Continuous variation of eigenvalues and Garding's inequality*, Differential Geom. Appl. **10** (1999), 187–189.

[129] I. Petrova and G. Stanilov, *A generalized Jacobi operator on the 4 dimensional Riemannian geometry*, Annuaire Univ. Sofia Fac. Math. Inform. **85** (1991), 55–64.

[130] Z. Rakić, *An example of rank two symmetric Osserman space*, Bull. Austral. Math. Soc. **56** (1997), 517–521.

[131] ———, *On duality principle in Osserman manifolds*, Linear Algebra App. **296** (1999), 183–189.

[132] B. A. Rozenfeld, *Noneuclidean geometries*, Gosudarstv. Izdat. Tehn.-Teor. Lit., Moscow, 1955. (Russian)

[133] H. S. Ruse, A. G. Walker T. J. Willmore, *Harmonic spaces*, Consiglio Nazionale delle Ricerche Monografie Matematiche, 8 Edizioni Cremonese, Rome, 1961.

[134] S. M. Salamon, *Quaternionic Kähler manifolds*, Invent. Math. **67** (1982), 143–171.

[135] ———, *Harmonic 4-spaces*, Math. Ann. **269** (1984), 169–178.

[136] K. Sekigawa and L. Vanhecke, *Volume preserving geodesic symmetries on four-dimensional Kähler manifolds*, Differential Geometry Peníscola, 1985, Proceedings (A. M. Naveira, A. Ferrández and F. Mascaró, eds.), Lecture Notes in Math., 1209, Springer, pp. 275–290.

[137] ———, *Four-dimensional almost Kähler Einstein manifolds*, Ann. Mat. Pura Appl. **157** (1990), 149–160.

[138] I. M. Singer, *Infinitesimally homogeneous spaces*, Comm. Pure Appl. Math. **13** (1960), 685–697.

[139] E. Spanier, *Algebraic Topology*, McGraw Hill, New York, 1966.

[140] G. Stanilov, *On the geometry of the Jacobi operators on 4-dimensional Riemannian manifolds*, Tensor (N.S.) **51** (1992), 9–15.

[141] ———, *Curvature operators based on the skew-symmetric curvature operator and their place in Differential Geometry*, Preprint 14.02.2000.

[142] G. Stanilov and V. Videv, *On a generalization of the Jacobi operator in the Riemannian geometry*, Annuaire Univ. Sofia Fac. Math. Inform. **86** (1992), 27–34.

[143] ———, *On Osserman conjecture by characteristical coefficients*, Algebras Groups Geom **12** (1995), 157–163.

[144] ———, *Four dimensional pointwise Osserman manifolds*, Abh. Math. Sem. Univ. Hamburg **68** (1998), 1–6.

[145] I. Stavrov, *A Note On Generalized Osserman Manifolds in the Lorentzian Setting*, preprint (2000).

[146] ———, *Ph. D. Thesis University of Oregon (anticipated).*

[147] N. E. Steenrod, *The Topology of Fiber Bundles*, Princeton University Press, vol. 14, 1951.

[148] ———, *Cohomology operations*, Ann. of Math. Studies #50 Lectures by N. E. Steenrod written and revised by D. B. A. Epstein, 1962.

[149] R. Stong, *Splitting the universal bundles over Grassmannians*, Algebraic and differential topology—global differential geometry, Teubner-Texte Math.,Teubner, Leipzig **70** (1984), 275–287.

[150] Z. I. Szabó, *A short topological proof for the symmetry of* 2 *point homogeneous spaces*, Invent. Math. **106** (1991), 61–64.

[151] F. Tricerri and L. Vanhecke, *Decomposition of a space of curvature tensors an a quaternionic Kähler manifolds and spectral theory*, Simon Stevin **53** (1979), 163–173.

[152] ———, *Curvature tensors on almost Hermitian manifolds*, Trans. Amer. Math. Soc. **267** (1981), 365–397.

[153] ———, *Geometry of a class of nonsymmetric harmonic manifolds*, Differential geometry and its applications (Opava, 1992), Math. Publ. 1, Silesian Univ. Opava, Opava (1993), 415–426.

[154] L. Vanhecke, *Geometry and symmetry*, Advances in Differential Geometry (F. Tricerri, ed.), World Scientific, Singapore, 1990, pp. 115–129.

[155] ———, *Geometry in normal and tubular neighbourhoods*, Proceedings Workshop on Differential Geometry and Topology, Cala Gonone (Sardinia), Rend. Sem. Fac. Sci. Univ. Cagliari, supp. al vol. 58 1988, pp. 73–176.

[156] L. Vanhecke and T. J. Willmore, *Interaction of tubes and spheres*, Math. Ann. **263** (1983), 31–42.

[157] A. G. Walker, *Canonical form for a Riemannian space with a parallel field of null planes*, Quart. J. Math. Oxford **1** (1950), 69–79.

[158] H. Weyl, *The Classical Groups*, Princeton Univ. Press, Princeton, 1946.

[159] G. Whitehead, *Homotopy Theory*, M.I.T. Press (1966).

[160] J. A. Wolf, *Spaces of constant curvature*, Univ. California (Berkeley, 1972).

[161] H. Wu, *Holonomy groups of indefinite metrics*, Pac. J. Math **20** (1967), 351–392.

[162] A. Yampolsky, *On the strong sphericity of the Sasaki metric of a spherical tangent bundle*, J. Math. Sci. **72** (1994), 3261–3266.

[163] T. Zhang, *Manifolds with indefinite metrics whose skew-symmetric curvature operator has constant eigenvalues*, Ph. D. thesis, University of Oregon (2000).

[164] ———, *Applications of algebraic topology in bounding the rank of the skew-symmetric curvature operator*, Topology and its applications (to appear).

Index